Communications in Computer and Information Science 2218

Rationale

The CCIS series is devoted to the publication of proceedings of computer science conferences. Its aim is to efficiently disseminate original research results in informatics in printed and electronic form. While the focus is on publication of peer-reviewed full papers presenting mature work, inclusion of reviewed short papers reporting on work in progress is welcome, too. Besides globally relevant meetings with internationally representative program committees guaranteeing a strict peer-reviewing and paper selection process, conferences run by societies or of high regional or national relevance are also considered for publication.

Topics

The topical scope of CCIS spans the entire spectrum of informatics ranging from foundational topics in the theory of computing to information and communications science and technology and a broad variety of interdisciplinary application fields.

Information for Volume Editors and Authors

Publication in CCIS is free of charge. No royalties are paid, however, we offer registered conference participants temporary free access to the online version of the conference proceedings on SpringerLink (http://link.springer.com) by means of an http referrer from the conference website and/or a number of complimentary printed copies, as specified in the official acceptance email of the event.

CCIS proceedings can be published in time for distribution at conferences or as post-proceedings, and delivered in the form of printed books and/or electronically as USBs and/or e-content licenses for accessing proceedings at SpringerLink. Furthermore, CCIS proceedings are included in the CCIS electronic book series hosted in the SpringerLink digital library at http://link.springer.com/bookseries/7899. Conferences publishing in CCIS are allowed to use Online Conference Service (OCS) for managing the whole proceedings lifecycle (from submission and reviewing to preparing for publication) free of charge.

Publication process

The language of publication is exclusively English. Authors publishing in CCIS have to sign the Springer CCIS copyright transfer form, however, they are free to use their material published in CCIS for substantially changed, more elaborate subsequent publications elsewhere. For the preparation of the camera-ready papers/files, authors have to strictly adhere to the Springer CCIS Authors' Instructions and are strongly encouraged to use the CCIS LaTeX style files or templates.

Abstracting/Indexing

CCIS is abstracted/indexed in DBLP, Google Scholar, EI-Compendex, Mathematical Reviews, SCImago, Scopus. CCIS volumes are also submitted for the inclusion in ISI Proceedings.

How to start

To start the evaluation of your proposal for inclusion in the CCIS series, please send an e-mail to ccis@springer.com.

Kang Li · Kailong Liu · Yukun Hu · Mao Tan ·
Long Zhang · Zhile Yang
Editors

Clean Energy Technology and Energy Storage Systems

8th International Conference on Life System Modeling and Simulation, LSMS 2024 and 8th International Conference on Intelligent Computing for Sustainable Energy and Environment, ICSEE 2024
Suzhou, China, September 13–15, 2024
Proceedings, Part III

Springer

Editors
Kang Li
University of Leeds
Leeds, UK

Kailong Liu
Shandong University
Jinan, China

Yukun Hu
University College London
London, UK

Mao Tan
Xiangtan University
Xiangtan, China

Long Zhang
University of Manchester
Manchester, UK

Zhile Yang
Chinese Academy of Sciences
Shenzhen, China

ISSN 1865-0929 ISSN 1865-0937 (electronic)
Communications in Computer and Information Science
ISBN 978-981-96-0231-5 ISBN 978-981-96-0232-2 (eBook)
https://doi.org/10.1007/978-981-96-0232-2

This Springer imprint is published by the registered company Springer Nature Singapore Pte Ltd.
The registered company address is: 152 Beach Road, #21-01/04 Gateway East, Singapore 189721, Singapore

Preface

This book series constitutes the proceedings of the 2024 International Conference on Life System Modeling and Simulation (LSMS 2024) and the 2024 International Conference on Intelligent Computing for Sustainable Energy and Environment (ICSEE 2024), which were held during September 13–15, in Suzhou, China. The LSMS and ICSEE international conference series aim to bring together international researchers and practitioners in the fields of advanced methods for life system modeling and simulation, as well as advanced intelligent computing theory, methodologies, and engineering applications in achieving net zero across all sectors to tackle the global climate change challenge. These events are built upon the success of previous LSMS conferences held in Shanghai, Wuxi, Nanjing and Hangzhou in 2004, 2007, 2010, 2014, 2017 and 2021, and ICSEE conferences held in Wuxi, Shanghai, Nanjing, Chongqing and Hangzhou in 2010, 2014, 2017, 2018 and 2021 respectively, and are based on large-scale UK-China collaboration projects on sustainable energy.

At LSMS 2024 and ICSEE 2024, technical exchanges within the research community took the form of keynote speeches, panel discussions, an early career researcher innovation competition, an industrial innovation forum, as well as oral and poster presentations. The LSMS 2024 and ICSEE 2024 conferences received 504 submissions from 17 countries and regions. All papers went through a rigorous peer review procedure and each paper received at least three single blind review reports. Based on the review reports, the Program Committee finally selected 158 high-quality papers for presentation at LSMS 2024 and ICSEE 2024. These papers cover 34 topics, and are included in five volumes of CCIS proceedings published by Springer.

The organizers of LSMS2024 and ICSEE 2024 would like to acknowledge the enormous contribution of the Program Committee and the referees for their efforts in reviewing and soliciting the papers, and the Publication Committee for their editorial work. We would also like to thank the editorial team from Springer for their support and guidance. Particular thanks go to all the authors; without their high-quality submissions and presentations the conferences would not have been successful.

Finally, we would like to express our gratitude to our sponsors and organizers, listed on the following pages.

September 2024

Kang Li
Kailong Liu
Yukun Hu
Mao Tan
Long Zhang
Zhile Yang

Organization

Sponsors

China Instrument and Control Society (CIS)
China Simulation Federation (CSF)
IEEE CC Ireland Chapter
IEEE Systems, Man & Cybernetics Society Technical Committee on Systems Biology

Organizers

Suzhou University of Science and Technology, China
Shanghai University, China
Shanxi University, China
University of Leeds, UK
Life System Modeling and Simulation Technical Committee of CSF, China
Embedded Instrument and System Technical Committee of China Instrument and Control Society, China

Co-sponsors

IEEE Industrial Electronics Society
UK China University Consortium on Engineering Education and Research
International Scientific Research Alliance on Intelligent Measurement Control and Applications of Complex Networked Systems
Shanghai Association for System Simulation
Shanghai Instrument and Control Society
Shanghai Association of Automation

Co-organizers

Queen's University Belfast, UK
Nanjing University of Posts and Telecommunications, China
University of Essex, UK
Queensland University of Technology, Australia
Tsinghua University, China
Peking University, China
Nantong University, China

Shenzhen Institute of Advanced Technology, CAS, China
Cardiff University, UK
Xiangtan University, China
Shanghai Key Laboratory of Power Station Automation Technology, China
Shanghai International Joint Laboratory of Intelligent Automation and Networked Control, China
Discipline Innovation and Talent Introduction Base for Intelligent Measurement & Control for Complex Networked Systems and Applications, Ministry of Science and Technology, China
Joint Laboratory for Intelligent Measurement and Control of Low-Carbon Energy Systems, China-UK
Key Laboratory of Complex Systems and Data Science, Ministry of Education, China
Jiangsu Industrial Intelligent and Low-Carbon Technology Engineering Center, China
Suzhou Key Laboratory of Intelligent Low-Carbon Technology Application, China
Suzhou Blockchain Data Privacy Protection Innovation Application Laboratory, China
Anhui Key Laboratory of Electric Drive and Control, Wuhu, China

International Advisory Committee Chairs

Chengshan, Wang	Tianjin University, China
Huosheng, Hu	University of Essex, UK
Pedrycz, Witold	University of Alberta, Canada
Xinzhou, Dong	Tsinghua University, China
Xudong, Zhao	University of Hull, UK
Rosei, Federico	University of Trieste, Italy

Honorary Chairs

Minrui, Fei	Shanghai University, China
Qinglong, Han	Swinburne University of Technology, Australia
Spurgeon, Sarah	University College London, UK
Umezu, Mitsuo	Waseda University, Japan

General Chairs

Chen, Peng	Shanghai University, China
Kang, Li	University of Leeds, UK
Juping, Gu	Suzhou University of Science and Technology, China
Jianzhong, Wu	Cardiff University, UK

International Industrial Advisory Committee Chairs

James, Yu	SP Energy Networks, UK
Yun, Li	CRRC, China
Kai, Chen	SAIC, China

International Program Committee Chairs

Dajun, Du	Shanghai University, China
Shiwei, Ma	China Simulation Federation, China
Xinchun, Jia	Shanxi University, China
Junfei, Qiao	Beijing University of Technology, China
Fang, Fang	North China Electric Power University, China
Luonan, Chen	University of Tokyo, Japan
Coombs, Tim	University of Cambridge, UK
McLoone, Sean	Queen's University Belfast, UK
Lei, Ma	Southwest Jiaotong University, China
Bin, Li	Tianjin University, China
Yuchu, Tian	Queensland University of Technology, Australia
Rakić, Aleksandar	University of Belgrade, Serbia
Zhiqiang, Yao	Xiangtan University, China

International Advisory Committee

Chee Khiang, Pang	Singapore Institute of Technology, Singapore
Frangi, Alejandro	University of Manchester, UK
Hongjie, Jia	Tianjin University, China
Jihong, Wang	Warwick University, UK
Jinghan, He	Beijing Jiaotong University, China
Keyou, You	Tsinghua University, China
Ling, Wang	Tsinghua University, China
Morris, Kevin	University of Leeds, UK
Varga, Liz	University College London, UK
Wen, Yu	National Polytechnic Institute, Mexico
Wenhui, Fan	Tsinghua University, China
Wenjun, Zhang	University of Saskatchewan, Canada
Yan, Jin	Queen's University Belfast, UK
Zhengtao, Ding	University of Manchester, UK
Zhiwei, Gao	Northumbria University, UK

IPC Local Chairs

Athanasopoulos, Nikolaos	Queen's University Belfast, UK
Azizi, Sadegh	University of Leeds, UK
Congzhi, Huang	NCEPU, China
Deqing, Huang	Southwest Jiaotong University, China
Fei, Teng	Imperial College, UK
Fuqiang, Li	Henan Agricultural University, China
Fuwen, Yang	Griffith University, Australia
Fuyuan, Hu	Suzhou University of Science and Technology, China
Guangya, Yang	Danmarks Tekniske Universitet, Denmark
Huiyu, Zhou	University of Leicester, UK
Jahromi, Amir	University of Leeds, UK
Jianhua, Zhang	NCEPU, China
Juan, Yang	Anhui University, China
Kailong, Liu	Shandong University, China
Kun, Zhang	Nantong University, China
Lin, Jiang	University of Liverpool, UK
Lisheng, Wei	Anhui Polytechnic University, China
Long, Cheng	Institute of Automation, Chinese Academy of Sciences, China
Long, Zhang	University of Manchester, UK
Mao, Tan	Xiangtan University, China
Menhas, Muhammad Ilyas	Mirpur University of Science and Technology, Pakistan
Naeem, Wasif	Queen's University Belfast, UK
Pfaender, Fabien	Université de technologie de Compiègne, France
Qing, Fang	Yamagata University, Japan
Qi, Wang	Shanxi University
Shaojun, Gan	Beijing University of Technology, China
Shan, Gan	Southeast University, China
Shuai, Li	University of Oulu, Finland
Stefanovski, Jovan D.	JP "Streževo", North Macedonia
Teng, Long	University of Cambridge, UK
Tengfei Zhang	Nanjing University of Posts and Telecommunications, China
Uddin, Mohammad Monir	North South University, Bangladesh
Wei, Jiang	Southeast University, China
Wei, Wang	NCEPU, China
Weiyan, Hou	Zhengzhou University, China
Wenju, Zhou	Shanghai University, China

Xiaoou, Li	National Polytechnic Institute, Mexico
Xiandong, Xu	Tianjin University, China
Xing, Zhang	University of Sheffield, UK
Xinli, Wang	Shandong University, China
Yang, Liu	Harbin Institute of Technology, China
Yongliang, Li	University of Birmingham, UK
Yuanjun, Guo	SIAT, CAS, China
Yukun, Hu	University College London, UK

Organization Committee

Chairs

Hongjie, Wu	Suzhou University of Science and Technology, China
Yulong, Wang	Shanghai University, China
Ni, Li	CSF, China
Mao, Tan	Xiangtan University, China
Yanpeng, Guan	Shanxi University, China
Ling, Wang	Shanghai University, China
Zhengtian, Wu	Suzhou University of Science and Technology, China
Shenghui, Guo	Suzhou University of Science and Technology, China

ROC Members

Qi, Kang	Shanghai Association for System Simulation, China
Weihua, Wu	Shanghai Instrument and Control Society, China
Wenbin, Dai	Shanghai Association of Automation, China
Xiaobo, Chi	Shanxi University, China
Xianming, Zhang	Swinburne University of Technology, Australia

Net-Zero Innovation Workshop Chairs

Zhile, Yang	SIAT CAS, China
Li, Zhang	MediaTek, China
Yanli, Liu	Tianjin University, China
Nan, Ma	Beijing University of Technology, China

Special Session Chairs

Sen, Wang	Imperial College London, UK
Qiang, Liu	University of Bristol, UK
Nan, Ma	Beijing University of Technology, China
Xiaolin, Mou	Shenzhen Technology University, China
Yihuan, Li	NCEPU, China

Publication Chairs

Qun, Niu	Shanghai University, China
Wanqing, Zhao	Newcastle University, UK

Publicity Chairs

Erfu, Yang	University of Strathclyde, UK
Zixiang, Fei	Shanghai University, China
Haochen, Hua	Hohai University, China
Weiming, Shao	China University of Petroleum, China

Registration Chairs

Yang, Song	Shanghai University, China
Shuchen, Ding	Suzhou University of Science and Technology, China

Secretary-Generals

Xin, Li	Shanghai University, China
Xin, Sun	Shanghai University, China
Zhi, Chen	Shanghai University, China
Xiaoliang, Wang	Suzhou University of Science and Technology, China

International Program Committee

Aolei, Yang	Shanghai University, China
Athanasopoulos, Nikolaos	Queen's University Belfast, UK
Azizi, Sadegh	University of Leeds, UK
Bin, Zhang	Shanghai University, China

Chee Khiang, Pang	Singapore Institute of Technology, Singapore
Congzhi, Huang	NCEPU, China
Deqing, Huang	Southwest Jiaotong University, China
Erfu, Yang	University of Strathclyde, UK
Fei, Teng	Imperial College London, UK
Frangi, Alejandro	University of Manchester, UK
Fuwen, Yang	Griffith University, Australia
Fuyuan, Hu	Suzhou University of Science and Technology, China
Guangya, Yang	Danmarks Tekniske Universitet, Denmark
Haochen, Hua	Hohai University, China
Hongjie, Jia	Tianjin University, China
Hongjie, Wu	Suzhou University of Science and Technology, China
Huiyu, Zhou	University of Leicester, UK
Jahromi, Amir	University of Leeds, UK
Jianhua, Zhang	NCEPU, China
Jihong, Wang	Warwick University, UK
Jinghan, He	Beijing Jiaotong University, China
Juan, Yan	Anhui University, China
Kailong, Liu	Shandong University, China
Kun, Zhang	Nantong University, China
Lan, Ma	Beijing University of Technology, China
Li, Zhang	MediaTek, China
Lin, Jiang	University of Liverpool, UK
Ling, Wang	Shanghai University, China
Ling, Wang	Tsinghua University, China
Lisheng, Wei	Anhui Polytechnic University, China
Long, Cheng	Institute of Automation, Chinese Academy of Sciences, China
Long, Zhang	University of Manchester, UK
Lu, Wang	Shanghai University, China
Mao, Tan	Xiangtan University, China
Menhas, Muhammad Ilyas	Mirpur University of Science and Technology, Pakistan
Naeem, Wasif	Queen's University Belfast, UK
Ni, Li	CSF, China
Peng, Zhou	Shanghai University, China
Pfaender, Fabien	Université de technologie de Compiègne, France
Qi, Kang	Shanghai Association for System Simulation, China
Qi, Wang	Shanxi University, China

Qiang, Liu	University of Bristol, UK
Qing, Fang	Yamagata University, Japan
Qun, Niu	Shanghai University, China
Sen, Wang	Imperial College London, UK
Shaojun, Gan	Beijing University of Technology, China
Shan, Gao	Southeast University, China
Shenghui, Guo	Suzhou University of Science and Technology, China
Shuchen, Ding	Suzhou University of Science and Technology, China
Teng, Long	University of Cambridge, UK
Tengfei, Zhang	Nanjing University of Posts and Telecommunications, China
Uddin, Mohammad Monir	North South University, Bangladesh
Wanqing, Zhao	Newcastle University, UK
Wei, Jiang	Southeast University, China
Wei, Wang	NCEPU, China
Weihua, Wu	Shanghai Instrument and Control Society, China
Weiming, Shao	Shanxi University, China
Weiyan, Hou	Zhengzhou University, China
Wenbin, Dai	Shanghai Association of Automation, China
Wenhui, Fan	Tsinghua University, China
Wenju, Zhou	Shanghai University, China
Wenjun, Zhang	University of Saskatchewan, Canada
Wen, Yu	National Polytechnic Institute, Mexico
Xiaobo, Chi	Shanxi University, China
Xiaoou, Li	National Polytechnic Institute, Mexico
Xiaoliang, Wang	Suzhou University of Science and Technology, China
Xiaolin, Mou	Shenzhen Technology University, China
Xiandong, Xu	Tianjin University, China
Xianming, Zhang	Swinburne University of Technology, Australia
Xin, Li	Shanghai University, China
Xin, Sun	Shanghai University, China
Xing, Zhang	University of Sheffield, UK
Xinli, Wang	Shandong University, China
Yan, Jin	Queen's University Belfast, UK
Yang, Song	Shanghai University, China
Yanpeng, Guan	Shanxi University, China
Yanli, Liu	Tianjin University, China
Yihuan, Li	NCEPU, China
Yongliang, Li	University of Birmingham, UK

Yuanjun, Guo	SIAT, CAS, China
Yukun, Hu	University College London, UK
Yulong, Wang	Shanghai University, China
Zhengtao, Ding	University of Manchester, UK
Zhengtian, Wu	Suzhou University of Science and Technology, China
Zhi, Chen	Shanghai University, China
Zhile, Yang	SIAT, CAS, China
Zixiang, Fei	Shanghai University, China

中国仿真学会
China Simulation Federation

CIS
中国仪器仪表学会
CHINA INSTRUMENT & CONTROL SOCIETY

IEEE SMC
Systems, Man, and Cybernetics Society

IEEE
CC Ireland Chapter

Contents – Part III

Clean and Smart Energy Technology

Climatic Efficiency Analysis of Ten Leading UK Offshore Wind Farms: A Data-Driven Approach

Liyang Ma, Chenyi Li, and Long Zhang(✉)

Department of Electrical and Electronic Engineering, University of Manchester, Manchester, UK
{liyang.ma,liyang.ma}@postgrad.manchester.ac.uk

Abstract. As wind energy becomes increasingly important in global energy systems, accurately evaluating the performance of wind farms is essential for understanding wind generation efficiency and fostering sustainable development. The aim of this paper is to assess wind power generation efficiency under different climatic wind conditions. A data-driven approach is used to simulate the power generation of ten leading wind farms in the UK, using real wind data and geographic information while considering wake effects. Furthermore, to evaluate whether these wind farms perform optimally when climate-related wind patterns change, we modify the layouts of the wind farms and compare their performance with the original designs. Extensive results from 15 years (2008 to 2023) of wind data show that the originally designed layouts of these ten leading wind farms will no longer be optimal when the wind resource changes.

Keywords: Wind Energy · Wind Farm Simulation · Data-driven Approach

1 Introduction

Over the last three decades, there have been remarkable advancements in new energy technologies, particularly in wind energy and solar energy. It is crucial to ensure the efficient and stable operation of these power sources. In the UK, the rapid expansion of wind farms faces several challenges, primarily due to the variable and unpredictable nature of wind energy. Wind farm outputs fluctuate significantly due to climatic patterns, leading to notable variations in different seasons and times. Furthermore, accurate simulation and forecasting of wind farm efficiency is crucial due to the complexities involved in ideal wind farm siting, installation, and maintenance costs, as well as the impact of policy and technology.

Researchers have developed various data-driven methods for evaluating wind farm performance, including computational fluid dynamics (CFD) simulation

K. Li et al. (Eds.): LSMS/ICSEE 2024, CCIS 2218, pp. 3–13, 2025.
https://doi.org/10.1007/978-981-96-0232-2_1

[1–3] and machine learning methods [4–6]. These tools facilitate decision-making on critical wind farm design and operation, such as optimising turbine layout, analysing wind farm characteristics, and predicting wind farm power. This evaluation is crucial to comprehending how wind farms actually perform under different climate conditions, particularly given the climatic variability of wind energy resources. The original field wind farm layout may no longer be suitable for the changing climate conditions in the future. In this study, we employ a data-driven approach to analyze the power and efficiency of real wind farms in the UK using actual wind data and environmental information. By simulating and making small adjustments to the field wind farm layout, we can observe changes in wind farm power and efficiency when weather conditions change slightly.

This paper is organized as follows. The second section describes the collection of input factors, including data on the geographic locations of real wind farms and data on real wind resources in these areas from 2008 to 2023. The third section describes the experimental process, including the construction of the framework, the selection of models and parameters, and the simulation design. The fourth section describes the simulation results. The fifth section describes the conclusions.

2 Data Collection and Preprocessing

2.1 Collection and Transformation of Real Geographical Data

This study analyses 10 wind farms located on the east and west coasts of the UK, which are shown in Fig. 1, including the Burbo Bank Offshore Wind Farm, the London Array Offshore Wind Farm, the North Hoyle Offshore Wind Farm, etc. The turbine layout and latitude and longitude information for these wind farms were obtained through Open Street Maps and Wikipedia Maps.To build the performance simulation model, the latitude and longitude coordinates were converted into 2D plane coordinates. This process converts coordinates using standard geographic conversion methods and considers the impact of the earth's curvature on measured distances. Furthermore, we collected turbine data, include body height, rotor diameter, sea level altitude, turbine rated power and power efficiency, and nameplate capacity of the wind farm, which can be utilised to input model parameters and ensure accuracy.

2.2 Collection and Preprocessing of Real Wind Data

This study utilises ERA5 hourly wind data from the European Centre for Medium-Range Weather Forecasts (ECMWF) [7] to analyse the 100-meter altitude horizontal wind speed (u, v components) across 15 years (Jan 2008-Jan 2024) at specific latitudes and longitudes. These two components represent the horizontal components of wind speed, while the vertical wind speed is so small that it can be roughly ignored. The monthly wind data is classified for each farm, creating 12 subsets for climatic analysis. Because the u and v components are

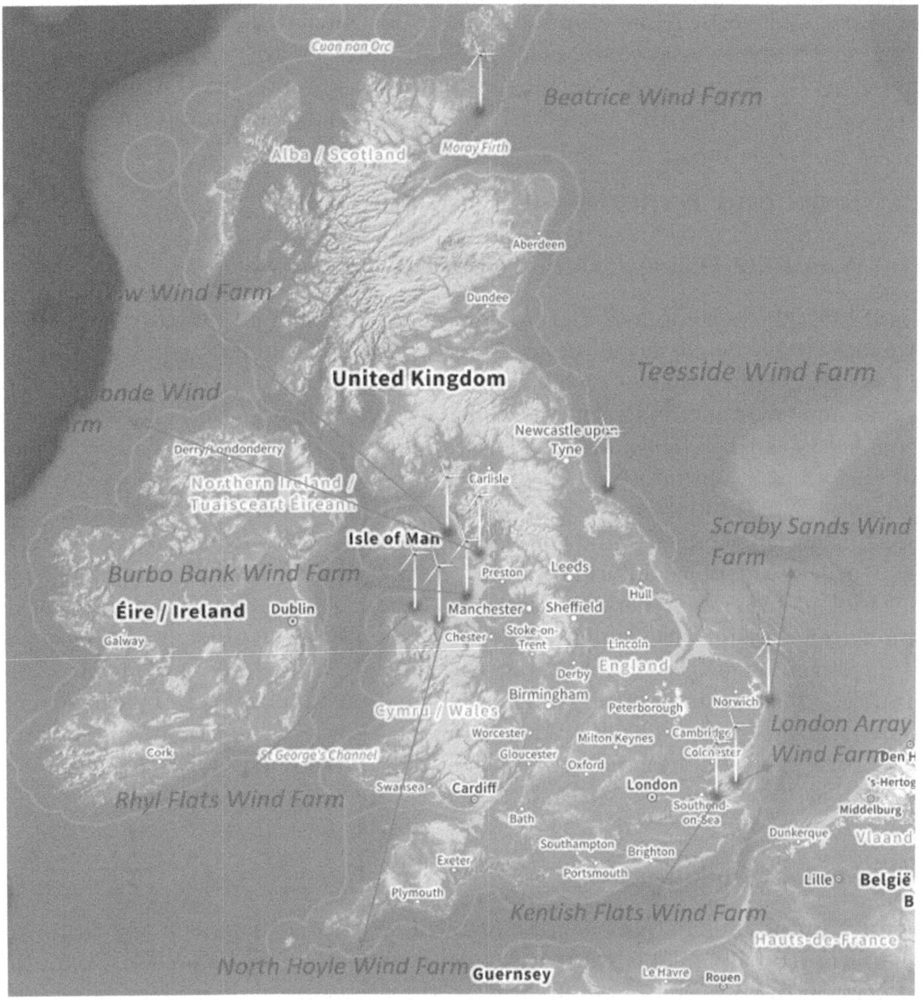

Fig. 1. The geographical locations of the 10 wind farms included in this study are marked in red on a map of the United Kingdom. (Color figure online)

directional, we cannot calculate their arithmetic mean directly. To address this, we first convert these components into wind speed V and corresponding wind direction angles θ, given by

$$V = \sqrt{u^2 + v^2} \tag{1}$$

$$\theta = arctan(u/v) \tag{2}$$

Wind speed is represented by the magnitude of the vector and is always non-negative. The wind direction angle varies from 0 to 360°. By averaging these values, we can obtain the approximate average wind speed, $\overline{V}$, and the average wind direction angle, $\overline{\theta}$. The required average wind speed components in the horizontal and vertical directions can then be calculated directly from these averages. The horizontal component, $\overline{u}$, is given by $\overline{V} \cdot \cos(\theta)$, and the vertical component, $\overline{v}$, is given by $\overline{V} \cdot \sin(\theta)$. This allows us to represent the wind speed's directional influence without explicitly resorting to equation notation.

3 Model and Experimental Simulation

3.1 WindFLO Framework and Script Development

WindFLO [8] is a computational framework used to simulate wind turbine performance within wind farms. It accurately simulates ambient wind conditions and evaluates and optimizes turbine layout by integrating multiple physical models and numerical methods, such as boundary layer theory [9]. The framework is publicly available [10] and is widely used in research on wind farm design and performance optimisation. In our study, we developed a customized script within the WindFLO framework to simulate the performance of wind farms. By inputting specific ambient wind conditions and adjusting various parameters, the script provides detailed information on power output and operating efficiency. This tool enables researchers to assess different turbine configurations and make data-driven decisions to simulate wind farm yield. One significant modification we implemented in our script is the elimination of the need for layout optimization adjustments. Instead, we concentrated on simulating the performance of pre-existing turbine layouts. This approach allows our script to more accurately reflect the conditions of actual wind farms, providing realistic performance simulations.

3.2 Model Configuration

Wind Model. Simulating accurate ambient wind conditions is a key factor in wind farm modelling. This study uses a log-law model [11] to precisely capture the vertical distribution of wind speed in the atmospheric boundary layer and describe it with a functional description. The log-law distribution is expressed by

$$U(z) = \frac{u^*}{k} \ln\left(\frac{z}{z_0}\right) \tag{3}$$

where k is the Von Karman constant, u^* is the friction speed, and z_0 represents the ground surface roughness. As all wind farms in this study are offshore, a uniform roughness input of 0.1 is selected due to the sea surface. In contrast, alternative models such as the power-law model or more sophisticated computational fluid dynamics (CFD) models also aim to describe wind speed distribution. While the power-law model is simpler, it lacks accuracy near the ground and may lead to significant errors in wind speed predictions at different heights. On the other hand, CFD models offer high precision by solving the Navier-Stokes fluid flow equations but demand substantial computational resources and time.

Wake Model. Given the uniform and predictable wind conditions in offshore wind farms and the even arrangement of turbines without complex terrain, the Jensen model is a suitable choice for assessing wind farm layout and turbine interactions. This model is low in complexity and conservative [12] and has demonstrated considerable reliability in predicting performance in early experiments. The model exhibiting a top-hat shape for the velocity deficit in the wake is given by

$$\frac{\Delta U}{U_\infty} = \frac{U_\infty - U_{\text{wake}}}{U_\infty} \tag{4}$$

where C_T is the thrust coefficient of the turbine, k_{wake} is the wake expansion rate which has chosen as 0.04, d_0 is the diameter of the wind turbine and the x is the distance from two turbines. $\Delta U / \Delta U_\infty$ is the filledized velocity deficit, where ΔU_∞ is the incoming wind speed and U_{wake} is the wake velocity, which can be defined as

$$\frac{\Delta U}{U_\infty} = \frac{1 - \sqrt{1 - C_T}}{\left(1 + \frac{2k_{\text{wake}}x}{d_0}\right)^2} \tag{5}$$

At the same time, Dynamic Wake Meandering (DWM) was selected as the wake merging mode to provide more accurate predictions by modelling the dynamic offset and spreading of the wake. The wake velocity distribution is usually assumed to be Gaussian, and its velocity loss can be expressed by

$$u_w(r, x, t) = \frac{\Delta U}{\sqrt{2\pi\sigma_w^2(x)}} \exp\left(-\frac{r^2}{2\sigma_w^2(x)}\right) \tag{6}$$

where ΔU is the peak velocity deficit, $\sigma_w(x)$ is the lateral wake diffusion width, r is the radial distance from the wake centre, x is the downstream distance, and t is the time. When wake merging occurs in a wind farm, the velocity deficits of multiple wind turbine wakes combine. The combined wake velocity deficit can be approximated using a superposition principle.

3.3 Simulation

Field Layout Simulation. To ensure realism and consistency in our simulations, we adjust parameters to reflect actual conditions before initiating simulations for each wind farm. Our custom script requires the input of steam turbine coordinates that align with the real-world environment. It is crucial to verify the accuracy of the steam turbine's rated power and quantity configuration information. The model then calculates turbine power P using the input wind speed components u and v for each month. This calculation involves the air density ρ, the swept area A of the turbine blades, the wind speed V, and the power coefficient C_p.

$$P = \frac{1}{2}\rho A V^3 C_p \tag{7}$$

The operating efficiency is calculated by simulating the total generated power P_{Farm} of the wind farm and dividing it by the rated power of the farm P_{Rated}, which ensure an objective evaluation of the power generated by the wind farm, are given by

$$\overline{\eta} = \frac{P_{\text{Farm}}}{P_{\text{Rated}}} \tag{8}$$

Modified Layout Simulation. In this section, the transformation is performed by simply rotation of the wind farm, making minimal changes to the layout, to simulate future conditions with small changes in wind resources. We employ a rotated layout strategy to modify the turbine positions defined by (x_i, y_i) within the same area. This involves rotating the layout 1° clockwise, 5° clockwise and 10° clockwise around a central point (c_x, c_y), resulting in a new configuration. Given that such rotation may lead some coordinates to extend beyond the first quadrant, we apply principles from linear transformation and coordinate rotation. Specifically, we initially shift each coordinate point relative to the center point, effectively repositioning the centre point at the origin. The rotation is facilitated by a rotated matrix with the angle set. Subsequent to the rotation, we apply an inverse translation concerning the original center point to the rotated coordinates, thereby deriving the new coordinate set, which can be expressed by

$$x_i' = c_y - (y_i - c_y) \tag{9}$$

$$y_i' = (x_i - c_x) + c_y \tag{10}$$

where (x_i', y_i') is the new coordinate after rotation. Figure 2 lists the rotation strategy for the London Array wind farm as an example. The London Array, once the largest wind farm in the UK, has 175 wind turbines and covers an area of more than 120 square kilometres. The rotation and translation operations are used to determine the 2D position coordinates of all wind turbines in the first quadrant.

4 Simulation Results

This section presents the results of the wind speed and wind farm performance assessment for the 10 wind farm areas from 2008 to 2023. Figure 3 shows the climatic pattern of average wind speeds for all wind farm areas, with higher average wind speeds in winter than in summer. Figure 4 shows the wind rose diagrams of the 15-year average annual data of the 3 selected wind farms, in which the wind resources of these areas are the most representative and obvious. In these 10 wind farm areas we studied, the wind rose diagrams show that southwest wind is the main wind source. Table 1 shows the correlation between the power output of the wind farms and the trend of wind speeds in a field layout. The efficiency of the wind farm (shown in Fig. 5) exhibits a positive correlation with power, with a U-shaped distribution.

When the modified layout is implemented, the power and efficiency of the wind farm may change, as demonstrated in Fig. 3. Climatic variations can cause different wind farms to exhibit irregular changes, highlighting the uncertainty

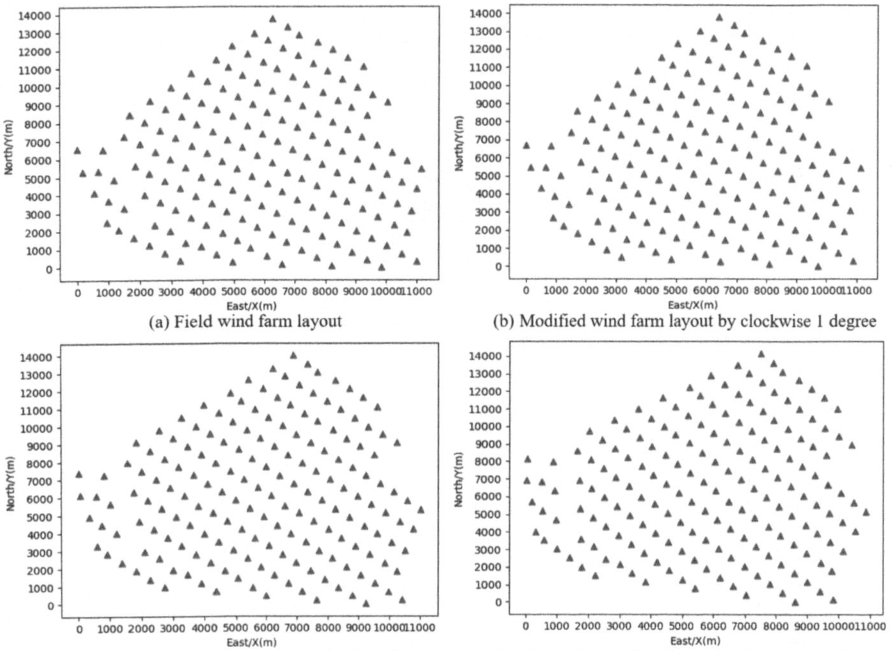

(a) Field wind farm layout

(b) Modified wind farm layout by clockwise 1 degree

(c) Modified wind farm layout by clockwise 5 degrees

(d) Modified wind farm layout by clockwise 10 degrees

Fig. 2. The wind farm layout rotation strategy used in this study is simulated, taking the London array as an example. (a) shows the field layout, (b) is the turbine layout after rotating 1°C clockwise, (c) shows the turbine layout after rotating 5°C clockwise and (d) is the turbine layout after rotating 10°C clockwise.

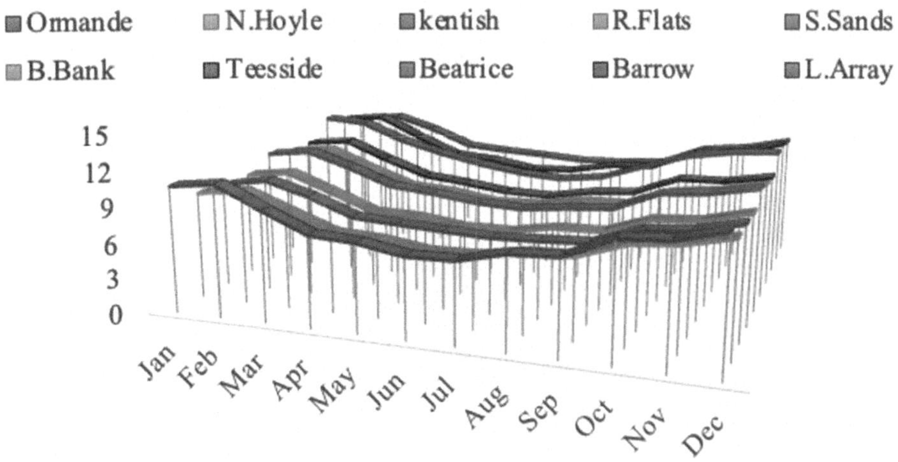

Fig. 3. Monthly average wind speed in the 10 wind farm areas from 2008 to 2023 (Unit: m/s).

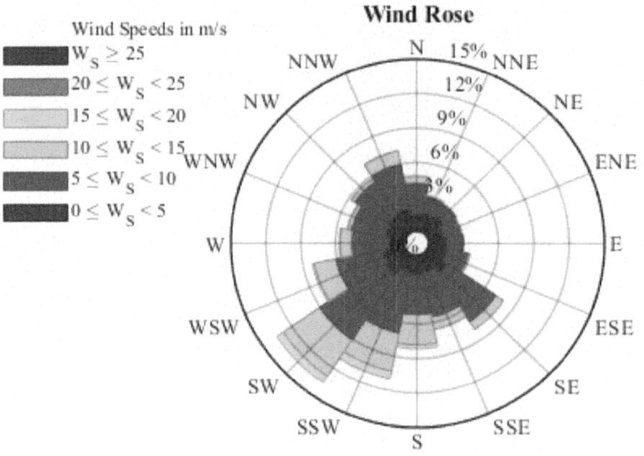

(a) The Ormonde Wind Farm area 15-year annual average windrose diagram.

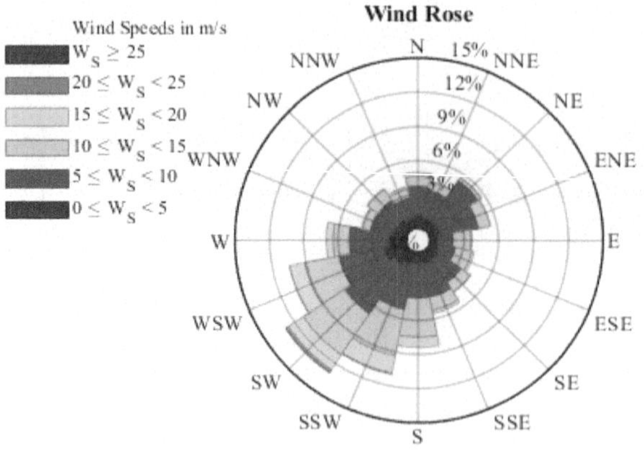

(b) The Kentish Wind Farm area 15-year annual average windrose diagram.

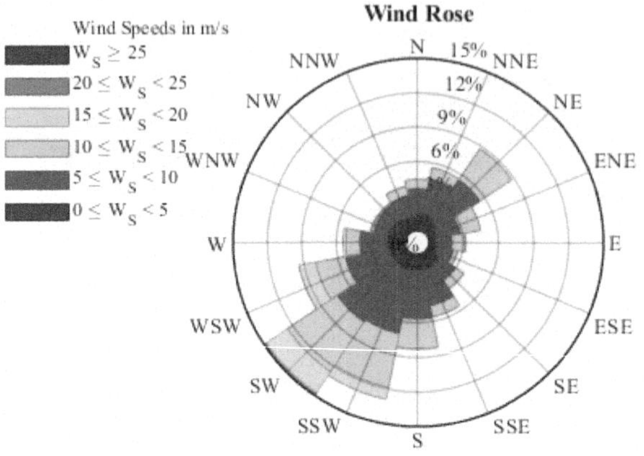

(c) The London Array Wind Farm area 15-year annual average windrose diagram.

Fig. 4. 15-year average windrose diagram of the 3 wind farms with the most representative wind conditions.

Table 1. Simulated Power Output of field 10 Wind Farms (Unit: MW)

Farms	Jan	Feb	Mar	Apr	May	Jun	Jul	Aug	Sep	Oct	Nov	Dec
B.Bank	122.2	137.9	82.5	47.6	44.2	32.1	35.5	43.3	49.1	88.1	99.3	130.9
Barrow	55.7	58.9	42.1	22.7	20.8	15.4	17.2	23.7	24.8	41.2	51.1	64.2
Beatrice	549.9	525.6	356.5	262.3	182.3	140.4	136.7	166.5	261.6	417.9	461.2	482.5
Kentish	69.4	42.1	58.9	35.1	30.3	27.8	25.5	26.6	29.6	51.6	57.9	72.3
N.Hoyle	50.8	54.8	34.5	19.7	17.7	13.5	14.1	18.9	20.9	35.4	41.8	56.3
L.Array	477.3	520.1	432.7	224.4	160.7	146.4	132.4	110.4	205.9	400.8	467.1	508.2
Ormonde	109.4	122.6	106.3	55.8	53.8	38.8	40.7	58.8	63.6	96.2	128.9	120.3
R.Flats	56.3	65.4	39.7	21.1	19.8	14.8	14.9	21	20.9	40.6	43.7	63.0
S.Sands	59.4	66.1	47.4	31.0	25.2	19.5	17.4	20.7	23.3	48.9	54.6	67.2
Teesside	48.9	51.1	35.2	20.6	96.7	13.4	12.8	15.3	19.8	34.3	35.6	50.1

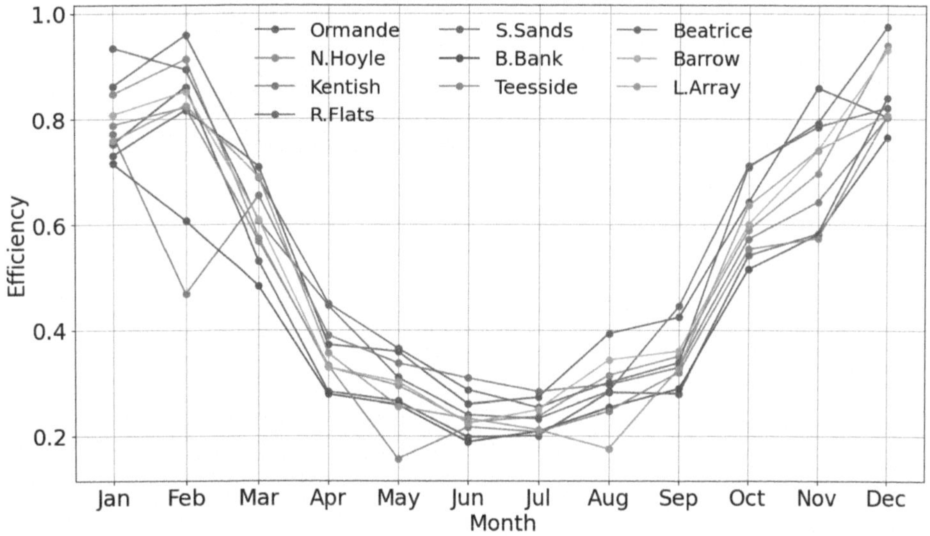

Fig. 5. Simulated power generation efficiency of 10 wind farms under field layout.

of the arrangement layout. At a rotated angle of 10°C, the North Hoyle Wind Farm (0.11%), the Burbo Bank Wind Farm (0.04%), the Teesside Wind Farm (0.025%), the Beatrice Wind Farm (0.02%), the Barrow Wind Farm (0.011%), and the London Array Wind Farm (0.06%) improved in the average efficiency. Efficiency improvements were observed at various wind farms when rotated at different angles. This validates the aim of our operation, where the rotated layout simulates the variability of the wind resource, such as wind direction. In uncertain climates, the field layout may not be optimal, whereas the rotated layout measures may be more applicable (Fig. 6).

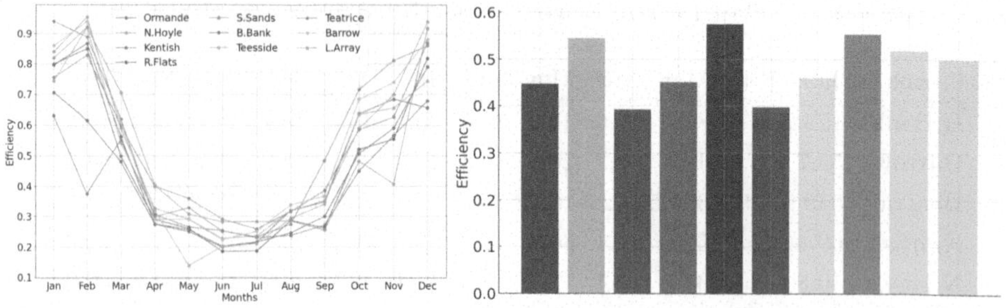

(a) The efficiency of 10 wind farms after rotating the layout by clockwise 1 degree. The left part is the efficiency line and the right part is the corresponding average efficiency histogram.

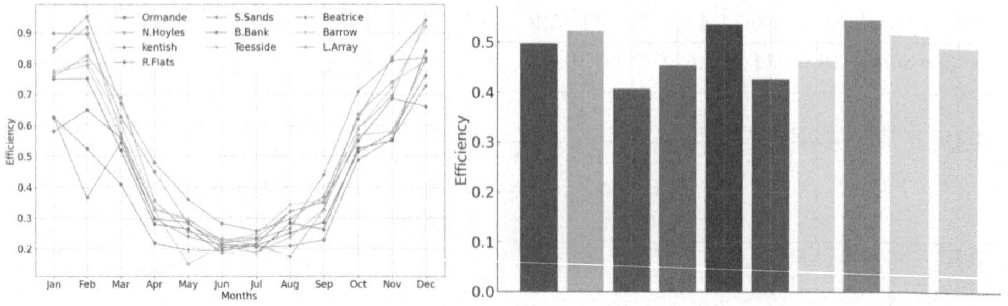

(b) The efficiency of 10 wind farms after rotating the layout by clockwise 5 degree. The left part is the efficiency line and the right part is the corresponding average efficiency histogram.

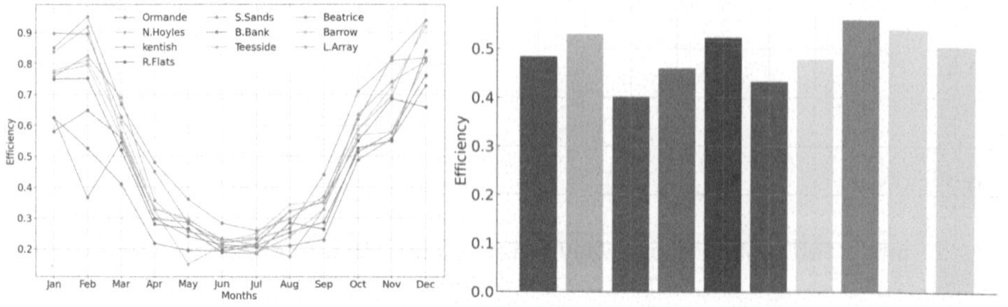

(c) The efficiency of 10 wind farms after rotating the layout by clockwise 10 degree. The left part is the efficiency line and the right part is the corresponding average efficiency histogram.

Fig. 6. Efficiency distribution of the wind farm after rotational layout: 1°C clockwise rotation (a), 5°C clockwise rotation (b) and 10°C clockwise rotation (c).

5 Conclusion

This study analyses the performance of 10 wind farms in the UK over a 15-year period, assessing power output and operational efficiency in different climatic wind patterns and with different layouts. The study uses a data-driven approach and the WindFLO framework to provide insights into wind farm design and simulation, as well as a tool for future wind energy prediction. The results indicate that a field wind farm layout may no longer be the optimal decision by changing climatic conditions and unpredictable future data. After slightly adjusting the field layout, the efficiency of the 6 wind farms was slightly improved. In future work, working with real data generated by the field wind farms instead of modeling data would be better.

References

1. Mohammadi, M., Gitizadeh, M., Roosta, A.: Dynamic stability improvement of a power system incorporating DFIG wind power plant using optimized control parameters of a SVC. In: IEEE International Power Engineering and Optimization Conference, pp. 416–421. IEEE, Melaka, Malaysia (2012)
2. Wang, Z., Tu, Y., Zhang, K., Han, Z., Cao, Y., Zhou, D.: An optimization framework for wind farm layout design using CFD-based kriging model. Ocean Eng. **293**, 116644 (2024)
3. Mittal, A., Sreenivas, K., Taylor, L.K., Hereth, L., Hilbert, C.B.: Blade-resolved simulations of a model wind turbine: effect of temporal convergence. Wind Energy **19**(10), 1761–1783 (2016)
4. Yang, S., Deng, X., Yang, K.: Machine-learning-based wind farm optimization through layout design and yaw control. Renew. Energy **224** (2024)
5. Yang, K., Deng, X.: Layout optimization for renovation of operational offshore wind farm based on machine learning wake model. J. Wind Eng. Ind. Aerodyn. **232**, 105280 (2023)
6. Bai, F., Ju, X., Wang, S., Zhou, W., Liu, F.: Wind farm layout optimization using adaptive evolutionary algorithm with Monte Carlo tree search reinforcement learning. Energy Convers. Manage. **252**, 115047 (2022)
7. Copernicus Climate Data Store Homepage. https://cds.climate.copernicus.eu/. Accessed 25 Feb 2024
8. Reddy, S.R.: Wind farm layout optimization (WindFLO): an advanced framework for fast wind farm analysis and optimization. Appl. Energy **269**, 115090 (2020)
9. Brauner, C.M., Gay, B., Mathieu, J.: Singular perturbations and boundary layer theory. In: Proceedings of the Conference Held at the École Centrale de Lyon, 8-10 December (1977)
10. WindFLO can be downloaded publicly. https://github.com/sohailrreddy/WindFLO. Accessed 19 Mar 2024
11. Sharma, P.K., Warudkar, V., Ahmed, S.: Effect of atmospheric stability on the wind resource extrapolating models for large capacity wind turbines: a comparative analysis of power law, log law, Deaves and Harris model. Energy Procedia **158**, 1235–1240 (2019)
12. Katic, I., Højstrup, J., Jensen, N.O.: A simple model for cluster efficiency. In: Proceedings of the European Wind Energy Association Conference and Exhibition (EWEC'86), vol. 1, pp. 407–410. Rome, Italy (1987)

A Novel Method for Wind Turbine 3D Reconstruction Using Quadrotor UAV

Yiming Xu[1], Hanming Sun[1], Dianhao Zhang[1(✉)], and Maoyang Chen[2]

[1] School of Electrical Engineering and Automation, Nantong University, Nantong, China
dzhang07@qub.ac.uk

[2] School of Zhangjian, Nantong University, Nantong, China

Abstract. Conducting a wind turbine inspection is crucial for wind-powered energy's overall performance and longevity. However, Wind turbine inspection is a tedious and dangerous process due to the extreme height. Additionally, the maintenance of these offshore or onshore wind turbines, especially in remote areas, remains a challenging task. To address this problem, accurate 3D object rendering approaches are adopted to enable real-time monitoring of the wind turbines performance and condition, which requires wind turbine 3D reconstruction before modeling. In this work, a vision-based autonomous wind turbine inspection framework using a quadrotor is designed based on neural radiance fields for image-based rendering. The simulation testing results and the real-world experiment on a wind turbine model have shown the effectiveness of the proposed approach.

Keywords: Wind Turbine Inspection · Neural Radiance Fields · Complex Scene Rendering · UAV Control

1 Introduction

Due to the clean and renewable properties of Wind energy, the number of wind farm installations all over the world has witnessed explosive growth in recent years. However, traditional maintenance and inspection require a well-trained group to climb the equipment, which requires high operation costs and is dangerous for the workers [1]. Therefore, tracking the turbine's real-time state and observing the small cracks using digital twins technology is a prerequisite. Towards this target, a quadrotor is applied to give a dense sampling of views and then use the samplings to reconstruct the wind turbine. The generated 3D model will then be imported into the digital twins environment.

There have been several studies using unmanned aerial vehicles (UAV) for inspection in the last decade [2,3]. The idea behind this is to navigate the UAV along a series of fixed inspection waypoints generated by vision-based path generation algorithms considering the length of the blades. However, there does not exist a work specifically handling the inspection trajectory for the UAV. In this

K. Li et al. (Eds.): LSMS/ICSEE 2024, CCIS 2218, pp. 14–24, 2025.
https://doi.org/10.1007/978-981-96-0232-2_2

work, we solve this problem by using object recognition technology. Then, the sampled images from different perspectives are imported into neural radiance fields (Nerf) to generate the 3D model.

Neural radiance field technology plays a key role in reproducing realistic scenes with complex geometry. Combined with the quadrotor-based visual servo, it can represent complex real-world geometry and appearance using the sampled images. Recently, [4] proposed a network called WTBNerf for wind turbine rendering and 3D reconstruction using conical truncated rays. To accurately obtain the visualization of the moving wind turbine blades, [5] presented a yet efficient algorithm to reconstruct the 3D model of blades combining image stitching and Nerf. Both works prove the applicability of employing Nerf for online wind turbine inspection.

Figure 1 plots the overall architecture of the inspection framework. At the beginning, the quadrotor takes an image of the turbine from the front side, and a YOLO-based approach is used to capture the parts containing the center actuator and blades. Then, we take the center of the bounding box as the center of the circular trajectory assuming the height of the turbine is known. Given the trajectory, we model the sampling task as a tracking problem, which enables the quadrotor to obtain the samples of the turbine from different perspectives around the center. Afterwards, the images are imported into Nerf to generate 3D rendering. The main contributions of the paper are as follows:

- We design a novel framework for wind turbine inspection which includes an automatical quadrotor trajectory generator and object 3D rendering modules.
- For the path generator, we employ an object detection-based method to generate the inspection trajectory based on the bounding box.
- We design an inspection scenario using the wind turbine model and prove the possibility of the proposed framework.

2 Related Works

2.1 Path Generation Based on Object Detection

Object detection-based path generation has been widely used in autonomous vehicles, which refers to the process of using object detection algorithms to identify obstacles, pedestrians, vehicles, and other relevant objects in the vehicles environment, and then generating a safe and efficient path for the vehicle to navigate through this environment. The idea behind is also suitable to the quadrotor UAVs and unmanned ground vehicles (UGVs), for example: [6] investigates a trajectory generation algorithm for dynamic target following and [7] presents an aerial online mapping system, which is updated by obstacle detection algorithm. In this work, a simplified path-generation method for quadrotors is presented considering the height of the wind turbine and the length of blades combining the object detection algorithm.

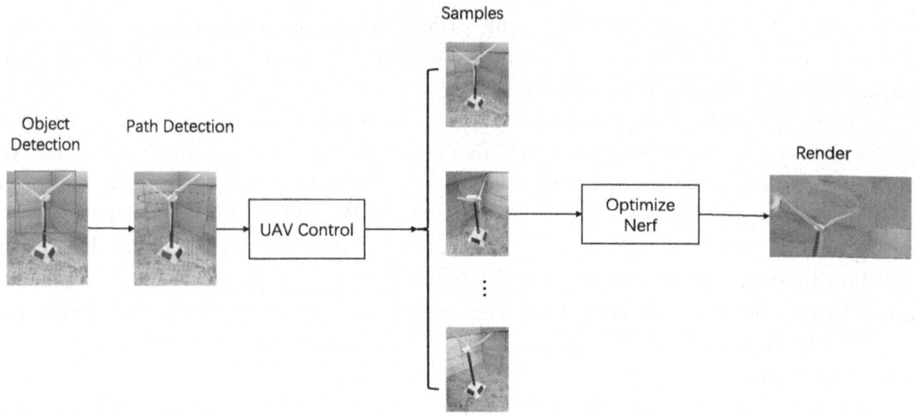

Fig. 1. The overall architecture of the wind turbine inspection architecture.

2.2 Neural Radiance Fields for Image-Based Rendering

Traditional Image stitching-based methods, such as Multi-View Stereo (MVS) [8], rely a lot on feature matching and image alignment. Additionally, they are sensitive to occlusion, lighting variations, and texture repetition, resulting in detail loss and reduced reconstruction accuracy [9]. To improve the quality of rendering, NeRF is able to generate higher-quality and more realistic views With a continuous 5D scene representation function, especially for cases with complex geometries and lighting variations. Given a dense sampling of views, photorealistic novel views can be reconstructed by simple light field sample interpolation techniques [9]. It is a sparser view sampling-based method, which predicts traditional geometry and appearance representations from images. There are two main categories for view synthesis, one is mesh-based representations and the other is volumetric representations. Mesh-based representations directly optimize mesh representations to reproduce a set of input images using gradient descent. However, due to local minima, poor conditioning of the loss landscape, and the initial topology problem, the mesh-based approach is unable to achieve satisfying performance during optimization [10].

Volumetric representations handle the task of high-quality photorealistic view synthesis based on a set of input RGB images. This method can represent complex shapes and materials, is well-suited for gradient-based optimization, and produces less visually distracting artifacts than mesh-based methods [10]. Although the existing volumetric techniques have achieved impressive results for novel view synthesis [11,12], the ability to scale to higher resolution imagery is limited by poor time and space complexity. In this work, we employ a volumetric-based method, which circumvents this problem by instead encoding a continuous volume within the parameters of a deep fully-connected neural network for better rendering performance.

3 UAV Platform

In this study, wind turbine inspection is performed using a DJI Phantom3 quadrator UAV equipped with a quad-core ARM CPU and a 256-core NVIDIA GPU, as shown in Fig. 1. Because of its outstanding speed and power economy, complicated control and vision algorithms may be executed in real time. GPS and IMU are ubiquitous sensing modules onboard, providing accurate location and attitude information via a data fusion technique. To perform vision-based inspection, the quadrotor carries a high-resolution camera placed on a gimbal as cargo. This camera is employed not only to generate control orders for the vision-based detecting system but also to record the blades. This camera will store high-quality photographs and videos on a microSD card (Fig. 2).

Fig. 2. The quadrotor used for inspection.

4 Autonomous Inspection Procedure

Autonomous inspection is designed with the following assumptions: (i) Wind turbine specifications are known, such as blade length and tower height. (ii) The approximate GPS coordinates of a wind turbine's hub position are known.

4.1 Object Detection

For the wind turbine detection module, we use WT-YOLO [13] to generate a bounding box that includes the hub and blades. The WT-YOLO approach outperforms previous methods for wind turbine detection and location on various land cover backgrounds, as well as multiresolution photos of the generated dataset. The dataset of wind turbine images consists of 353 tiles and 398 complete wind turbine instances collected from Google Earth.

4.2 Path Generation, UAV Tracking and Images Sampling

Once obtaining the bounding box of the wind turbine, we select the hyperplane containing the center of the bounding box. Then a circular trajectory is generated with the center of the hyperplane and the maximum radius (i.e. higher than the length of the blade). The generated 3D trajectory can be written as

$$((r + r_{\text{safe}})sin(\omega t) + x_h, (r + r_{\text{safe}})cos(\omega t) + y_h, z_h), \tag{1}$$

where (x_h, y_h, z_h) denotes the 3D position of the hub, the height of the hub z_h and the length of blades r are known. The angular velocity of the quadrotor is denoted as ω, r_{safe} is a safe distance to the end of the blade. The position state dynamics of the quadrotor are shown as follows

$$\ddot{x} = \frac{u_1}{m}(cos\phi sin\theta cos\psi + sin\phi sin\psi), \tag{2}$$

$$\ddot{y} = \frac{u_1}{m}(cos\phi sin\theta sin\psi - sin\phi sin\psi), \tag{3}$$

$$\ddot{z} = \frac{u_1}{m}cos\phi cos\theta - g, \tag{4}$$

where $u_1 = b(\omega_1^2+\omega_2^2+\omega_3^2+\omega_4^2)$. (ϕ, θ, ψ) is the angle rotated about the roll, yaw, and pitch axis, respectively. b is a constant value. (x, y, z) is the position of the quadrotor along 3D axes of the inertial frame. $(\omega_1, \omega_2, \omega_3, \omega_4)$ is the actuator's angular velocity of the quadrotor, g is the gravity value, m is the weight of the

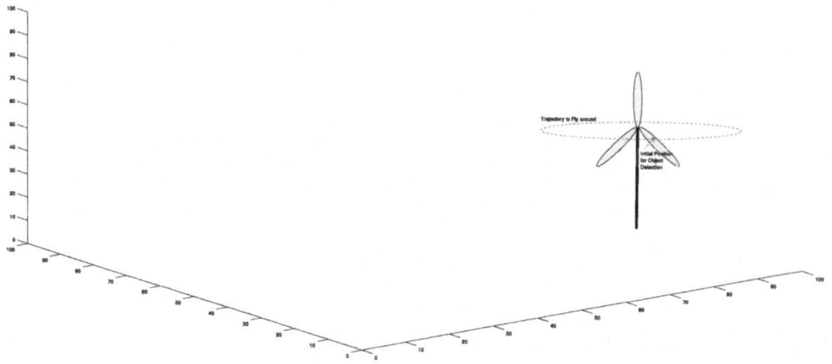

Fig. 3. The trajectory of the quadrotor UAV during wind turbine inspection. The red line denotes the generated trajectory for the quadrator to track. (Color figure online)

quadrotor. Afterward, a TVIBLF algorithm is adopted to track the generated trajectory [14], which enables the accuracy of the tracking and avoids the disturbance caused by wind. As shown in Fig. 7, the quadrotor performs the object detection task at the initial position. Once the position of the hub is determined, the quadrotor is required to move to the red dotted line and track the trajectory in speed ω (Fig. 3).

4.3 View Synthesis and Rendering

In this work, feature extraction, feature matching, sparse 3D reconstruction, and image pose estimation modules are implemented in COLMAP, which can efficiently process a large amount of image data and provide accurate pose estimation. Then the estimated pose is imported to the Nerf for the view synthesis and rendering, the detailed processing is shown in Fig. 4.

The continuous scene can be represented as a 5D vector-valued function whose input is a 3D location $\chi = (x, y, z)$ and 2D viewing direction (θ, ϕ). The output is an emitted color $c = (R, G, B)$ and volume density σ. In practice, we express direction as a 3D Cartesian unit vector C_d. We approximate this continuous 5D scene representation with an MLP network $F_\Theta : (\chi, C_d) \rightarrow (c, C_d)$ and optimize its weights Θ to map from each input 5D coordinate to its corresponding volume density and directional emitted color, the transformation is shown in Fig. 5. the MLP F_Θ takes the 3D coordinate with 8 fully-connected layers as

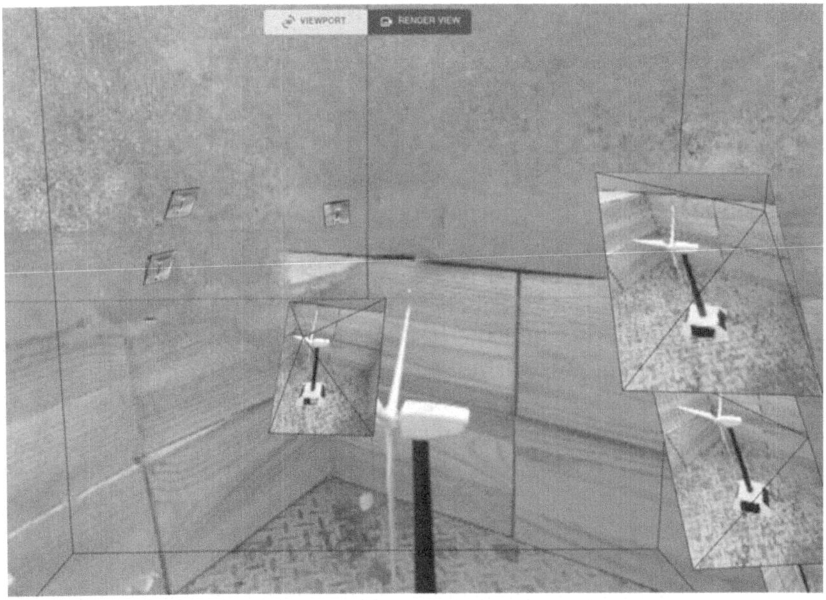

Fig. 4. 3D reconstruction in Nerf. It includes the photos of the turbine model during the flight and the generated 3D model.

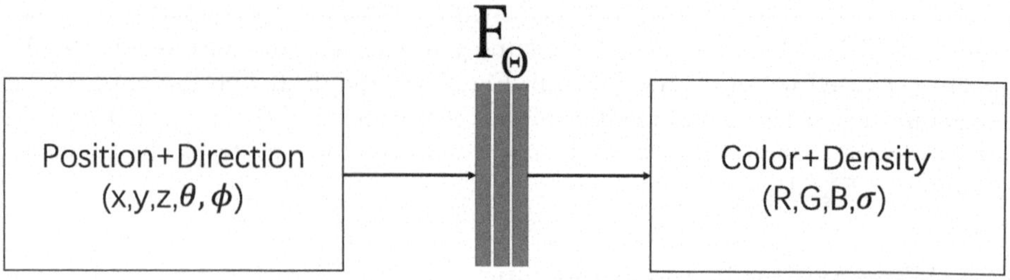

Fig. 5. The overall architecture of the wind turbine inspection architecture.

input (using ReLU activations and 256 channels per layer), and outputs σ and a 256-dimensional feature vector. The feature vector consists of the camera ray's viewing direction, then fed into an extra fully-connected layer with ReLU activation and 128 channels, generating a view-dependent RGB color. Figure 5 shows a 3D rendering of the NeRF reconstruction model, showing the structure and details of the wind turbine from multiple perspectives. From the reconstruction results, it can be seen that for images taken from different angles of the same large building, the NeRF model can reconstruct the 3D scene that is visually suppressed by high rollers, reducing the inconsistency of color and shadow caused by lighting changes. Coding the scale structure allows the model to recapture the overall contours and local details of large buildings with excellent performance. The model was able to accurately reconstruct large areas of smooth and detailed decorative areas, maintaining the geometric integrity and detail richness of the building.

4.4 Results

We show results on complex real-world scenes captured with a high-resolution camera all around. This dataset consists of 40 images (1279×1706 in pixel) and holds $\frac{1}{7}$ out of these for the test set. The output images from the Nerf are 1288×2048 pixels. According to Fig. 6, the Nerf method produces heavily smoothed geometry compared with the actual wind turbine model.

Figure 7 plots the actual and the desired trajectory for inspection using an IBLF-based path-tracking algorithm [14]. The generated trajectory proves the tracking performance of using IBLF-based algorithm. According to Fig. 8, the maximum error during path tracking of the x,y,z-axes are 8.7×10^{-4} m, 1.0×10^{-3} and 3.9×10^{-3}, respectively. It indicates that the path-tracking accuracy of the UAV is high.

Fig. 6. Nerf is specifically designed for this use case using the images from different perspectives. Our method is able to represent fine geometry, as shown in the turbines blades and the hub. Our method also correctly reconstructs partially occluded regions, such as the black pillar behind the blades.

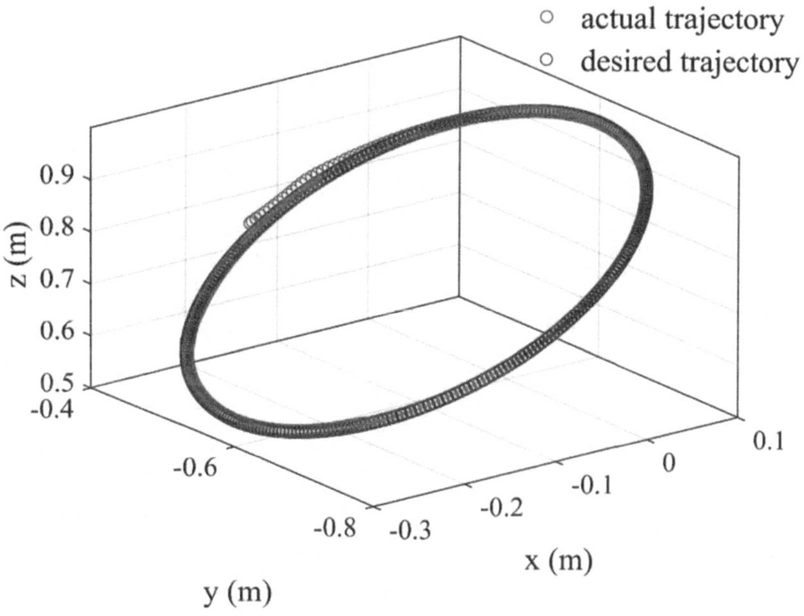

Fig. 7. The comparison between the real inspection trajectory and the desired trajectory. The Blue line indicates the desired trajectory and the red line indicates the real trajectory during flight. (Color figure online)

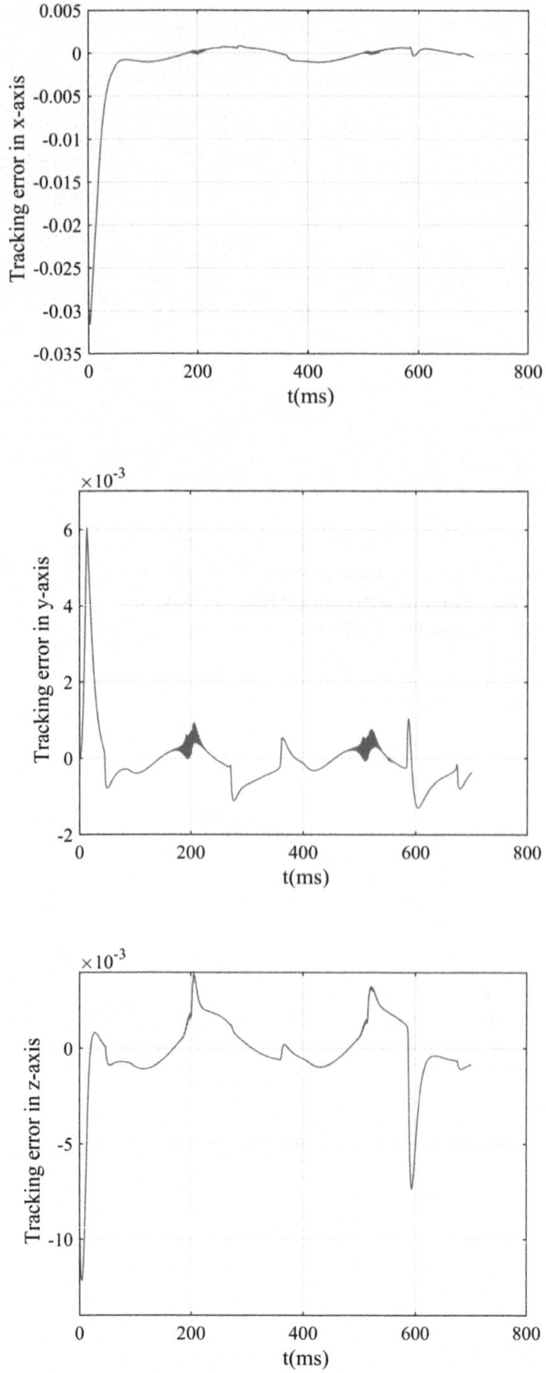

Fig. 8. The tracking error in x,y,z-axes.

5 Conclusion

Our work directly addresses the maintenance problem in wind turbine inspection, which presents the prior work for 3D object reconstruction in digital twins environments. The results and the attached video prove that representing scenes as 5D neural radiance fields produces fine renderings. Apart from this, the proposed framework is the first work to consider object detection in the quadrotor path generation, which enables automatically scene sampling. We believe that this framework makes progress towards online low-cost wind turbine inspection and 3D modeling of complex scenes.

References

1. Gu, W., Hu, D., Cheng, L., Cao, Y., Rizzo, A., Valavanis, K.P.: Autonomous wind turbine inspection using a quadrotor. In: 2020 International Conference on Unmanned Aircraft Systems (ICUAS), pp. 709–715. IEEE (2020)
2. Stokkeland, M., Klausen, K., Johansen, T.A.: Autonomous visual navigation of unmanned aerial vehicle for wind turbine inspection. In: 2015 International Conference on Unmanned Aircraft Systems (ICUAS), pp. 998–1007 (2015)
3. Schäfer, B.E., Picchi, D., Engelhardt, T., Abel, D.: Multicopter unmanned aerial vehicle for automated inspection of wind turbines. In: 2016 24th Mediterranean Conference on Control and Automation (MED), pp. 244–249. IEEE (2016)
4. Yang, H., Tang, L., Ma, H., Deng, R., Wang, K., Zhang H.: Wtbnerf: wind turbine blade 3d reconstruction by neural radiance fields. In: Zhang, H., Ji, Y., Liu, T., Sun, X., Ball, A.D. (eds.) Proceedings of TEPEN 2022, pp. 675–687. Springer, Cham (2023). https://doi.org/10.1007/978-3-031-26193-0_60
5. Yang, C., Liu, X., Zhou, H., Ke, Y., See, J.: Towards accurate image stitching for drone-based wind turbine blade inspection. Renew. Energy **203**, 267–279 (2023)
6. Jeon, B.F., Shim, D., Kim, H.J.: Detection-aware trajectory generation for a drone cinematographer. In: 2020 IEEE/RSJ International Conference on Intelligent Robots and Systems (IROS), pp. 1450–1457 (2020)
7. Lee, J., Lim, J., Pyo, S., Lee, J.: Aerial online mapping on-board system by real-time object detection for UGV path generation in unstructured outdoor environments. J. Field Rob. **40**(7), 1754–1765 (2023)
8. Smith, M.W., Carrivick, J.L., Quincey, D.J.: Structure from motion photogrammetry in physical geography. Prog. Phys. Geograhy-Earth Environ. **40**(2, SI), 247–275 (2016)
9. Shi, L., Hassanieh, H., Davis, A., Katabi, D., Durand, F.: Light field reconstruction using sparsity in the continuous fourier domain. ACM Trans. Graph. (TOG) **34**(1), 1–13 (2014)
10. Mildenhall, B., Srinivasan, P.P., Tancik, M., Barron, J.T., Ramamoorthi, R., Ng, R.: NeRF: representing scenes as neural radiance fields for view synthesis. In: Vedaldi, A., Bischof, H., Brox, T., Frahm, J.-M. (eds.) ECCV 2020. LNCS, vol. 12346, pp. 405–421. Springer, Cham (2020). https://doi.org/10.1007/978-3-030-58452-8_24
11. Zhou, T., Tucker, R., Flynn, J., Fyffe, G., Snavely, N.: Stereo magnification: learning view synthesis using multiplane images. arXiv preprint arXiv:1805.09817 (2018)

12. Tulsiani, S., Zhou, T., Efros, A.A., Malik, J.: Multi-view supervision for single-view reconstruction via differentiable ray consistency. In: Proceedings of the IEEE Conference on Computer Vision and Pattern Recognition, pp. 2626–2634 (2017)
13. Zhai, Y., Chen, X., Cao, X., Cui, X.: Identifying wind turbines from multiresolution and multibackground remote sensing imagery. Int. J. Appl. Earth Obs. Geoinf. **126**, 103613 (2024)
14. Zhang, D., Van, M., Mcllvanna, S., Sun, Y., McLoone, S.: Adaptive safety-critical control with uncertainty estimation for human-robot collaboration. IEEE Trans. Autom. Sci. Eng. 1–14 (2023)

A Method for Evaluating the Operating Health of Wind Turbines Under Wake Effect

Huixue Shao, Xiaolu Lai(✉), Hao Fu, Lei Huang, and Bitao Xiao

Guodian Nanjing Automation Co., Ltd., Nanjing 210032, China
532042116@qq.com

Abstract. This method uses the existing wind turbine operation data, environmental data and meteorological data to realize the prediction and quantitative evaluation of wake flow through non-intrusive mode. In this method, the characteristic factors affecting the wake are extracted and the wake prediction classification tree is constructed. The feature classification tree is used to determine whether the wind turbine is affected by the wake. This method uses the classification tree of benchmarking wind turbine to determine the latest benchmarking wind turbine and obtain health data. The wind speed under the influence of the wake is calculated according to the Gaussian model, and the wind speed loss rate is calculated to get the loss factor. According to the loss factor, the operation health degree of wind turbine under the influence of wake is quantified, which provides the decision basis for the yaw strategy of wind turbine.

Keywords: Non-invasive · Wake Prediction Classification Tree · Classification Tree for Benchmarking Wind Turbine · Gaussian Model · Loss Factor

1 Introduction

The power generation efficiency of wind turbine directly affects the amount of power generation and the level of economic benefits. When the efficiency of wind turbines is reduced, people often think that the performance of wind turbines is degraded, or sub-components are faulty. Therefore, on-site personnel generally adopt maintenance methods to find hidden trouble, and take corresponding measures to reduce the occurrence of faults and improve the operation efficiency of wind turbines.

However, because the wind turbine is affected by the topological arrangement of the wind farm, terrain, weather and other factors, when a wind turbine is operating in the wake area of the upstream wind turbine, its power generation will be greatly affected.

The wake is a kind of air flow structure, which has the characteristic of reducing the wind speed. The blades of a wind turbine absorb dynamic energy from the incoming air flow, and the wake flow is increased because the turbine blocks the air flow. Wind turbines in wind farms can affect the performance of other wind turbines downstream through the wake of wind turbines downwind. If a turbine is not completely covered by the free flow, but is located in the wake path, the reduction of wind speed will lead to the reduction of power generation of the turbine. On the other hand, the increased shear wind in the

K. Li et al. (Eds.): LSMS/ICSEE 2024, CCIS 2218, pp. 25–36, 2025.
https://doi.org/10.1007/978-981-96-0232-2_3

wake will cause the dynamic load of the wind turbine located downwind to increase. As shown in Fig. 1, the wake effect can cause serious energy loss and has a significant impact on the economic efficiency of wind field. Affected by the wake effect, the energy loss of the whole wind field can reach 10%–20%, and the power loss of a single turbine can reach 30%–40% [1]. In a Danish onshore wind farm, measurements have found that the wake of the first row of wind turbines can reduce the output of downstream wind turbines by 60%. At the same time, due to the increase of wake turbulence intensity, the fatigue load of wind turbines immersed in the wake will increase by 5%–15%.

Therefore, when the power generation efficiency of the wind turbine is reduced, it is difficult to distinguish between the wake effect and the performance reduction caused by the aging of the wind turbine itself, and accurate maintenance cannot be achieved. If the wind turbines with reduced power generation efficiency are uniformly repaired. First of all, for wind turbines in good condition, the maintenance function and dismantling test are unnecessary behaviors. Dismantling test may not only reduce the reliability of wind turbines, but also reduce the income of wind farms. Secondly, these maintenance work requires wind farm personnel to cooperate with cranes and other equipment for detailed inspection, consuming a lot of human and material resources, and the need to climb the wind turbine engine room, increasing the risk factor of personnel, once an accident causes greater losses.

Fig. 1. Wake effect of wind turbine.

Therefore, how to predict the wake effect of wind turbines and evaluate the operating health of wind turbines has become an urgent concern.

There are three main methods to study the wake effect: experimental measurement [1], computational fluid dynamics and wake theoretical model.

1.1 Experimental Measurement Method

The data obtained by the experimental measurement method is real and reliable, but the research period is long and consumes a lot of manpower and material resources [2]. At present, wind tunnel experiments are often used in research, which is costly

Fig. 2. Wake test of wind turbine.

and unrealistic under the current productivity conditions. Most of the existing research results are based on enumeration or empirical methods to determine the yaw Angle of wind turbines to simulate wake generation, and the wake monitoring methods have not been fully studied (Fig. 2).

1.2 Calculation of Wake Model

The generation of atmospheric boundary layer inflow and the calculation of the aerodynamic performance of the fan are based on the atmospheric boundary layer solver SOWFA developed by the Renewable Energy Laboratory of the United States Department of Energy. The accuracy of the software has been verified in most studies. The specific governing equation is as follows:

$$\frac{\partial \overline{u_i}}{\partial x_i} = 0 \tag{1}$$

$$\frac{\partial \overline{u_i}}{\partial t} + \frac{\partial}{\partial x_j}(\overline{u_j u_i}) = -\frac{1}{\rho_0}\frac{\partial}{\partial x_i}\overline{p_0}(x, y) - \frac{\partial \widetilde{p}}{\partial x_i} - 2\varepsilon_{i3k}\Omega_3\overline{u_k} - \frac{\partial}{\partial x_j}\left(\tau_{ij}^D\right) + g\left(\frac{\overline{\theta} - \theta_0}{\theta_0}\right)\delta_{i3} + \frac{1}{\rho_0}f_i^T \tag{2}$$

In the formula, the first term on the right side of the equal sign is the background pressure gradient term, which acts as the driving force to promote the formation of wind; The second term is the modified pressure gradient term. The third term is the Coriolis force term, ε_{i3k} is the alternating tensor, and Ω_3 is the Earth rotation vector; The fourth term is the fluid stress tensor term. The fifth item is buoyancy caused by space temperature difference; The sixth item is the force of the fan blade on the flow field, which can be modeled by the actuating line model. In the fifth term, $\overline{\theta}$ represents the potential temperature of the solvable scale, and θ_0 is the reference value of the potential temperature, which is 300 K. In addition, a potential temperature transport equation independent of the N-S equation needs to be solved to obtain the potential temperature field:

$$\frac{\partial \overline{\theta}}{\partial t} + \frac{\partial}{\partial x_j}(u_j \overline{\theta}) = -\frac{\partial}{\partial x_j}(q_j) \tag{3}$$

where q_j is the turbulent heat flux, modeled according to the following expression:

$$q_j = -\frac{\nu^{SGS}}{pr_t}\frac{\partial \widetilde{\theta}}{\partial x_j} \tag{4}$$

where pr_t is the turbulent Prandtl number and its value is 1/3.

The computational fluid dynamics method simulates the wake flow phenomenon by numerical modeling of wind turbines and solving the fluid control equation. It has high accuracy and comprehensive data [3–8], but it requires high computational resources. When used for wind turbine cluster prediction, the lengthy calculation time of computational fluid dynamics method is not acceptable for engineering applications. The wake offset between wind turbines is closely related to the thrust coefficient, wake expansion coefficient and the flow direction spacing of wind turbines. However, most of the existing studies are carried out under specific working conditions, and the generation of turbulent boundary layer is relatively fixed. Due to the high cost, no relevant projects have conducted the above studies completely.

1.3 Analytic Wake Model

The analytical wake model has the characteristics of simple form, fast calculation speed and can meet the requirements of certain engineering accuracy. Therefore, using the analytical wake model to calculate the output power of wind turbines can greatly improve the calculation efficiency.

There are three main analytical wake models:

(1) The Park wake model is derived based on conservation of mass. It assumes that the velocity loss in the wake region of the wind turbine meets the "top-cap" distribution, ignores the near wake region where the pressure returns to atmospheric pressure, assumes that the wake radius behind the disk is the wake radius where the pressure returns to atmospheric pressure, and the wake expands linearly. The wake velocity expression is:

$$\mathrm{U} = \mathrm{U}_{\infty}[1 - \left(1 - \sqrt{1 - \mathrm{C_T}}\right) * (\frac{R}{R + kx})^2] \tag{5}$$

where U_{∞} is the average inlet wind speed at the height of the wheel hub of the wind turbine. $\mathrm{C_T}$ is the thrust coefficient of wind turbine; R is the radius of wind turbine disk; k is the coefficient of wake expansion, which ranges from 0.04–0.05 for offshore wind field and 0.075 for onshore wind field. x is the downstream flow distance of the wind turbine disk.

(2) The Frandsen wake model is derived based on conservation of momentum. It is also assumed that the velocity loss in the wake region of wind turbines meets the "top hat" distribution, and the wake velocity expression is:

$$U = U\infty(\frac{1}{2} + \frac{1}{2}\sqrt{1 - 2 * \frac{A_0}{A}C_T}) \tag{6}$$

where, A_0 is the disk area of the wind turbine; A is the wake cross-sectional area at the downstream location.

(3) Bastankhah wake model is derived based on conservation of momentum and conservation of mass. It is a two-dimensional wake model assuming that the velocity distribution in the wake region meets the Gaussian distribution. The wake velocity loss expression is as follows:

$$\frac{\Delta U}{U\infty} = (1-\sqrt{1-\frac{C_T}{8(k^*x/D_0+\varepsilon)^2}}) * \exp\{-\frac{1}{2(k^*x/D_0+\varepsilon)^2} * [(\frac{z-z_h}{D_0})^2 + (\frac{y}{D_0})^2]\} \tag{7}$$

where, k^* is the change rate of the standard deviation of the Gaussian distribution with the downstream distance x; ε is the relative value of the standard deviation of the Gaussian distribution at the disk position of the wind turbine, which is obtained by taking the same mass flow at the disk as Frandsen's wake model. z is the vertical coordinate of any position behind the disk of the wind turbine, and z_h is the hub height of the wind turbine. y is the horizontal coordinate of any position behind the disk of the wind turbine.

Jensen model, AV wake model and Lissaman model can be used for full-field wake calculation. The non-viscous near field wake model and the simplified wake model are suitable for near field wake and far field wake respectively. At present, the theoretical model of fan wake is studied.

The problem is that, in addition to Jensen model, there is no widely accepted wind turbine wake model with practical theoretical basis. Although Jensen model has theoretical support, it contains a lot of idealized assumptions, and the calculation accuracy is low. Therefore, it is urgent to put forward a wind turbine wake theoretical model which has both practical physical significance and high precision and can be applied universally.

The existing wake theoretical model is a mathematical expression of the regional velocity distribution of wind turbine wake based on idealized hypothesis or experimental data fitting, which often can not meet the requirements of field engineering accuracy. Due to the limitations of operation and maintenance and site, it is difficult to deploy sensors in wind farms and wind turbines that have been put into operation to obtain model parameters and analyze the wake state of wind turbines.

This paper presents a method for evaluating the health of wind turbines under wake effect, which can reduce the calculation amount and improve the prediction accuracy of wake effect.

2 A Method for Evaluating the Operating Health of Wind Turbines Under Wake Effect

This paper provides a method to judge the health of wind turbines under the wake effect, including:

1. Obtain wind turbine operation data, environmental data and meteorological data, input the pre-built wake effect prediction classification tree for classification, and determine whether wind turbines are affected by wake;

2. If it is determined that the wind turbine is affected by the wake, the wake center location and wind speed loss rate are calculated according to the pre-constructed Gaussian model, and the wind speed loss rate is taken as the loss factor under the wake effect;
3. If it is determined that the wind turbine is not affected by wake flow, the data set of all wind turbines not affected by wake flow is established, and the data set is input into the pre-constructed classification tree model of the benchmark wind turbine to determine the latest benchmark wind turbine, and the wind speed - power curve of the benchmark wind turbine is taken as health data. The wind speed loss rate which is not affected by the wake effect is calculated as the loss factor.
4. The loss factor under the wake effect is compared with the loss factor of the wind turbine that is not affected by the wake effect, and the influence of the wake effect on the operation health of the wind turbine is judged.

2.1 Build Wake Effect Prediction Classification Tree

Decision tree is a supervised learning algorithm in machine learning methods, which is a tree structure that classifies samples based on features and can be used for classification and regression. The main steps of building a decision tree include:

Starting from the root node, the calculation is carried out according to each feature of the training data, and the training data is assigned to its child nodes according to the uncertainty of each feature. Along the branch, it may reach the leaf node or reach another internal node, and then the remaining features are recursively executed until a leaf node is reached. When we reach the leaf node, we get the final classification result. Therefore, for wake related data, the steps to establish a decision tree are as follows:

Obtain and discretize wind turbine operation historical data, environmental historical data and meteorological historical data;

In this method, the distribution probability characteristics of all power dispersions are calculated, so as to achieve accurate classification and identification of power dispersions. This method studies the probability characteristics of power scatter distribution, and uses probabilistic statistical methods to find out the reasonable power scatter distribution of a certain type of wind turbine, and precisely classify and label all power scatter points. Power scatter type identification is the core of wind turbine performance monitoring and the key to ensure accuracy. The power generation performance index and power scatter distribution shape index of wind turbines are calculated according to the marked scatter points (Fig. 3).

ID3 algorithm is used to determine each feature of the classification tree, and the initial wake effect prediction classification tree is constructed. When classifying features, discrete classification methods with different quantization ranges are used to get the group of wake prediction classification trees.

Suppose a random variable X takes the value $X = \{x_1, x_2, \ldots, x_n\}$, and the probabilities of each are $\{p_1, p_2, \ldots, p_n\}$, then the formula for the information entropy of X is:

$$\mathrm{Ent(X)} = -\sum_{i=1}^{n} p_i \log_2 p_i \tag{8}$$

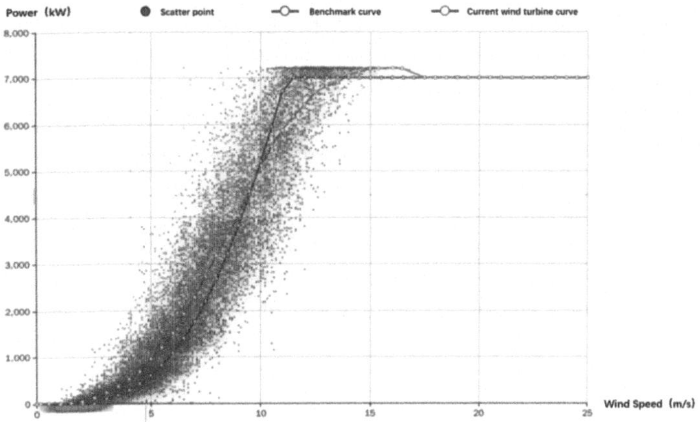

Fig. 3. Schematic diagram of discrete wind turbine operation data.

Ent() represents the information entropy function, X_i represents the value of the i th random variable X, p_i represents the probability of taking the value of the i th random variable X, i = 1, 2, …, n, n is the number of random variable X values;

The entropy of the whole data set is called the original entropy, and the entropy of data set *D* divided according to a certain feature *a* is the conditional entropy. Information gain = original entropy - conditional entropy. The specific method of information gain division is to calculate the information gain corresponding to each type of feature *a*, and then select the feature with the smallest information gain for division:

$$Gain(D, a) = Ent(D) - \sum_{v=1}^{m} p_v Ent(D^v) \tag{9}$$

Here, Gain(D,a) is the information gain of data set D after partition according to a certain feature a, v is a classification of feature a, m is the number of classification of feature a, p_v is the proportion of v classification in the total number of feature a, and D^v is the data set after partition according to v classification of feature a.

The wind direction at the present position in front of the downstream wind turbine is detected by LiDAR wind meter in real time, and the local wind direction at the location of the downstream wind turbine is detected by mechanical or ultrasonic wind meter. Calculate the deviation between the two wind direction detection results; With the local wind direction of the downstream wind turbine itself as the horizontal coordinate and the wind direction measurement deviation as the vertical coordinate, the statistical diagram of wind direction measurement deviation is drawn. If the wake of upstream wind turbines has a significant impact on local wind turbines in a certain wind direction range, the deviation in the statistical diagram will be presented as a positive and negative alternating waveform, and the amplitude and start-stop interval of the waveform will be obtained, and then the influence degree and sector interval of downstream wind turbines by the wake of upstream wind turbines will be determined. This is used to determine whether the wind turbine is affected by the wake of the upstream wind turbine.

The random tree algorithm is to use the training data to randomly calculate many decision trees and form a tree. Then use this tree cluster to predict the unknown data and select the classification tree with the highest accuracy. When classifying a feature, different classification methods are used. This method uses discrete classification methods with different quantization ranges to obtain a wake effect prediction classification tree. The classification tree whose classification accuracy is greater than 90% is selected as the final wake prediction classification tree.

The discrete wind turbine historical operation data, environmental historical data and meteorological historical data are input into the wake effect prediction classification tree for classification, and the classification tree whose classification accuracy rate is higher than the set value is selected as the final wake effect prediction classification tree.

The whole process is shown in Fig. 4:

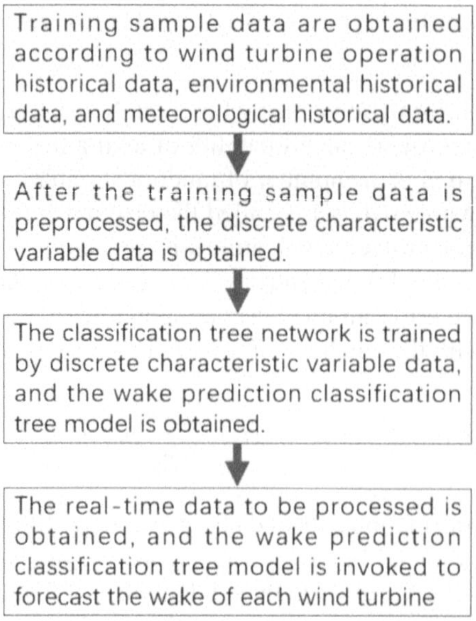

Fig. 4. Construction diagram of wake prediction classification tree.

2.2 Build a Classification Tree for Benchmarking Wind Turbines

Wind turbine operation data independent of wake flow is obtained and discretized. ID3 algorithm and random tree algorithm are used to classify the characteristics of wind turbine operation data. Taking the rated wind speed power curve as the reference basis, the classification tree whose classification results of the wind speed power curve and the rated wind speed power curve fitting rate of more than 90% in the classification tree group is used as the final benchmark wind turbine classification tree. The whole process is shown in Fig. 5:

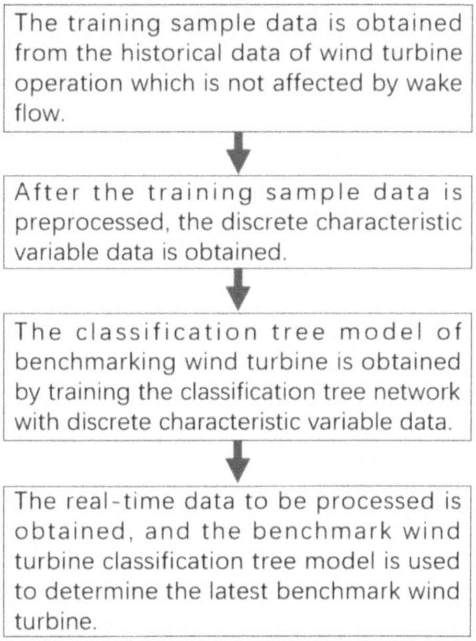

Fig. 5. Construction diagram of the benchmarking wind turbine classification tree.

2.3 Calculation of Loss Factors for Wind Turbines not Affected by Wake Effect

The operation data of all wind turbines not affected by wake were input into the classification tree of benchmark wind turbines to determine the latest benchmark wind turbines, and the wind speed of benchmark wind turbines was taken as health data, and the wind speed loss rate of wind turbines not affected by wake effects was calculated as a loss factor.

$$V_D' = \frac{u_0 - u_1}{u_0} * 100\% \tag{10}$$

Here, V_D' is the wind speed loss rate of the wind turbine unaffected by the wake effect, u_0 is the wind speed of the benchmark wind turbine, u_1 is the wind speed of the wind turbine unaffected by the wake effect (Fig. 6).

2.4 Calculation of Loss Factors of Wind Turbines Affected by Wake Effect

If the wind turbine is determined to be affected by the wake, the wake center location and wind speed are calculated according to the Gaussian model. The relationship between the radial wind speed measured by Doppler LiDAR and the real wind speed is:

$$u_{LOS}(y, r) = u_{actual}\cos(\theta - \varphi) = u_{actual}\left(\frac{\sqrt{r^2 - y^2}}{r}\cos\varphi + \frac{y}{r}\sin\varphi\right) \tag{11}$$

where r is the distance between the lidar and the measuring point, $u_{LOS}(y,r)$ is the radial wind speed, u_{actual} is the real wind speed, φ is the wind direction, θ is the azimuth of the

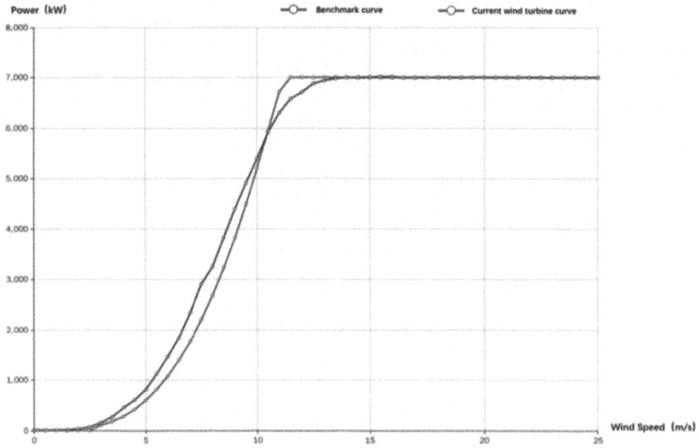

Fig. 6. Comparison of power and wind speed curves between benchmarking wind turbine and wind turbine not affected by wake effect.

lidar, y is the vertical axis in the rectangular coordinate system. Gaussian model is used to calculate the wake center position:

$$u_{LOS}(y, r) = \{u_{actual} - a * \exp[\frac{-(y - y_c)^2}{2s_w^2}]\}(\frac{\sqrt{r^2 - y^2}}{r}\cos\varphi + \frac{y}{r}\sin\varphi) \quad (12)$$

In the formula, y_c corresponds to the wake center, and S_w is the parameter controlling the wake width. a is the Gaussian fitting parameter, which is obtained by fitting the field engineering experiment results.

The wake center position of the wind turbine is calculated, and the wake wind speed at the center position is obtained, and the wind speed loss rate is:

$$V_D = \frac{u - u_{wake}}{u} * 100\% \quad (13)$$

Here, V_{D} is the loss rate of wind speed, u is the ambient wind speed, u_{wake} is the wake wind speed, and V_D is the loss factor under the wake effect (Fig. 7).

2.5 Wind Turbine Health Evaluation

By comparing the loss factors of wind turbine affected by wake with those of wind turbine not affected by wake, the health of wind turbine operation under wake effect can be evaluated.

This method can well reflect the influence of wake effect on wind turbine operation. According to the actual operation data, the invention deals with the wind turbine group with strong wake correlation and the wind turbine group with weak correlation respectively, and has certain engineering application value. The health evaluation process of wind turbines is shown in Fig. 8.

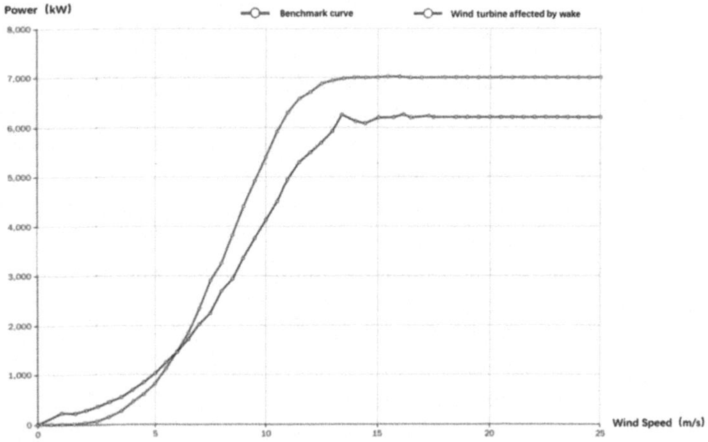

Fig. 7. Comparison of power and wind speed curves between benchmarking wind turbines and wind turbines affected by wake effect.

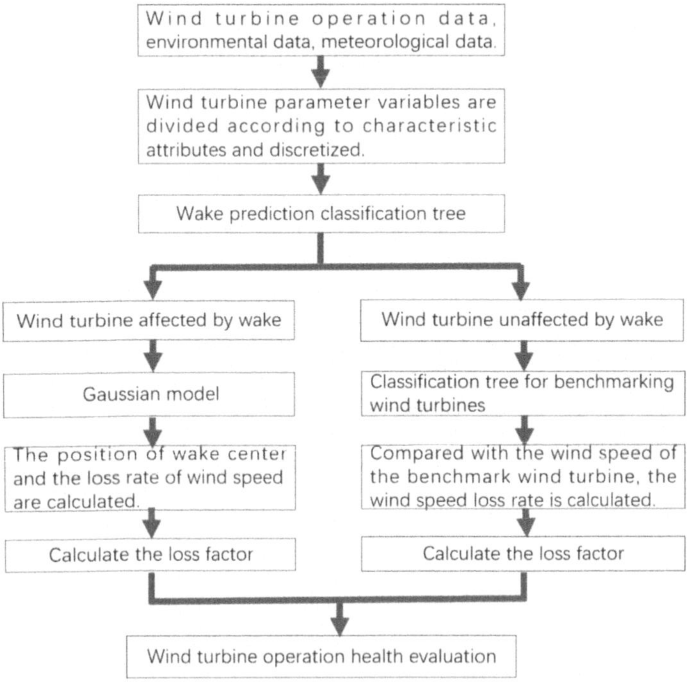

Fig. 8. Health evaluation diagram of wind turbine operation.

3 Conclusion

This paper introduces a method for evaluating the operating health of wind turbines under the wake effect. The method is simple, effective and has clear physical meaning by predicting the wake effect through the existing operating data, environmental data and meteorological data of wind turbines.

First of all, the method fully considers the factors affecting the wake, such as wind speed, wind direction, wind turbine size, blade sweep area, wind field topology, topography, weather, etc., and takes into account the data information reflected by each variable and the interaction between variables, so as to improve the wake prediction accuracy.

Secondly, for the determination of benchmarking wind turbines, it takes into account that during the operation of the wind turbines, the original benchmarking wind turbines may be aging and their operating status deteriorates as time accumulates. The classification tree of benchmarking wind turbines is used to dynamically determine the latest benchmarking wind turbines.

Third, the implementation of this method does not need to add additional sensor arrays and other equipment, is not subject to operation and maintenance and site restrictions, and is suitable for use in wind farms that have been put into operation, and has certain engineering application value.

References

1. Fleming, P., King, J., Dykes, K.: Initial results from a field campaign of wake steering applied at a commercial wind fam. Wind Energy Sci. **4**(2), 273–285 (2019)
2. Dijk, M.T., Wingerden, J.W., Ashuri, T.: Wind farm multi-objective wake redirection for optimizing power production and loads. Energy **5**(2), 561–569 (2017)
3. Sun, H., Wu, S.W., Wang, C.: Impact analysis of wake effect on wind farm output. J. North China Electr. Power Univ. **21**(4), 45–46 (2015)
4. Gu, B., Hu, H., Liu, Y.Q., et al.: Study of wind farm optimal control technology considering wake effect. Acta Energ. Sol. Sin. **39**(2), 359–368 (2018)
5. Xie, S.B., Archer, C.: Self-similarity and turbulence characteristics of wind turbine wakes via large-eddy simulation. Wind Energy **18**(10), 1815–1838 (2014)
6. Cao, X.X., Yang, H.X., Lu, L.: Optimization of wind turbine layout position in a wind farm using a newly developed two-dimensional wake model. Appl. Energy **174**(1), 192–200 (2016)
7. Tian, L.L., Zhu, W.J., Shen, W.Z., et al.: Development and validation of a new two-dimensional wake model for wind turbine wakes. J. Wind Eng. Ind. Aerodyn. **3**(7), 90–99 (2015)
8. Wu, Y., Porte, F.: Atmospheric turbulence effects on wind-turbine wakes. Energies **5**(12), 5340–5362 (2012)

Toward Next-Generation Buildings: A Review of Renewable Energy Integration and Intelligent Building Energy Management Technologies

Zhengrun Zhao and Kang Li(✉)

University of Leeds, Leeds LS2 9JT, UK
k.li1@leeds.ac.uk

Abstract. Buildings have significant potential to reduce energy use and increase the share of renewable energy, and the building sector is currently experiencing a significant transformation, driven by the urgent need for decarbonization and the increased use of renewable energy sources. Countries worldwide are pursuing policies to move towards the next generation of more energy-efficient and low-carbon green buildings. In order to achieve net-zero energy buildings or plus-zero energy buildings, renewable energy integration is considered to be an important approach. However, the complexity of building energy systems, the variability of renewable energy production, as well as the increasing prevalence of electrified equipment introduce significant challenges to energy management in buildings. This paper presents a brief review of the new trends in building energy developments and various technological considerations that should be taken into account in the decision-making process for the development of next-generation buildings. The paper particularly focuses on intelligent building energy technologies that enhance energy efficiency, and some future research directions in building energy studies are also discussed.

Keywords: Building energy · Energy management · Renewable energy · Intelligent algorithms

1 Introduction

Rapid growth in energy demand, driven by the needs of households and industries, is placing significant stress on the global environment. The increasing prevalence of electrified products, the pursuit of enhancing living standards, and the swift expansion of large-scale, energy-intensive computing data centers over the past few years have made our energy demand reach unprecedented levels. However, the reliance on traditional energy sources has led to increasing energy costs and substantial greenhouse gas emissions [1]. A number of countries and regions are actively developing strategies to achieve the ambitious goal

K. Li et al. (Eds.): LSMS/ICSEE 2024, CCIS 2218, pp. 37–51, 2025.
https://doi.org/10.1007/978-981-96-0232-2_4

of net-zero emissions [2] in response to these challenges. The building sector, as the predominant energy consumer, its operational energy usage accounts for approximately 30% of the final energy consumption globally. This share further increases to about 34% when considering the energy used in the production of building materials used in construction [3]. Furthermore, nearly one-third of global CO_2 emissions are attributed to the building sector, including indirect emission which is defined as the CO_2 produced by the power and heat conversion and processing sector to satisfy buildings' electricity and heat demand [4]. The detailed percentage of global building energy consumption and CO_2 emissions are shown in Fig. 1. Promoting highly performing and efficient buildings is believed to be one of the promising solutions to the energy and environmental challenges, and has been adopted by many countries and regions. For example, the EU is promoting high-performance buildings, intending to achieve a decarbonized building stock by 2050 [5]. The United States also aims to realize a net-zero-emissions economy by the same year [6,7], with the decarbonization of buildings forming a crucial part of this strategy [8]. However, according to the United Nations Environment Programme [9], the trajectory of building decarbonization efforts is progressing on a less optimistic path: the current pace of advancement towards these targets is proven to be inadequate. Consequently, it is essential to frame and implement more effective strategies that can expedite the transition toward the net-zero objective.

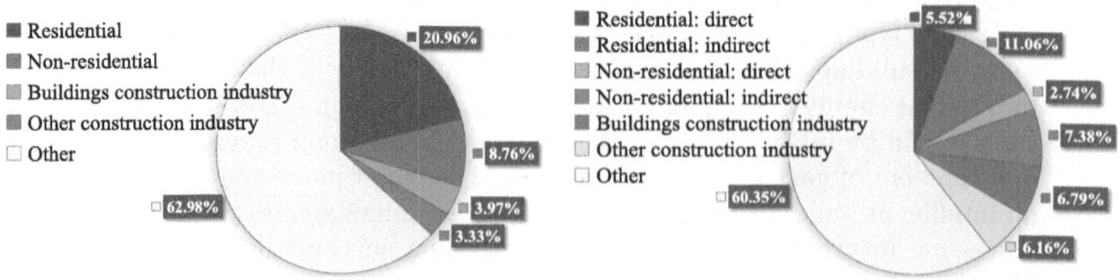

Fig. 1. Global Energy Comsumption (Left) [3] and Global CO_2 emissions (Right) [4]

In fact, net-zero energy buildings (NZEBs), whose on-site renewable energy and demand are balanced over a period (typically yearly), along with plus-zero energy buildings (PZEB), whose energy supply exceeds the demand [10], are targets for building renovation and development. The adoption of renewable energy sources (RES) and energy storage systems (ESS), together with the implementation of energy management strategies to optimize RES-integrated buildings are essential for achieving the NZEBs and even PZEBs targets, and for rational coordination of various components in the energy system, an effective building energy management system (BEMS) is crucial. Without proper energy management, even if RES are integrated into the building, residual power may be wasted, which can significantly diminish the economic and environmental benefits brought by these renewable sources, and even lead to the instability of the power grid rather than the expected grid auxiliary. Thus, comprehensive gathering and analysis of building energy profiles, and accurate estimates of energy demand and generation are crucial for effective building energy management.

This paper first presents a brief review of the commonly used RES in buildings, as well as the equipment that can be used in building energy management. Then, various technologies that contribute to the development of high-performance buildings are surveyed. Finally, this paper gives a brief discussion of future challenges and trends before concluding the paper.

2 Building Energy: Consumer, Producer and Prosumer

There are many devices related to building energy. This paper categorizes these devices into three types based on their functions: Consumers, Producers, and Prosumers. Figure 2 illustrates the relationship between these categories.

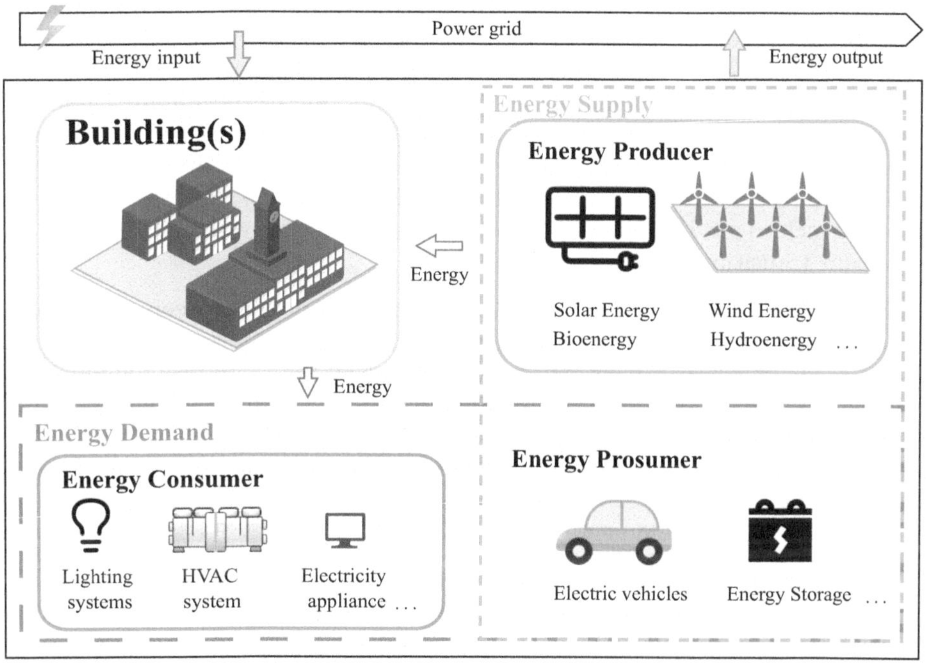

Fig. 2. Building Energy Consumer, Producer and Prosumer

2.1 Building Energy Consumer

Heating, Ventilation and Air Conditioning (HVAC) System are prevalent in buidlings and account for about 50% of the building energy consumption

[11]. They have a significant impact on the indoor environment and are important in striking a balance between indoor thermal comfort and energy consumption. Therefore, the control strategies of HVAC systems are crucial for building energy management. In some large buildings located in areas with high year-round temperatures, the chiller plant acts as the core of the HVAC system and is composed of many subsystems such as chillers, cooling towers, pumps, etc. [12,13]. These components work collaboratively to deliver the expected thermal performance. In addition, heat pumps (HPs) are more favored in milder climates areas [14] as a combination of cooling and heating devices, which are an important research object that is considered essential for achieving low-carbon buildings [15]. HPs can transfer energy from heat sources into useful heat by moving heat from a higher temperature to a lower temperature area. The excess RES can be transformed into thermal energy, and the transferred heat can be used for space heating or hot water supply in buildings by using HPs [16]. There are many types of HPs, such as air source heat pumps (ASHPs), water source heat pumps (WSHPs), and ground source heat pumps (GHSPs). ASHPs transfer heat from the ambient air, but can have seasonal variations in their coefficient of performance due to air temperature fluctuations [17]. WSHPs exchange heat with bodies of water such as lakes, rivers, or industrial wastewater, this type of HP is considered more environmentally friendly due to the minimal temperature fluctuation of water compared to air [16]. GSHPs use underground pipes around a building to exchange heat with underground soil which is relatively stable in temperature [15]. In addition, the integration of traditional heating systems, such as gas boilers, with HPs to form hybrid HP systems [18] is also an important transition strategy towards NZEB.

Lighting Systems constitute a substantial portion of energy consumption in commercial buildings, representing approximately 30% or more of office building energy usage. It is worth mentioning that lighting systems can provide illumination and dissipate heat to the surrounding environment simultaneously, both of which should be taken into consideration in building energy management [19]. Sensing technologies have enabled the development of occupancy-based [20,21] and illuminance-based [22,23] lighting system control strategies, making lighting systems more flexible and energy-efficient.

Other Energy-Consuming Appliances are also important to maintain the normal operation of buildings or the comfort of occupants. Different buildings have unique energy consumption patterns, for example, in data centers, the energy consumption of servers can be significant. In general, energy usage for water heating, cooking, refrigeration, and other electrical appliances [19] can be considered when highly precise energy management is required.

2.2 Building Energy Producer

Solar Energy is usually collected by the photovoltaic (PV) system, which is an important component of the energy generation system in modern buildings, with its energy production highly dependent on solar radiation and the installation footprint [24]. It is worth noting that, photovoltaic-thermal (PV/T) is also a popular technology, which can use solar radiation to generate electricity and harvest the heat generated from the PV panels simultaneously. Due to the decrease in PV cell efficiency at higher temperatures, this operational principle also ensures the electricity generation efficiency of PV/T systems [25]. Building-integrated photovoltaic (BIPV) is another type of PV system that acts as an alternative to traditional building envelope materials, and can be used as a part of the roofs, skylights, or facades [26], but it should be noticed that the cost and product development issues of BIPV are major concerns before a significant technical breakthrough is achieved [27,28].

Wind Energy constitutes a significant portion of the RES [29]. Wind turbines convert kinetic energy from the wind into usable energy [30], and the production of such energy is highly dependent on the wind speed and turbine blade design, which makes it challenging to maintain a stable, nominal power output for a long time due to the stochastic nature of wind [31]. Wind turbines have different installation types, which can be installed in the stand-alone area, on the roof of the building [32–34], or even integrated into the building architecture [35]. Various installation types provide more flexibility and research opportunities for wind energy-integrated buildings.

Power Grid is a common energy source for almost every building right now. It acts as the main or only power source in the non-RES integrated buildings and can fill the electricity supply gaps when the RES cannot meet the demand, or even receive surplus electricity from the building if possible [36]. The rise of smart grid technologies enables the power grid to deliver electricity more efficiently, enhances the capability of the grid to interact with buildings, and allows households to participate in the electricity market actively [37].

Other Energy Sources can also be considered in the building energy system besides the above-mentioned common energy sources. For instance, bioenergy is a reliable, stable, and sustainable source that can provide a relatively stable power output and has a low reliance on weather conditions [38], at the same time, woody biomass, agricultural residues, and waste, etc. [39] provide a wide range of choices for bioenergy sources. Hydropower is a leading source of RES [40], by converting water pressure into mechanical shaft power, hydro-turbines can generate electricity that is clean and renewable [41]. In addition, Geothermal power, hydrogen energy, etc. are also available clean energy sources, Christopher *et al.* [42] concluded comprehensive conversion factors of renewable source technologies, which can be used during the RES evaluation.

2.3 Building Energy Prosumers

Some devices play a dual role as they can both consume and produce energy for the building energy system, this paper refers to these devices as prosumers:

Energy Storage Systems are essential to integrate multiple energy sources, as they provide a buffer between energy generation and consumption thus reducing the impact of the intermittency and enhancing the value of RES [43]. There is a growing demand for ESS solutions that are both cost-competitive and effective, particularly in the current context where the integration of RES is on the rise. Besides storage of the energy generated by RES, ESS can also be used to store the power grid electricity during the valley period, and then release the energy during the peak period. Chatzivasileiadi *et al.* [43] compared multiple energy storage technologies for building applications, and identified lithium-ion as a promising option. Schmidt *et al.* [44] also conducted an assessment of the levelized cost of storage for nine different ESS technologies, including Lithium-Ion, Hydrogen, Sodium Sulfur, and Flywheel energy storage, among others. They compared their suitability for various applications. and also concluded that lithium-ion technology is likely to be the most cost-efficient option in the foreseeable future after a comprehensive evaluation. Given these findings, research on lithium-ion batteries holds paramount importance. In addition, thermal energy storage systems (TESS) provide an important means of energy savings by absorbing and releasing heat, thus enabling load shifting [45]. The most common types of TESS include sensible heat storage (SHS) and latent heat storage (LHS) [46]. SHS stores energy in the form of sensible heat by raising the temperature of the storage medium, which requires the medium to have a high specific heat capacity, long-term stability under thermal cycling and low cost. Phase change materials (PCM) based LHS offer high energy storage density and are considered to have significant development potential. PCMs store heat as latent heat of fusion at the phase change temperature [47], making them particularly efficient for thermal energy storage. LHS [46,48] is highly versatile. They are capable of actively storing energy by charging and discharging PCMs on demand, and can also be passively subject to the ambient temperature. [49] provided a comprehensive review of the modeling techniques and simulation methods for PCM-based TESS, guiding the integration of TESS in building energy simulations.

Electric Vehicle (EV) is a new type of building energy system load, which has gained widespread adoption in recent years and is anticipated to become a prevalent load in the foreseeable future. An EV can serve as a load during charging and as an energy source during discharging. As charging stations proliferate across commercial and residential buildings, the bidirectional energy flow from EVs is set to become an important component of smart building management. However, the large stochasticity of times when EVs arrive and depart, the variability in energy requirements, and the contingency on EV owners' willingness to engage in coordinated energy programs, present formidable challenges to the integration of EVs in building energy management. Moreover, when accounting for EVs, it is imperative to incorporate the battery degradation costs due to the power exchange [50,51].

3 Building Energy Management: Intelligent Transformation

3.1 Building Digitalization

Internet of Things (IoT) serves as the backbone of smart buildings, enabling users to communicate wirelessly with a wide range of devices within the building. Due to the aim of maximizing occupant productivity while minimizing costs and reducing environmental impact [52], smart buildings need to monitor and control various building equipment, and the IoT provides a platform for the information exchange and data sharing between different devices, thus facilitating more advanced service [53] and providing the basis for building an intelligent building management system. An effective IoT platform should have capabilities for data collection, storage, processing, and visualization [54]. To achieve these functions, the integration of various data transmission protocols [55], the development of edge and cloud computing platforms, the construction of intelligent algorithms and the development of visualization software are all essential components of a complete IoT platform. These elements work together to enhance the functionality and efficiency of smart building systems, enabling real-time data management and advanced operational capabilities.

Digital Twin(DT) 's definition has evolved with the growing of understanding. Reference [56] analyzed 46 different definitions and summarised DT as *'a virtual representation of a physical system (and its associated environment and processes) that is updated through the exchange of information between the physical and virtual systems'*. With the support of IoT technologies, the use of DT has become more feasible. This technology bridges the visual and physical worlds, allowing information to be displayed in real time. Such capabilities provide building managers with detailed and intuitive data, enabling more effective energy management and timely decision-making [57]. Hosamo *et al.* [58] proposed a DT-based air handling unit predictive maintenance framework, and the results show that the framework can effectively detect faults and bring economic benefits. Therefore, DT can also be used for predictive maintenance. By monitoring the equipment status, failures may be predicted and prevented before they get worse [58].

Blockchain is a shared and distributed data structure that can store digital transactions securely [59]. These transactions are grouped into blocks and cryptographically linked in sequence [60]. In addition, the integration of transparent, digitized and automated smart contracts within the blockchain allows for the automatic execution of subsequent logic when predetermined conditions are met [60]. This capability significantly reduces the costs associated with intermediaries traditionally required to process transactions. In the context of building energy management, buildings as prosumers can use blockchain technology to engage in decentralized and secure energy trading [61]. Smart contracts can encapsulate transaction details, such as the sources and quantities of energy exchanged, etc. and therefore enhance the efficiency of these transactions [54].

3.2 Building Energy Estimation

Traditional Physical-Based Methods: Estimating or predicting building energy consumption is an important aspect of building energy management [62,63]. However, the increasing integration of diverse electrical appliances into the electric network [64] makes the estimation task more complex. Traditional approaches often rely on physical-based models, they are based on fundamental principles such as energy-mass balance and heat transfer and attempt to describe the energy transfer process mathematically and quantitatively. To facilitate the calculation process, simulation software, such as EnergyPlus, TRNSYS, and ESP-r, has become indispensable tools for modeling building energy patterns in line with physical principles [65]. However, as these tools require comprehensive data on building structures and their surroundings, the simulation process is both time-consuming and labor-intensive. Furthermore, these simulations are heavily dependent on the user's expertise, which may raise the costs of their widespread adoption. Simplified methodologies have been researched to make the calculation easier: Al-Homoud *et al.* [66] introduced the degree-day method to roughly estimate small buildings' heating energy consumption, and the bin method for buildings where internal loads are significant or cooling loads do not linearly correlate with outdoor/indoor temperature differences. Westphal *et al.* [67] introduced a simple calculation method based on weather data to estimate the heating and cooling loads of non-residential buildings. While these simplified models are easy to use, their simple basic structure often results in less accurate estimation or predictions.

Intelligent Approaches: With the development of computing power and the increasing availability of data, the data-driven method has been proven an effective approach for complex estimation and prediction tasks. In recent years, research works in these areas have surged, and many studies attempt to apply various data-driven methods to building energy estimation and prediction. The diversity of building construction typologies, end-uses (e.g., residential or non-residential), and other factors in the building stock often result in a high degree of heterogeneity in design and operating conditions. Manfren *et al.* [68] tried to develop a generalizable and accurate building energy model to bridge the 'performance gap' of different buildings. This research employs a flexible regression-based approach to model energy consumption data for a passive house. This approach combines multiple sub-models into one using dummy variables and introduces a semi-automated model performance comparison and selection process. Idowu *et al.* [69] used various machine learning methods, including support vector machine, regression tree, multiple linear regression, etc. to predict the thermal load in district heating substation, and results show that the support vector machine has a better performance. Chen *et al.* [12] proposed a knowledge-embedded associative-temporal dynamic graph convolutional network, by incorporating empirical knowledge graph into the association graph, this method achieved a better performance in cooling load prediction. It is foreseeable that more and more mature data-driven energy consumption estimation and prediction methods will appear in the future. Its architecture design and data processing methods are the focus of this field and deserve in-depth study.

3.3 Building Energy Optimization

Diverse energy management strategies are essential for achieving energy conservation in buildings due to the variability in building materials, energy usage habits of occupants, and regional climatic conditions. Research in this area is focused on achieving a balance between meeting the energy demands of buildings, minimizing energy costs, and ensuring eco-friendliness [70].

Traditional Control: Early building controls focused on mechanisms such as proportional-integral (PI) and proportional-integral-derivative (PID) controllers [71,72]. Soyguder *et al.* [73] designed a PID-controller-based HVAC system, whose two damper gap rates are controlled according to the indoor temperature and humidity. Fuzzy sets were used to obtain the optimal PID parameters, and fuzzy adaptive control was used to improve the system's performance. In addition, event- or rule-driven control methods are also popular in building energy management due to their simplicity and ease of implementation: Doukas *et al.* [70] presented a rule-set-based energy management system for buildings that utilizes logical judgment and expert knowledge for decision-making. Dai *et al.* [74] proposed an event-driven air conditioning control strategy for rapid demand response in buildings. The strategy aims to address the issue of disordered water distribution caused by the shutdown of some cooling units.

Intelligent Approaches: Traditional energy management methods have limited scope to adapt to the complex dynamic environment, and such strategies are highly dependent on the expertise of the designer, which may restrict their flexibility and adaptability [75]. To overcome these challenges, research is increasingly turning towards intelligent control methods, which offer a promising avenue for developing more intelligent and adaptive control strategies: Baldi *et al.* [75] proposed a building optimization and control system design method based on parametric cognitive adaptive optimization, which provides efficient strategies in two different scenarios: one using building dynamics simulation models to assist in the control strategy design, and another directly leveraging actual operational data from buildings for optimal control. The results from both scenarios indicate a significant reduction in building energy consumption and an improvement in comfort levels. Jia *et al.* [76] contended that promoting a friendly interaction between buildings and grids is beneficial: they suggested that buildings should adjust their power consumption strategies based on the grid's load conditions. They introduced a grid-friendly interaction indicator to measure the degree of this positive interaction and used Non-dominated Sorting Genetic Algorithm II to optimize the energy system. Furthermore, their research highlights the potential of energy storage devices to reduce CO_2 emissions, albeit with the trade-off of increased total expenses. Ghafoori *et al.* [77] developed a machine learning model to predict the electricity demand profile of buildings. Their approach integrates the scheduling of EV trips and state-of-charge requirements into the building's energy management strategy and employs binary linear programming to schedule the charging and discharging of EVs, thereby helping to reduce peak energy demand in buildings. To achieve NZEB, the integration of RES in buildings is likely to increase in the future, coupled with the widespread adoption of smart grids. Consequently, comprehensive BEMS are set to become a focal point, offering significant economic, environmental, and social benefits.

4 Conclusions and Recommendations for Future Development

The purpose of this review is to outline current directions in building energy management. This paper has reviewed several key issues associated with achieving high-energy-performance buildings. It has first discussed the widely used energy equipment in buildings and promising building energy management technologies. The equipment is classified as energy producers, consumers, and prosumers based on their functionalities. Then, this paper has given a review of building energy management methods. In view of the fact that the current energy efficiency of most buildings and the penetration rate of new energy sources still have a great deal of room for improvement, it is expected that the RES and advanced building energy management technologies will continue to evolve, so the following research prospects are suggested:

1) Enhancing the energy efficiency of existing equipment: This strategy can be advanced through technical innovations and policy interventions. In terms of technical innovations, the development of high-efficiency electricity appliances and reducing their production costs are important. On the policy side, incentives and regulations can be used to encourage the widespread adoption of these technologies.
2) Integrating RES effectively: To facilitate easier use of renewable energy in buildings, smart grid technologies should be developed to lower the RES entry barriers. Policy support is also essential to attract more households and businesses to access renewable energy.
3) Developing intelligent algorithms: Intelligent management systems are almost an inevitable trend, especially in the current exponential growth of new artificial intelligence technologies and applications. Therefore, developing efficient, adaptive, and more flexible artificial intelligence-empowered BEMSs, and employing optimization methods to minimize energy waste and maximize returns is a promising direction.

4) Unleashing the potential of building-grid interaction: Buildings can actively participate in load management of the grid, and enhance the flexibility of the grid. Considering the interaction between the power grid and buildings to achieve grid-friendly building energy management is also an important goal.

Acknowledgments. Mr. Zhengrun Zhao would like to thank the Chinese Scholarship Council for sponsoring his research.

Disclosure of Interests. The authors declare that they have no known competing financial interests or personal relationships that could have appeared to influence the work reported in this paper.

References

1. Hadjipaschalis, I., Poullikkas, A., Efthimiou, V.: Overview of current and future energy storage technologies for electric power applications. Renew. Sustain. Energy Rev. **13**(6–7), 1513–1522 (2009)
2. Guterres, A.: United Nations: carbon neutrality by 2050: the world's most urgent mission. https://www.un.org/sg/en/content/sg/articles/2020-12-11/carbon-neutrality-2050-the-world's-most-urgent-mission. Accessed 10 April 2024
3. IEA: Final energy consumption of buildings relative to other sectors, 2022, IEA, Paris, licence: CC BY 4.0. https://www.iea.org/data-and-statistics/charts/final-energy-consumption-of-buildings-relative-to-other-sectors-2022. Accessed 15 April 2024
4. IEA: Global CO2 emissions from buildings, including embodied emissions from new construction, 2022. IEA, Paris, licence: CC BY 4.0. https://www.iea.org/data-and-statistics/charts/global-co2-emissions-from-buildings-including-embodied-emissions-from-new-construction-2022. Accessed 15 April 2024
5. European Commission: Energy performance of buildings directive. https://energy.ec.europa.eu/topics/energy-efficiency/energy-efficient-buildings/energy-performance-buildings-directive. Accessed 18 April 2024
6. Arent, D.J., et al.: Challenges and opportunities in decarbonizing the us energy system. Renew. Sustain. Energy Rev. **169**, 112939 (2022)
7. National Academies of Sciences, Engineering, and Medicine: Accelerating Decarbonization of the U.S. Energy System. The National Academies Press, Washington, DC (2021)
8. Langevin, J., et al.: Decarbonizing the U.S. Economy by 2050: A National Blueprint for the Buildings Sector, Technical report, Lawrence Berkeley National Laboratory (2024)
9. United Nations Environment Programme: 2022 Global Status Report for Buildings and Construction: Towards a Zero-emission. Efficient and Resilient Buildings and Construction Sector, Nairobi (2022)
10. Ha, J.W., Park, K.S., Kim, H.Y., Song, Y.H.: A study of zero energy building verification with measuring and model-based simulation in exhibition building. Archit. Res. **20**(3), 93–102 (2018)
11. Costa, A., Keane, M.M., Torrens, J.I., Corry, E.: Building operation and energy performance: Monitoring, analysis and optimisation toolkit. Appl. Energy **101**, 310–316 (2013)

12. Chen, Z., et al.: A knowledge embedded graph neural network-based cooling load prediction method using dynamic data association. Energy Build. **278**, 112635 (2023)
13. Zhao, Z., Chen, Z., Deng, Q., Tang, P., Peng, T.: Cooling load prediction based on correlative temporal graph convolutional network. In: 2022 4th International Conference on Industrial Artificial Intelligence (IAI), pp. 1–6. IEEE, Shenyang, China (2022)
14. Ahmadi, M.H., Ahmadi, M.A., Sadaghiani, M.S., Ghazvini, M., Shahriar, S., Alhuyi Nazari, M.: Ground source heat pump carbon emissions and ground-source heat pump systems for heating and cooling of buildings: a review. Environ. Progress Sustain. Energy **37**(4), 1241–1265 (2018)
15. Yunna, W., Ruhang, X.: Green building development in China-based on heat pump demonstration projects. Renew. Energy **53**, 211–219 (2013)
16. Gaur, A.S., Fitiwi, D.Z., Curtis, J.: Heat pumps and our low-carbon future: a comprehensive review. Energy Res. Soc. Sci. **71**, 101764 (2021)
17. Dongellini, M., Naldi, C., Morini, G.L.: Seasonal performance evaluation of electric air-to-water heat pump systems. Appl. Therm. Eng. **90**, 1072–1081 (2015)
18. D'Ettorre, F., Conti, P., Schito, E., Testi, D.: Model predictive control of a hybrid heat pump system and impact of the prediction horizon on cost-saving potential and optimal storage capacity. Appl. Therm. Eng. **148**, 524–535 (2019)
19. Harish, V., Kumar, A.: A review on modeling and simulation of building energy systems. Renew. Sustain. Energy Rev. **56**, 1272–1292 (2016)
20. de Bakker, C., Aries, M., Kort, H., Rosemann, A.: Occupancy-based lighting control in open-plan office spaces: a state-of-the-art review. Build. Environ. **112**, 308–321 (2017)
21. Guo, X., Tiller, D., Henze, G., Waters, C.: The performance of occupancy-based lighting control systems: a review. Lighting Res. Technol. **42**(4), 415–431 (2010)
22. Ding, X., Yu, J., Si, Y.: Office light control moving toward automation and humanization: a literature review. Intell. Build. Int. **12**(4), 225–256 (2020)
23. Wagiman, K.R., Abdullah, M.N., Hassan, M.Y., Radzi, N.H.M., Kwang, T.C., et al.: Lighting system control techniques in commercial buildings: current trends and future directions. J. Build. Eng. **31**, 101342 (2020)
24. Park, M., Wang, Z., Li, L., Wang, X.: Multi-objective building energy system optimization considering EV infrastructure. Appl. Energy **332**, 120504 (2023)
25. Good, C., Andresen, I., Hestnes, A.G.: Solar energy for net zero energy buildings-a comparison between solar thermal, PV and photovoltaic-thermal (PV/T) systems. Sol. Energy **122**, 986–996 (2015)
26. Peng, C., Huang, Y., Wu, Z.: Building-integrated photovoltaics (BIPV) in architectural design in China. Energy Build. **43**(12), 3592–3598 (2011)
27. James, T., Goodrich, A., Woodhouse, M., Margolis, R., Ong, S.: Building-integrated Photovoltaics (BIPV) in the Residential Sector: An Analysis of Installed Rooftop System Prices. Tech. rep, National Renewable Energy Laboratory (2011)
28. Kylili, A., Fokaides, P.A.: Investigation of building integrated photovoltaics potential in achieving the zero energy building target. Indoor Built Environ. **23**(1), 92–106 (2014)
29. Sadorsky, P.: Wind energy for sustainable development: driving factors and future outlook. J. Clean. Prod. **289**, 125779 (2021)
30. Manyonge, A., Ochieng, R., Onyango, F., Shichikha, J.: Mathematical modelling of wind turbine in a wind energy conversion system: power coefficient analysis. Appl. Math. Sci. **6**(91), 4527–4536 (2012)

31. Khan, A., Memon, S., Said, Z.: Predictive permanent magnet synchronous generator based small-scale wind energy system at dynamic wind speed analysis for residential net-zero energy building. Int. J. Solar Thermal Vacuum Eng. **3**(1), 29–49 (2021)
32. Rezaeiha, A., Montazeri, H., Blocken, B.: A framework for preliminary large-scale urban wind energy potential assessment: roof-mounted wind turbines. Energy Convers. Manage. **214**, 112770 (2020)
33. Balduzzi, F., Bianchini, A., Carnevale, E.A., Ferrari, L., Magnani, S.: Feasibility analysis of a Darrieus vertical-axis wind turbine installation in the rooftop of a building. Appl. Energy **97**, 921–929 (2012)
34. Mithraratne, N.: Roof-top wind turbines for microgeneration in urban houses in New Zealand. Energy Build. **41**(10), 1013–1018 (2009)
35. Mertens, S.: Wind energy in urban areas: concentrator effects for wind turbines close to buildings. Refocus **3**(2), 22–24 (2002)
36. Wu, W., Skye, H.M.: Residential net-zero energy buildings: review and perspective. Renew. Sustain. Energy Rev. **142**, 110859 (2021)
37. Kolokotsa, D.: The role of smart grids in the building sector. Energy Build. **116**, 703–708 (2016)
38. Ahmed, A., Ge, T., Peng, J., Yan, W.C., Tee, B.T., You, S.: Assessment of the renewable energy generation towards net-zero energy buildings: a review. Energy Build. **256**, 111755 (2022)
39. Nzihou, A.: Handbook on Characterization of Biomass, Biowaste and Related By-Products. Springer (2020)
40. Moran, E.F., Lopez, M.C., Moore, N., Müller, N., Hyndman, D.W.: Sustainable hydropower in the 21st Century. Proc. Natl. Acad. Sci. **115**(47), 11891–11898 (2018)
41. Yuksek, O., Komurcu, M.I., Yuksel, I., Kaygusuz, K.: The role of hydropower in meeting Turkey's electric energy demand. Energy Policy **34**(17), 3093–3103 (2006)
42. Christopher, S., et al.: Renewable energy potential towards attainment of net-zero energy buildings status-a critical review. J. Clean. Prod. **405**, 136942 (2023)
43. Chatzivasileiadi, A., Ampatzi, E., Knight, I.: Characteristics of electrical energy storage technologies and their applications in buildings. Renew. Sustain. Energy Rev. **25**, 814–830 (2013)
44. Schmidt, O., Melchior, S., Hawkes, A., Staffell, I.: Projecting the future levelized cost of electricity storage technologies. Joule **3**(1), 81–100 (2019)
45. Bastani, A., Haghighat, F., Kozinski, J.: Designing building envelope with PCM wallboards: design tool development. Renew. Sustain. Energy Rev. **31**, 554–562 (2014)
46. Romdhane, S.B., Amamou, A., Khalifa, R.B., Said, N.M., Younsi, Z., Jemni, A.: A review on thermal energy storage using phase change materials in passive building applications. J. Build. Eng. **32**, 101563 (2020)
47. Hasnain, S.: Review on sustainable thermal energy storage technologies, part i: heat storage materials and techniques. Energy Convers. Manage. **39**(11), 1127–1138 (1998)
48. Tyagi, V.V., Buddhi, D.: PCM thermal storage in buildings: a state of art. Renew. Sustain. Energy Rev. **11**(6), 1146–1166 (2007)
49. Al-Saadi, S.N., Zhai, Z.J.: Modeling phase change materials embedded in building enclosure: a review. Renew. Sustain. Energy Rev. **21**, 659–673 (2013)
50. Beer, S., et al.: An economic analysis of used electric vehicle batteries integrated into commercial building microgrids. IEEE Trans. Smart Grid **3**(1), 517–525 (2012)

51. Wu, D., Zeng, H., Lu, C., Boulet, B.: Two-stage energy management for office buildings with workplace EV charging and renewable energy. IEEE Trans. Transp. Electrification **3**(1), 225–237 (2017)
52. Kumar, A., Singh, A., Kumar, A., Singh, M.K., Mahanta, P., Mukhopadhyay, S.C.: Sensing technologies for monitoring intelligent buildings: a review. IEEE Sens. J. **18**(12), 4847–4860 (2018)
53. Jia, M., Komeily, A., Wang, Y., Srinivasan, R.S.: Adopting internet of things for the development of smart buildings: a review of enabling technologies and applications. Autom. Constr. **101**, 111–126 (2019)
54. Li, J., Herdem, M.S., Nathwani, J., Wen, J.Z.: Methods and applications for artificial intelligence, big data, internet of things, and blockchain in smart energy management. Energy AI **11**, 100208 (2023)
55. Verma, A., Prakash, S., Srivastava, V., Kumar, A., Mukhopadhyay, S.C.: Sensing, controlling, and IoT infrastructure in smart building: a review. IEEE Sens. J. **19**(20), 9036–9046 (2019)
56. VanDerHorn, E., Mahadevan, S.: Digital twin: generalization, characterization and implementation. Decis. Support Syst. **145**, 113524 (2021)
57. Himeur, Y., et al.: AI-big data analytics for building automation and management systems: a survey, actual challenges and future perspectives. Artif. Intell. Rev. **56**(6), 4929–5021 (2023)
58. Hosamo, H.H., Svennevig, P.R., Svidt, K., Han, D., Nielsen, H.K.: A digital twin predictive maintenance framework of air handling units based on automatic fault detection and diagnostics. Energy Build. **261**, 111988 (2022)
59. Andoni, M., et al.: Blockchain technology in the energy sector: a systematic review of challenges and opportunities. Renew. Sustain. Energy Rev. **100**, 143–174 (2019)
60. Van Cutsem, O., Dac, D.H., Boudou, P., Kayal, M.: Cooperative energy management of a community of smart-buildings: a blockchain approach. Int. J. Electr. Pow. Energy Syst. **117**, 105643 (2020)
61. Ghorbanian, M., Dolatabadi, S.H., Siano, P., Kouveliotis-Lysikatos, I., Hatziargyriou, N.D.: Methods for flexible management of blockchain-based cryptocurrencies in electricity markets and smart grids. IEEE Trans. Smart Grid **11**(5), 4227–4235 (2020)
62. Afzal, S., Ziapour, B.M., Shokri, A., Shakibi, H., Sobhani, B.: Building energy consumption prediction using multilayer perceptron neural network-assisted models; comparison of different optimization algorithms. Energy **282**, 128446 (2023)
63. Li, X., Wen, J.: Review of building energy modeling for control and operation. Renew. Sustain. Energy Rev. **37**, 517–537 (2014)
64. Wu, D., Lin, W.: Efficient residential electric load forecasting via transfer learning and graph neural networks. IEEE Trans. Smart Grid **14**(3), 2423–2431 (2022)
65. Zhao, H.X., Magoulès, F.: A review on the prediction of building energy consumption. Renew. Sustain. Energy Rev. **16**(6), 3586–3592 (2012)
66. Al-Homoud, M.S.: Computer-aided building energy analysis techniques. Build. Environ. **36**(4), 421–433 (2001)
67. Westphal, F.S., Lamberts, R.: The use of simplified weather data to estimate thermal loads of non-residential buildings. Energy Build. **36**(8), 847–854 (2004)
68. Manfren, M., James, P.A., Tronchin, L.: Data-driven building energy modelling-an analysis of the potential for generalisation through interpretable machine learning. Renew. Sustain. Energy Rev. **167**, 112686 (2022)
69. Idowu, S., Saguna, S., Åhlund, C., Schelén, O.: Applied machine learning: forecasting heat load in district heating system. Energy Build. **133**, 478–488 (2016)

70. Doukas, H., Patlitzianas, K.D., Iatropoulos, K., Psarras, J.: Intelligent building energy management system using rule sets. Build. Environ. **42**(10), 3562–3569 (2007)
71. Kasahara, M., Matsuba, T., Kuzuu, Y., Yamazaki, T., et al.: Design and tuning of robust PID controller for HVAC systems. ASHRAE Trans. **105**, 154 (1999)
72. Salsbury, T.: A temperature controller for vav air-handling units based on simplified physical models. HVAC&R Res. **4**(3), 265–279 (1998)
73. Soyguder, S., Alli, H.: Fuzzy adaptive control for the actuators position control and modeling of an expert system. Expert Syst. Appl. **37**(3), 2072–2080 (2010)
74. Dai, M., Li, H., Wang, S.: Event-driven demand response control of air-conditioning to enable grid-responsive buildings. Autom. Constr. **150**, 104815 (2023)
75. Baldi, S., Michailidis, I., Ravanis, C., Kosmatopoulos, E.B.: Model-based and model-free plug-and-play building energy efficient control. Appl. Energy **154**, 829–841 (2015)
76. Jia, S., Sheng, K., Huang, D., Hu, K., Xu, Y., Yan, C.: Design optimization of energy systems for zero energy buildings based on grid-friendly interaction with smart grid. Energy **284**, 129298 (2023)
77. Ghafoori, M., Abdallah, M., Kim, S.: Electricity peak shaving for commercial buildings using machine learning and vehicle to building (V2B) system. Appl. Energy **340**, 121052 (2023)

Improving the Construction of a Multi-modal Underwater Image Recognition System for Anomaly Detection in Nuclear Power Plant Cooling Source Areas

Yu Huang[1], Kai Wang[2], Yuntao Gu[2], Chaofan Zhou[1], Yuanjun Guo[3], and Zhile Yang[3](✉)

[1] China Nuclear Industry Maintenance Co., Ltd., Shenzhen Branch, Shenzhen, China
[2] Taiyuan University of Technology, Taiyuan 030024, China
[3] Shenzhen Institute of Advanced Technology, Chinese Academy of Sciences, Shenzhen 518055, China
zl.yang@siat.ac.cn

Abstract. The cooling facilities of nuclear power plants are crucial components for transferring the accumulated heat from the reactor to the external environment. However, the complex structure of these facilities makes traditional manual anomaly detection methods inefficient, time-consuming, and poses certain risks to human safety. Therefore, this paper proposes a system for anomaly detection in the cooling source areas of nuclear power plants based on improved multi-modal underwater image recognition. The system utilizes deep learning and image processing techniques to denoise, compensate for illumination, and correct water flow in collected images of the cooling source areas. Wavelet-based noise removal techniques are applied to further enhance image clarity, providing accurate inputs for machine vision algorithms. By training an improved convolutional neural network model, real-time automatic identification and classification of underwater anomalies such as foreign objects and pipeline cracks are achieved, significantly improving the monitoring efficiency and safety of the cooling source areas in nuclear power plants. Comparative analysis with actual cases validates the excellent stability and accuracy of the proposed method in handling underwater operation images in the cooling source areas of nuclear power plants.

Keywords: Machine vision · Cooling source · Anomaly detection

1 Introduction

Nuclear power generation is valued for its clean, sustainable energy contribution amid rising global demand. Its efficiency and low-carbon profile make it pivotal in the energy transition. However, safety remains a key focus, given the potential consequences of even minor issues in nuclear plants. Strict safety measures and effective risk management are essential to ensure safe operations.

K. Li et al. (Eds.): LSMS/ICSEE 2024, CCIS 2218, pp. 52–66, 2025.
https://doi.org/10.1007/978-981-96-0232-2_5

1) Traditional anomaly detection methods in nuclear power plants, relying on manual inspections and monitoring, ensure operational integrity to some degree. Yet, they are time-consuming, prone to human error, and lack real-time, 24/7 monitoring, especially amidst complex equipment inspections, data collection, and analysis. As a result, they often struggle to promptly detect and respond to subtle, rapidly evolving anomalies.
2) Machine vision-based automatic anomaly detection systems have gained significant attention due to their real-time extraction of key information from large datasets. These systems utilize advanced technologies including deep learning, image processing, and pattern recognition to identify various anomalies such as pipe blockages, leaks, and abnormalities in temperature and flow rate. Despite their success in other fields, applying machine vision systems to the cooling source area of nuclear power plants encounters several key challenges:
3) The environmental conditions in the cooling source area of nuclear power plants are extremely harsh, including high radiation, extreme humidity, and complex and variable lighting conditions. These factors severely affect the image quality and the accuracy of algorithms. Obtaining high-quality images in such environments is challenging, as existing machine vision algorithms, such as those based on Convolutional Neural Networks (CNNs) and Support Vector Machines (SVMs), possess advanced image analysis capabilities but have high demands for image quality. For instance, images captured by underwater cameras may suffer from blurring, distortion, or uneven lighting due to water flow disturbances, suspended particles, or reflected light.
4) The visual characteristics of each anomaly in the cooling source area are often extremely subtle, presenting a wide variety of anomaly types. For instance, extremely fine cracks or early corrosion may only manifest as faint color differences or subtle texture changes in an image, making them difficult to comprehensively and accurately identify using existing pattern recognition techniques. Despite the use of techniques like deep learning algorithms and image segmentation in current machine vision systems, their effectiveness in recognizing anomalies in these extreme environments is limited.
5) Nuclear power plants adhere to stringent operational safety standards due to the severe consequences that may arise from any misjudgment. Although the integration of machine vision systems can help mitigate errors resulting from manual monitoring, the current real-time image processing and automatic anomaly detection systems are not without their risks, such as the potential for identification errors and missed detections.

This paper introduces an innovative approach to detect anomalies in the cooling source area of nuclear power plants. It employs an abnormal recognition system based on enhanced multimodal underwater image recognition (as shown in Fig. 1). Utilizing deep learning and image processing techniques, the system enhances image quality by addressing noise, adjusting lighting conditions, and correcting water flow distortions in acquired images. Additionally, wave removal techniques are implemented to further improve image clarity, enhancing inputs for machine vision algorithms. Through training an enhanced convolutional neural network model, the system achieves real-time automatic recognition and classification of underwater foreign objects, pipeline cracks, and other irregularities. These advancements significantly boost monitoring efficiency and safety in the cooling source area of nuclear power plants.

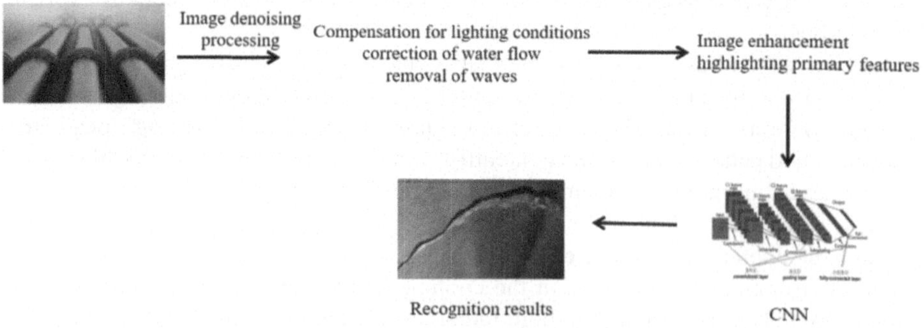

Fig. 1. Overall system framework

The main contributions of this paper include:

(1) The processing and optimization of underwater images in the cooling source area of nuclear power plants involve utilizing denoising, lighting compensation, and water flow correction operations to address the challenging lighting and water quality conditions.
(2) The introduction of wave removal technology to further improve the clarity of underwater images, providing a solution specifically designed to address visual interference in dynamic underwater environments.
(3) The development of a specially tuned convolutional neural network model to more effectively identify and classify underwater foreign objects and structural damage.

2 Development of Machine Vision Algorithms in Aquatic Environments

2.1 Image Denoising Processing

Compensation for Lighting Conditions. In machine vision anomaly detection systems for nuclear power plant cooling areas, lighting compensation is critical. Complex lighting conditions, such as uneven intensity, reflections, and refractions, can result in overexposed or underexposed areas in images, leading to information loss or increased noise. This hampers precise feature extraction and analysis by machine vision algorithms. Day-night changes and weather variations further exacerbate these challenges, causing instability in target detection and recognition. Underwater environments face additional visibility and clarity issues, making anomaly detection even more challenging. These environmental factors significantly affect image quality and target recognition accuracy.

Proper lighting compensation in underwater operations ensures brightness and color consistency in images, reducing noise and enhancing target detection algorithms' robustness. Accurate display of texture and color information of target objects, such as pipelines and valves, aids in timely safety hazard detection. Traditional methods like global histogram equalization or linear adjustments are ineffective underwater due to dynamic lighting changes, resulting in unstable pixel values and reduced target recognition accuracy. An improved approach involves nonlinear lighting compensation, which better simulates lighting effects and effectively corrects uneven lighting. The specific implementation is as follows:

$$I_{compensated}(x, y) = a \cdot I(x, y) + b \tag{1}$$

I(x,y) represents the pixel grayscale or brightness value of the original image at coordinates (x, y). a is the scaling factor used to control the overall brightness of the image. If $a > 1$, the image becomes brighter, while if $0 < a < 1$, the image darkens. b is the translation factor used to change the black level of the image. If $b > 0$, all pixel values increase accordingly, thus brightening the overall image; conversely, if $b < 0$, the overall image darkens.

In the nuclear power plant cooling source area, linear lighting compensation is ineffective in filtering image noise. Nonlinear compensation, mimicking human color perception under varying lighting intensities, outperforms linear methods. This paper employs a nonlinear composite neighborhood lighting compensation approach, with the formula as follows:

$$I_{compensated}(x, y) = log(a \cdot I(x, y) + b) + c \tag{2}$$

Using translation and scaling coefficients (a, b, and c), tailored to different areas within the nuclear power plant, this method incorporates a logarithmic transformation and constant addition to the original pixel values. This flexible approach accommodates diverse lighting conditions, providing detailed image compensation. Nonlinear compensation accurately restores original visual features, especially when pixel values exhibit nonlinear distributions under varying lighting conditions.

Correction of Water Flow. In machine vision and image processing, dynamic water environments like flowing rivers can distort, blur, or displace targets in consecutive frame images, challenging accurate detection and tracking. Optical flow estimation methods correct these effects caused by water flow in images.

Optical flow estimation calculates pixel motion velocity in image sequences, depicting relative motion between consecutive frames caused by target movement or water flow. Recent advancements include deep learning-based algorithms, which employ neural networks to estimate motion vectors (optical flow). These models surpass traditional methods in accuracy and robustness, particularly under complex scenes and large motion backgrounds.

Deep learning methods utilize extensive training data, including annotated real optical flow datasets, enabling neural networks to learn image sequence variations and derive pixel-level motion trajectories. This paper employs the PWC-Net optical flow learning network (Fig. 2), accurately estimating two-dimensional motion vectors of pixels in each frame by analyzing relationships between consecutive images. It discerns water flow speed and direction, correcting its effects on images.

Wave Removal. In image processing, wave removal targets the elimination of ripple and distortion caused by water surface fluctuations, enhancing target detection and recognition accuracy in aquatic environments. In nuclear power plant cooling source areas,

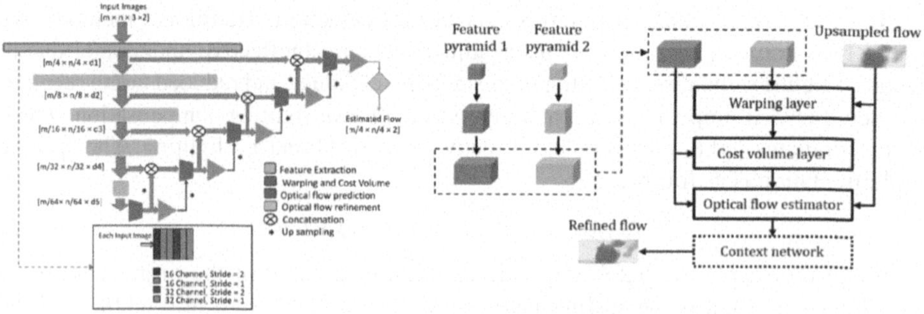

Fig. 2. PWC-Net Optical Flow Estimation Network Structure

waves can degrade image quality, obscuring devices and structures and complicating analysis and judgment in machine vision anomaly detection systems.

To tackle this, the paper utilizes adaptive filtering and image restoration techniques to eliminate wave noise and enhance image quality. Initially, it computes the noise estimate within each pixel neighborhood. This method involves padding the image pixel matrix with zeros and selecting a window size to calculate the average of all pixel values within the neighborhood centered on the pixel being processed. This average reflects the expected noise level in that neighborhood:

$$\mu_p = \frac{1}{N}\sum\nolimits_{(i,j\in Neighborhood(p))}^{N} I(p + (i,j)) \tag{3}$$

N represents the number of pixels in the neighborhood, and $I(p + (i,j))$ denotes the pixel value at coordinates (i,j) within the neighborhood. The standard deviation of the pixels within the neighborhood is computed as follows:

$$\sigma_p = \sqrt{\frac{1}{N-1}\sum\nolimits_{(i,j)\in Neiborhood(p)}^{N} (I(p + (i,j)) - \mu_p)^2} \tag{4}$$

where σ_p is the noise estimate within the neighborhood of the target pixel p. When noise is predominantly Gaussian, its standard deviation approximates the noise's standard deviation. By considering the variation of pixel values within the neighborhood and the noise level, filtering weights for each pixel are dynamically adjusted. Ultimately, adaptive filtering effectively eliminates image noise.

Lastly, inverse filtering techniques are applied to further enhance image restoration. Given the intricate nature of underwater noise in nuclear power plant cooling areas, the paper employs a total variation regularization method based on sparse representation. This approach minimizes the sum of squared errors between the observed image and the ideal image while controlling the total variation of the target image, reducing noise and preserving image edge continuity:

$$min_{I\_true}||H \cdot I_{True} - I_{observed}||_2^2 + \lambda \cdot TV(I_{true}) \tag{5}$$

where I_{true} is the true image we want to recover, $I_{observed}$ is the observed image containing noise or distortion, H is the system matrix representing degradation processes such as blurring and downsampling, $||\cdot||_2^2$ denotes the squared L2 norm, used to measure the discrepancy between the true and observed images, $\mathrm{TV}(\mathrm{I_{true}})$is the total variation of the ideal image used for regularization to avoid excessive smoothing that leads to loss of detail, and λ is the regularization parameter used to balance the importance between the data fidelity term and the regularization term. For a two-dimensional image signal, total variation can be defined by the following equation:

$$TV(I_{true}) = \sum\nolimits_{x,y} \sqrt{(D_x I_{true}(x,y))^2 + \left(D_y I_{true}(x,y)\right)^2} \tag{6}$$

where D_x and D_y are the differences between adjacent pixels along the x-axis and y-axis of the image, respectively:

$$D_x I_{true}(x,y) \approx I_{true}(x+1,y) - I_{true}(x,y) \tag{7}$$

$$D_y I_{true}(x,y) \approx I_{true}(x,y+1) - I_{true}(x,y) \tag{8}$$

To address the specific noise of underwater environments, including wave noise, and considering the optical properties of water, this paper employs the Red Dark Channel Prior (RDCP) method for dehazing and removing wave patterns. The RDCP method identifies the dark channel of the image and then corrects the transmission rate of the image based on prior information, thus restoring a clear image:

$$J(x) = (I(x) - A) \times T(x) + A \tag{9}$$

where $J(x)$ is the image after dehazing processing, $I(x)$ is the original hazy image, A is the estimate of the global atmospheric light, and $T(x)$ is the transmission rate.

2.2 Image Denoising Processing

In recent years, deep learning has seen widespread use in underwater image dehazing and target recognition. These approaches involve training neural networks on image datasets to enhance clarity or extract target features for intelligent recognition. In nuclear power plant cooling water systems, compromised water quality due to factors like algae growth or mechanical failures can lead to turbidity, impacting visual system effectiveness. Precise identification and prediction of abnormal changes in water, such as foreign objects or structural damage, are crucial. However, conventional visual detection struggles with low contrast, unclear features, and image blurring caused by suspended particles and color variations. This section explores the applicability of deep learning networks in detecting water quality anomalies and proposes methods to enhance their performance and accuracy in complex environments.

Traditional object detection methods require manual extraction of features from the target region, which is time-consuming and lacks robustness. With the emergence of deep learning convolutional neural networks, object detection algorithms have entered a new stage. Existing object detection algorithms are mainly divided into two-stage algorithms and one-stage α-stage algorithms. Two-stage algorithms mainly include the RCNN series algorithms (RCNN, Fast RCNN, Faster RCNN). These algorithms first generate region proposals and then perform classification and regression tasks on these proposals. Therefore, detection has been improved, but processing time has also increased accordingly. This type of algorithm is more suitable for detecting static underwater objects such as seabed rocks and corals. Single-stage algorithms mainly include SSD algorithm and YOLO series algorithms (YOLO, YOLOv2, YOLOv3). These algorithms improve detection speed while maintaining detection effectiveness as much as possible, and adopt a direct regression method to predict the category and position of the target. Therefore, this type of algorithm is suitable for frequent object detection scenarios due to frequent aquatic activities (e.g., fish movements).

Effective target feature detectors and classifiers provide advantages for deep learning methods in murky water environments, which is why computers must adapt to the blurry features of murky water and accurately identify target features. Lakshmi and Santhanam [8] proposed two types of classifiers: one for distinguishing between targets and backgrounds, a binary classification convolutional neural network, and another for predicting backgrounds or a type of target (such as shoes and ropes), a multi-class classification convolutional neural network. A convolutional neural network (CNN) was trained to classify 64 × 64 inputs. Images were classified, and the classifiers were used as target feature detectors. The accuracy of the binary convolutional neural network was 93.9%, and the target detection accuracy of the multi-class convolutional neural network was 90.1%, higher than existing multi-class detectors (88%).

During underwater operations in nuclear power plant cooling zones, low-quality underwater videos or images are common due to water characteristics like suspended solids, poor visibility, and complex light effects. This affects real-time monitoring, defect detection, and precision operations due to issues like contrast reduction, color distortion, and detail loss. To tackle this, the study proposes an image processing algorithm that combines physical modeling with deep learning. Initially, to address underwater image optical characteristics, an enhanced light transmission model is introduced to simulate water absorption and scattering for various light wavelengths. This model can be represented as:

$$I(x, y, \lambda, z) = I_0(\lambda)e^{-\mu_a(z)z}T(z) + B(\lambda, z) \tag{10}$$

I(x, y, z, λ) represents the light intensity received at position (x, y) and depth z underwater, at wavelength λ. $I_0(\lambda)$ denotes the initial light intensity of the light source, $\mu_a(z)$ is the absorption coefficient along the vertical direction, T(z) is the transmittance function varying with depth, and B(λ, z) represents the contribution of background light.

Based on the aforementioned physical model, a reverse restoration algorithm based on deep neural networks was designed and implemented. The network architecture consists of an encoder-decoder structure, supplemented with residual connections and attention mechanisms, aimed at extracting potential clear image features from the degraded underwater images. The objective loss function for network optimization can be formulated as:

$$L(\theta) = \sum\nolimits_{(i=1)}^{n} ||F_\theta(I_i^{underwater}) - (I_i^{(\text{ground\_truth})})||^2 + \alpha R(\theta) \tag{11}$$

F_θ is the neural network model to be trained, parameterized by θ, $I_i^{underwater}$ is the input underwater image, $I_i^{ground\_truth}$ is the corresponding ground truth clear image, N is the number of samples, $R(\theta)$ is the regularization term used to prevent overfitting, and α is the coefficient balancing the regularization term weight.

Semantic constraints, guided by prior knowledge, are integrated into the deep learning framework using Generative Adversarial Networks (GANs) to enhance image restoration. Specific scene features like pipeline structures and equipment signage are defined and embedded into the model. The generator produces complete images based on conditional input, while the discriminator distinguishes between real and generated images, ensuring that generated images adhere to specified semantic constraints. This approach enhances both visual realism and semantic accuracy in image content.

3 Image Segmentation

3.1 Local Information Entropy Optimization

Information entropy is a measure of uncertainty in image information. For underwater images, especially under complex lighting and low contrast conditions, using local information entropy as an auxiliary feature can help improve segmentation accuracy. When constructing the segmentation energy function, adding an optimization term based on local information entropy encourages more subdivisions in regions with higher entropy. This is because these regions often contain rich details and edge information. Adding the local information entropy optimization term to the original energy functional refines and optimizes the original image boundaries:

$$E(\phi) = \ldots - \gamma \int_{\Omega} H(\phi(x)) \cdot Entropy(f(x))dx \tag{12}$$

where $Entropy(f(x))$ represents the local information entropy at pixel x, and γ is the coefficient adjusting the importance of the local entropy term.

3.2 Combining Depth Information

The anomaly detection system designed in this study is equipped with a depth sensor, which can utilize depth information to guide the segmentation process, particularly beneficial for 3D scene understanding. Depth information serves as an additional constraint to direct the segmentation algorithm's focus towards objects located within specific depth ranges. The depth sensor provides depth values corresponding to each pixel from the acquired depth map, transforming the 2D image into 3D information. During segmentation, integrating depth information into the algorithm involves techniques such as depth-guided clustering, depth-guided region growing, and depth-constrained convolutional neural networks (DC-CNN). This approach allows for more precise differentiation of objects at different depth levels, mitigating issues like occlusion and overlap due to depth. The study employs a more sophisticated Depth-Constrained Convolutional Neural Network (DC-CNN) to optimize underwater image recognition. For a standard 2D convolutional layer, its forward propagation computation is represented as:

$$z^{[l]}(i,j) = b^{[l]} + \sum_{m=0}^{M-1} \sum_{n=0}^{N-1} W^{[l]}(m,n) \cdot A^{[l-1]}(i+m, j+n) \tag{13}$$

where $z^{[l]}$ is the output feature map of the l-th layer, $A^{[l-1]}$ is the input feature map of the l-1-th layer, $W^{[l]}$ is the convolutional kernel of the current layer, (m, n) is the bias of the convolutional kernel, (i, j) is the position index on the output feature map, $b^{[l]}$ is the bias term of that layer, and M and N are the width and height of the convolutional kernel, respectively (Fig. 3).

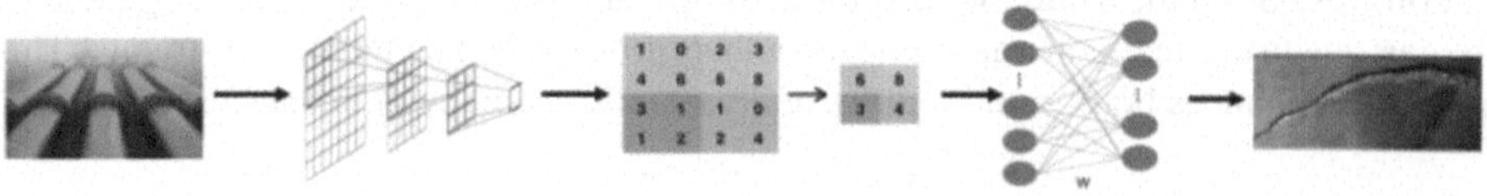

Fig. 3. Color Feature Extraction in Convolutional Neural Networks, CFE-CNN, From left to right: input image; convolution operation; pooling operation; fully connected layer; output image

After the convolutional layer, introduce a regularization term based on depth information D, and incorporate a loss term that establishes a relationship between certain predicted feature responses of the model and the given depth values.

$$L_D\left(z^{[l]}, D\right) = \frac{1}{|\Omega|} \sum\nolimits_{(i,j)\in\Omega} \rho\left(D(i, j), z^{[l]}(i, j)\right) \tag{14}$$

Here Ω is the set of positions in the feature map that need to be considered, $D(i, j)$ is the depth value corresponding to the location (i, j), and ρ $(\cdot)$ is a distance metric, which is the Euclidean distance used to measure the difference between feature response and depth. The overall goal of the depth constraint is to minimize this additional loss function while optimizing other network objective functions.

In the energy function $E(\phi)$, incorporate depth information as an additional constraint. For a depth-aware level set model, a constraint can be added to ensure that segmentation boundaries only occur within a specified depth range. If the depth information is represented as D(x, y), and we know the approximate depth range of the target object is $[\boldsymbol{d_{min}}$⍰$\boldsymbol{d_{max}}]$, the segmentation energy function can be modified as follows:

$$E(\phi) = \ldots - \eta \int\limits_{\Omega} H(d_{min} - D(x, y)) \cdot H(D(x, y) - d_{max}) \cdot g(\phi(x, y))dx \tag{15}$$

$$E() = \ldots - \eta H(d_{min} - D(x, y)) \cdot H(D(x, y) - d_{max}) \cdot g((x, y))dx \quad (15)$$

where ϕ is the level set function, $g(\phi)$ represents the energy term related to segmentation, H is the Heaviside function, and η is the weight of the depth constraint term. This constraint encourages segmentation boundaries to appear within the depth range $[d_{min}, d_{max}]$.

Depth information plays a crucial role in underwater image recognition and segmentation. By integrating it into image processing and machine learning algorithms, the adaptability of algorithms to the complexity of underwater environments can be significantly enhanced.

3.3 Adaptive Feature Fusion

Adaptive feature fusion is a method for integrating multiple features in image segmentation to enhance segmentation performance. It is particularly useful in scenarios with severe color distortion, such as images of the cooling source area in nuclear power plants, where advantages of color features, texture features, local brightness features, among others, can be fully utilized. By employing learning mechanisms to dynamically adjust the weights of each feature, the algorithm can exhibit stronger adaptability and robustness across different environments. The use of deep learning neural networks to fuse adaptive features mainly manifests in network architecture design and loss function optimization. In this study, a multi-path network structure is designed, with each path handling a specific feature. In the color feature branch, RGB color features are directly extracted using convolutional layers. In the texture feature branch, a specially designed texture feature extraction layer within the convolutional network is utilized for texture feature extraction. For local brightness features present in the images, feature maps are constructed based on local brightness information in the images.

Designing a convolutional kernel of size $f_h \times f_w$ performs an inner product operation on each RGB channel value within a small window $f_h \times f_w$ of the input image, and the results from the three channels are summed to obtain a pixel value on the output feature map for that window. Additionally, a bias term b is included. The formula for computing the output feature map can be represented as:

$$O_{i,j} = \sum\nolimits_{m=0}^{f_h-1} \sum\nolimits_{n=0}^{f_w-1} \sum\nolimits_{c=0}^{2} W_{m,n,c} \cdot I_{i+m,j+n,c} + b \tag{16}$$

where $O_{i,j}$ is the value at the position (i, j) on the output feature map. $W_{m,n,c}$ represents the weight of the convolutional kernel at position (m, n) and channel c (for RGB images, c takes values 0, 1, 2 corresponding to the weights of red, green, blue). $I_{i+m,j+n,c}$ is the value of the input image at position $(i + m, j + n)$ and channel c. b is the corresponding bias term. Then, after multiple convolutional layers and ReLU activation functions are stacked, advanced color features are extracted.

For the texture feature branch, this paper utilizes filters related to texture extraction—Gabor filters. A convolutional layer is designed to simulate filtering behavior, creating a layer with symmetric convolutional kernels that are orientation-sensitive, similar to the sine wave component of Gabor filters.

$$K(x, y; \theta, \lambda, \sigma, \gamma) = \\ exp\left(-\frac{\left(x' \cos\theta + y' \sin\theta\right)^2 + \left(\gamma^2 x' \sin\theta - y' \cos\theta\right)^2}{2\sigma^2}\right) \cos\frac{2\pi\left(x' \sin\theta + y' \cos\theta\right)}{\lambda} \tag{17}$$

where (x, y) represents the position in the original coordinate system, $\left(x', y'\right)$ represents the position in the new coordinate system after rotation, θ is the angle of the filter (i.e., the rotation angle), λ is the spatial frequency of the filter, σ is the standard deviation of the Gaussian kernel, controlling the spatial locality of the filter, γ is the aspect ratio of the elliptical Gaussian kernel, reflecting orientation selectivity. In the convolutional layers of the neural network, a set of convolutional kernels resembling the shape of Gabor filters is initialized to begin training. This allows the network to learn filters sensitive to texture by itself. This paper designs an image recognition system that adopts multiple such convolutional kernels, each with different orientation and scale parameters, forming a complete filter bank to extract texture features of different orientations and frequencies.

For extracting local brightness features from images, this paper utilizes image gradient operators (Laplacian operator) to compute the brightness change rate at each pixel of the image. The resulting gradient magnitude and direction can be used to construct feature maps:

$$G(x, y) = \sqrt{I_x^2(x, y) + I_y^2(x, y)} \tag{18}$$

$G(x, y)$ represents the gradient magnitude, where I_x and I_y are the gradient components in the horizontal and vertical directions, respectively. Then, Gaussian filters are applied to the image to smooth noise and emphasize local brightness contrast:

$$G\left(x', y'; \sigma\right) = \frac{1}{2\pi\sigma^2} e^{-\frac{x'^2+y'^2}{2\sigma^2}} \tag{19}$$

$G\left(x', y'; \sigma\right)$ is a two-dimensional Gaussian function, followed by Gaussian filtering of the image:

$$I_{filtered}(x, y) = (I*G)(x, y) = \sum_{x'} \sum_{y'} I\left(x - x', y - y'\right) G\left(x', y'; \sigma\right) \tag{20}$$

After applying the Gaussian filter, $I_{filtered}(x, y)$ represents the image. By using the Gaussian filter, high-frequency noise components in the image are effectively suppressed. Moreover, because the Gaussian filter preserves the overall shape and low-frequency information of the image, edges and significant brightness contrast areas in the image are relatively highlighted. To further enhance local brightness contrast, the Laplacian operator is applied. The Laplacian operator, being a second-order differential operator, can emphasize edges and brightness change points in the image. Applying the Laplacian operator on the image enhances local brightness contrast:

$$LoG(x, y) = \nabla^2 I_{filtered}(x, y) \tag{21}$$

where ∇^2 represents the Laplacian operator. By applying a Gaussian filter to the original image followed by computing the Laplacian, it's termed as "Laplacian of Gaussian" (LoG). This process effectively removes noise while highlighting edges and contrast. The combined description of the above two processes in mathematical terms is as follows:

$$LoG(x, y) = \nabla^2 (I * G)(x, y) \tag{22}$$

$$LoG(x, y) = {}^2 (I * G)(x, y) \tag{22}$$

The above three network branches are fused together through a simple concatenation operation.

$$F_{fusion} = Concatenate(F_{color}, F_{texture}, F_{brightness}) \tag{23}$$

The fused features are then fed into a fully connected layer or another feature fusion layer for the final classification or segmentation prediction.

4 Experimental Analysis

The experimental section validates the effectiveness of the proposed underwater image processing algorithm, combining deep learning and physical models, in nuclear power plant cold zones. It includes algorithm implementation, performance evaluation, and comparison with existing technologies, showcasing its superiority over traditional methods.

The experiment is based on the Python programming language, utilizing the PyTorch framework to construct the deep learning model for underwater image processing. The final experimental results are obtained using the Matplotlib image analysis tool. The deep learning model training and inference processes are accelerated using an Nvidia GeForce RTX3050 Laptop GPU server. The underwater image acquisition equipment includes a high-sensitivity underwater camera, a high-definition imaging system mounted on an underwater robot, and necessary lighting equipment to adapt to the low-light underwater environment in the cold zone.

The experimental data consists entirely of RGB images, comprising two parts: one collected from the underwater operation site in the cold zone of the nuclear power plant, covering various time periods, lighting conditions, and water quality conditions, representing real data. The other part consists of images acquired non-underwater, which are then degraded using a known underwater optical model to simulate image degradation under different depths, suspended particle concentrations, and light intensities. The combination of real and simulated data enriches the diversity of training samples and improves the generalization ability of the deep learning model (Fig. 4).

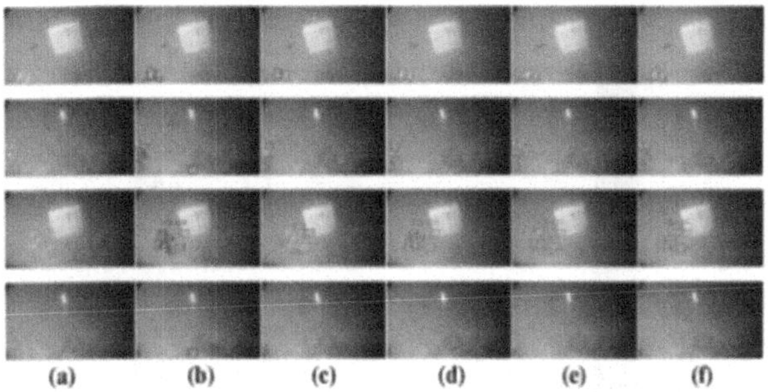

Fig. 4. The underwater images captured by the underwater image acquisition equipment in the cold zone of the nuclear power plant.

The image dehazing performance is evaluated using Peak Signal-to-Noise Ratio (PSNR) to assess the image quality of the DCP and physics-based dehazing methods (Fig. 5):

$$\mathrm{PSNR} = 10 \cdot \log_{10}\left(\frac{\mathrm{MAX}_I^2}{\mathrm{MSE}}\right) \tag{24}$$

PSNR stands for Peak Signal-to-Noise Ratio, where MSE represents the variance between an original m × n image and its corresponding image with added noise:

$$\mathrm{MSE} = \frac{1}{mn}\sum\nolimits_{i=0}^{m-1}\sum\nolimits_{i=0}^{n-1}\left[I(i,j) - K(i,j)\right]^2 \tag{25}$$

K(i, j) represents the pixel value at position (i, j) after adding noise. A higher signal-to-noise ratio indicates a smaller proportion of noise signal in the image, indicating better noise removal effectiveness. The results demonstrate that the dehazing effect of the dark channel prior method surpasses that of entirely physics-based dehazing methods.

Using precision in object detection determines the probability of the system correctly identifying the detected target (Fig. 6).

$$\mathrm{P} = \frac{TP}{TP + FP} \tag{26}$$

P represents precision, TP represents the number of correctly detected and identified targets, and FP represents the number of incorrectly detected targets. A higher precision indicates better performance in target detection. The experimental section compared the precision of object detection between the Faster R-CNN and YOLOv5 algorithms.

Methods based on deep convolutional neural networks (CNNs) have significantly improved the accuracy of underwater image recognition. These models can automatically learn complex feature representations, enabling high-precision target recognition even in complex underwater environments with low light, low visibility, and suspended particulate matter. The loss function can converge in a relatively short time.

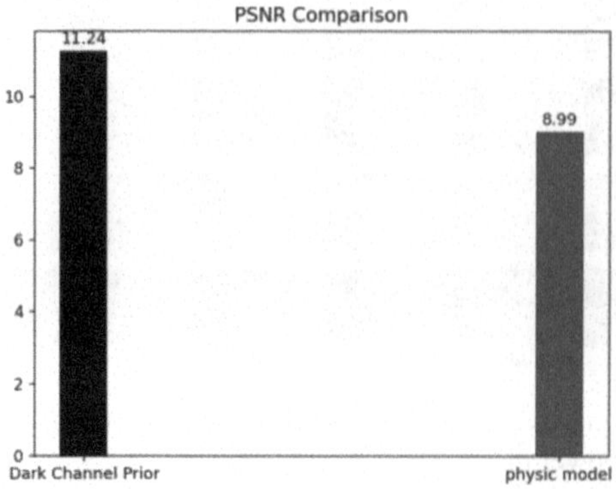

Fig. 5. Comparison of Peak Signal-to-Noise Ratio between Dark Channel Prior Dehazing and Physics-Based Dehazing Methods

Comparing the recognition performance of images in non-underwater environments (ideal conditions), underwater environments, and after being processed by the image recognition algorithm, it is evident that the algorithm designed in this paper maximally restores the true values of the images, effectively filtering out the noise effects overlaid by the underwater environment (Fig. 7).

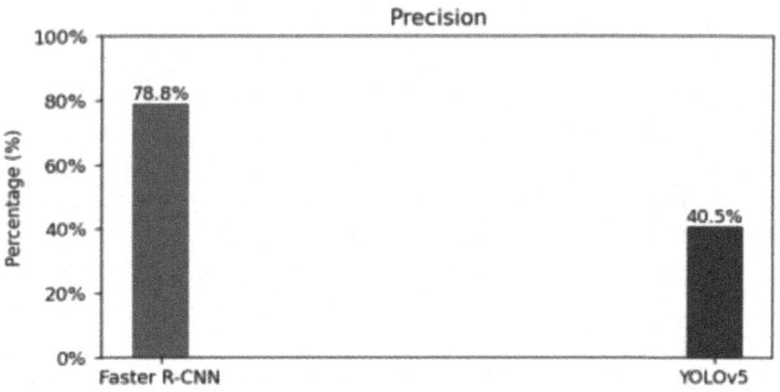

Fig. 6. Precision comparison graph between Faster R-CNN (in red) and YOLOv5 algorithm (in blue). (Color figure online)

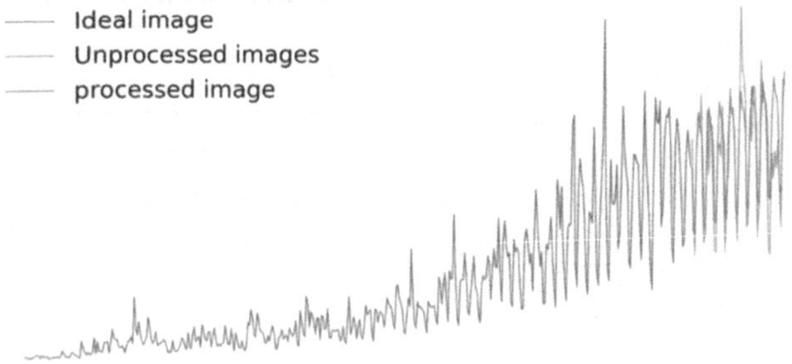

Fig. 7. Comparison chart of pixel values in images

5 Conclusion

The developed underwater image processing algorithm integrates physical modeling, deep learning, and attention mechanisms to effectively address image quality issues in nuclear power plant operations. By enhancing image clarity and detail presentation, this algorithm reduces misjudgment risks and improves operational safety. Overall, the algorithm demonstrates promising performance in processing underwater images in the challenging conditions of nuclear power plant environments.

Acknowledgments. This study was funded by the Science and Technology Project of Shenzhen (grant numbers GJHZ20210705141401005, CJGJZD20220517141405012 and KCXST20221021111402006), the "Nanling Team Project" of Shaoguan City, the Shenzhen Excellent Innovative Talents (grant number RCYX20221008093036022) and the Shenzhen-Hong Kong Joint Funding Project (A) (grant number SGDX20230116092053005).

Disclosure of Interests.. The authors have no competing interests to declare that are relevant to the content of this article.

References

1. Sun, J.: New technologies and equipment for coal mine monitoring. Ind. Min. Autom. **41**(01), 1–5 (2015). https://doi.org/10.13272/j.issn.1671-251x.2015.01.001
2. Liu, X.: Condition monitoring system for mine rotating equipment based on vibration signals. Coal Mine Mach. **41**(01), 140–144 (2020). https://doi.org/10.13436/j.mkjx.202001048
3. Yuan, H., Wu, Q., Huang, B.: Analysis of smart control systems for urban street lighting in the context of communication technology evolution. Light Light (S1), 100–103 (2022)
4. Lu, Y., Liu, Y., Hu, C., et al.: Research on IoT sensing technology for overhead lines based on LoRa. Electr. Appl. **38**(07), 68–72 (2019)
5. Zhu, J., Guo, L., Wang, Q., et al.: Design and implementation of an agricultural greenhouse monitoring system based on LoRa. Commun. Technol. **51**(10), 2430–2435 (2018)
6. Huo, Z.: Application analysis of LoRa technology in mine wireless communication. Ind. Min. Autom. **43**(10), 34–37 (2017). https://doi.org/10.13272/j.issn.1671-251x.2017.10.006
7. Gu, H.: Statistical distribution and cause analysis of titanium tube deformation defects in nuclear power plant condensers. Sci. Technol. Innov. **01**, 7–8 (2021)
8. Men, C., Han, R.: Research on improvement strategies for water intake engineering design in China's nuclear power plants. Water Supply Drain. **58**(S2), 264–268 (2022). https://doi.org/10.13789/j.cnki.wwe1964.2022.04.01.0002
9. Perezyabov, O.A., Maltseva, N.K., Ilinski, A.V.: Measuring the modulation-transfer function of radiation-tolerant machine-vision system using the sum of harmonic components of different frequency. In: Proceedings of the SPIE, vol. 10231, Optical Sensors, 102311R (2017). https://doi.org/10.1117/12.2265612
10. Volodin, V.S., Tolokonskij, A.O.: Application of machine learning for solving problems of nuclear power plant operation. In: Klimov, V.V., Kelley, D.J. (eds.) Biologically Inspired Cognitive Architectures 2021, BICA 2021. SCI, vol. 1032, pp. 585–589. Springer, Cham (2022). https://doi.org/10.1007/978-3-030-96993-6_65

A Novel Anti-noise Fault Diagnosis Method for Nuclear Energy Steam Turbine Based on VCformer

Yu Huang[1], Jiajing Zhou[2], Zhao An[2], Mo Li[1], Zhile Yang[2], and Yuanjun Guo[2](✉)

[1] China Nuclear Industry Maintenance Co., Ltd., Shenzhen Branch, Shenzhen 518120, China
[2] Shenzhen Institutes of Advanced Technology, Chinese Academy of Sciences, Shenzhen 518055, Guangdong, China
yj.guo@siat.ac.cn

Abstract. This paper introduces the VCformer, a Variational Mode Decomposition-Convolutional Neural Network-Transformer model, optimized for fault diagnosis in nuclear steam turbines amidst noisy environments. The VCformer employs VMD to break down vibration data into five distinct modal signals, and these signals, along with the original noisy input, are individually processed through a dual-layer CNN-Transformer framework, enabling precise feature extraction and fault classification for each modal component. Following this individual analysis, the VCformer utilizes an ensemble learning strategy, where a voting system dynamically integrates the outcomes from the dual-layer CNN-Transformer framework of each signal. This process effectively adjusts the voting weights in real-time, culminating in an optimized final output that leverages the comprehensive insights gained from both the original and decomposed signals. Empirical testing has demonstrated the VCformer's robust performance, with a minimal decrease in accuracy of only 0.16% even at a 20% noise level. This innovative methodology highlights the VCformer's capability to provide accurate diagnostics in environments with significant noise, representing a novel advancement in intelligent monitoring technology for the energy sector.

Keywords: VMD · Nuclear Energy · Anti-noise · Fault Diagnosis

1 Introduction

The onset of Industry 4.0 has catalyzed a transformative shift in traditional manufacturing, heralding a new era replete with both unprecedented opportunities and significant challenges [1]. This transformative phase is characterized by the escalating scarcity of fossil fuels and a growing cognizance of their environmental impacts, prompting a global pivot towards renewable energy sources. Such a shift is not only a response to depleting fossil resources but is also propelled by international endeavors to achieve carbon neutrality and reduce emissions. The assimilation of renewable energies into the existing power grids necessitates the development of advanced management and energy storage systems, essential for addressing the variability and intermittency intrinsic to these energy sources.

K. Li et al. (Eds.): LSMS/ICSEE 2024, CCIS 2218, pp. 67–78, 2025.
https://doi.org/10.1007/978-981-96-0232-2_6

Central to the ethos of Industry 4.0 is the comprehensive integration of digital technologies into the fabric of the economy, with the aim of embedding intelligence and automation throughout the production, management, and services sectors. In this evolving landscape, steam turbines, integral to energy conversion processes, are compelled to adapt to the demands of this new era. The integration of nuclear energy with the principles of Industry 4.0 further underscores the urgency for enhanced fault diagnosis mechanisms in nuclear steam turbines. Given their pivotal role in nuclear power generation systems, maintaining the security and reliability of these turbines is paramount for ensuring sustainable energy production and advancing environmental stewardship [2].

Current methodologies for fault diagnosis in nuclear steam turbines can be broadly categorized into four developmental trajectories: Those grounded in physical models, those relying on signal processing, approaches leveraging machine learning, and hybrid methodologies amalgamating these paradigms [3]. Physical model-based methods, however, grapple with inherent limitations, necessitating precise knowledge, incurring high development costs, exhibiting limited adaptability, and struggling to capture the nuanced nonlinear interactions and unidentified failures within nuclear steam turbines [4]. Conversely, signal processing-based methods mandate extensive data preprocessing, susceptible to the influence of measurement noise and various forms of interference. These methods may also encounter challenges in comprehending the intricate nonlinear interactions inherent in nuclear steam turbines, impeding their ability to discern faults not easily distinguishable in signal data [5].

Recent years have witnessed a burgeoning interest among researchers in the application of deep learning methodologies for nuclear steam turbine fault diagnosis [6]. An et al. introduced a comprehensive deep learning framework for fault detection and prediction, merging principal component analysis (PCA) and Informer to address the complexities of online monitoring for nuclear power valves [7]. Yin et al. addressed the challenge of sample imbalance in fault diagnosis for rotating machinery by employing adaptive synthetic sampling. They coupled this with ensemble empirical mode decomposition (EEMD) and continuous wavelet transform (CWT) for time-frequency image conversion of vibration data, culminating in the construction of a deep residual neural network for precise fault diagnosis in rotating machinery within nuclear power plants [8]. Chen et al. proposed a deep learning model anchored in long short-term memory (LSTM) and dropout mechanisms to predict the remaining time until automatic emergency shutdown of nuclear power plants under abnormal conditions [9]. Yi et al. introduced an unsupervised anomaly detection model based on Transformer, augmenting its robustness through the utilization of a modified loss function grounded in the maximum cross-correlation criterion for model training. This model aimed to achieve digital and intelligent status monitoring of nuclear power plants, thereby enhancing its resilience [10]. Yang et al. proposed a new method for extracting the status features of rolling bearings using Variational Mode Decomposition (VMD) and Improved Envelope Spectrum Entropy (IESE) to accurately identify bearing conditions and address bearing fault issues [11].

In this paper, we introduce the VCformer, a Variational Mode Decomposition-Convolutional Neural Network-Transformer model, meticulously engineered for anti-noise classification. The VCformer employs VMD to segment vibration data into distinct modes effectively, significantly reducing noise and enhancing signal clarity. This segmented data then undergoes precise feature extraction by a Convolutional Neural Network (CNN), which, in tandem with the Transformer architecture, creates a powerful two-layer CNN-Transformer framework. A distinctive feature of the VCformer is its innovative decision-making process, which leverages a voting system across six different outcomes, including the original data and five modal components obtained from VMD decomposition. This approach not only guarantees a comprehensive and robust diagnostic evaluation but also optimizes sample representation, thereby substantially enhancing the fault diagnosis accuracy and generalizability of the model. The results demonstrate the VCformer's outstanding diagnostic capabilities, establishing it as a revolutionary tool in the energy sector for enhancing machinery maintenance and reliability. Its novel integration of advanced methodologies positions the VCformer as a significant innovation in machinery fault diagnosis.

2 Method

2.1 Diagnostic Signal Decomposition

VMD is a signal processing technique that decomposes a complex signal into a predefined number of modes with specific bandwidths. It operates by solving a variational problem to iteratively extract each mode, characterized by a central frequency and bandwidth, ensuring the modes are both sparse in the frequency domain and have limited bandwidth. The mathematical formulas and concepts involved in this function mainly include:

Fourier Transform: Used to convert a time-domain signal to its frequency-domain representation, the formula is:

$$\hat{f} = \mathcal{F}(f) = \int_{-\infty}^{\infty} f(t)e^{-j\omega t}dt \tag{1}$$

where $\hat{f}$ is the Fourier transform of the signal f, and ω represents the angular frequency.

Wiener Filter Update: In the VMD algorithm, the spectrum of each mode $\hat{u}_k$ is updated using a Wiener filter approach:

$$\hat{u}_k^{n+1} = \frac{\hat{f}^{+} - \sum\limits_{i \neq k} \hat{u}_i^n - \frac{\lambda^n}{2}}{1 + \alpha(\omega - \omega_k^n)^2} \tag{2}$$

Here, $\hat{u}_k^{n+1}$ is the spectrum of the k th mode at iteration $n+1$, $\hat{f}^{+}$ is the positive frequency part of the signal spectrum, λ is the dual variable, α is the balance parameter for data fidelity, and ω_k^n is the center frequency of the k th mode at iteration n.

Center Frequency Update: The center frequency ω_k is updated using the formula:

$$\omega_k^{n+1} = \frac{\sum \omega|\hat{u}_k^{n+1}(\omega)|^2}{\sum |\hat{u}_k^{n+1}(\omega)|^2} \tag{3}$$

This is calculated as the weighted average frequency of the spectrum $\hat{u}_k^{n+1}$, where the weights are $|\hat{u}_k^{n+1}(\omega)|^2$.

Dual Variable Update: The dual variable λ is updated through a gradient ascent method:

$$\lambda^{n+1} = \lambda^n + \tau\left(\sum_k \hat{u}_k^{n+1} - \hat{f}^+\right) \tag{4}$$

where τ is the time-step of the dual ascent.

Using VMD with parameters alpha = 3000, tau = 0, K = 5, DC = 0, init = 1, and tol = 1e-7 on noise data from a normally functioning system allows for the precise decomposition into five distinct modes, offering clear insights into the noise structure. This analysis, as shown in Fig. 1, aids in understanding the system's operational characteristics and identifying any underlying issues.

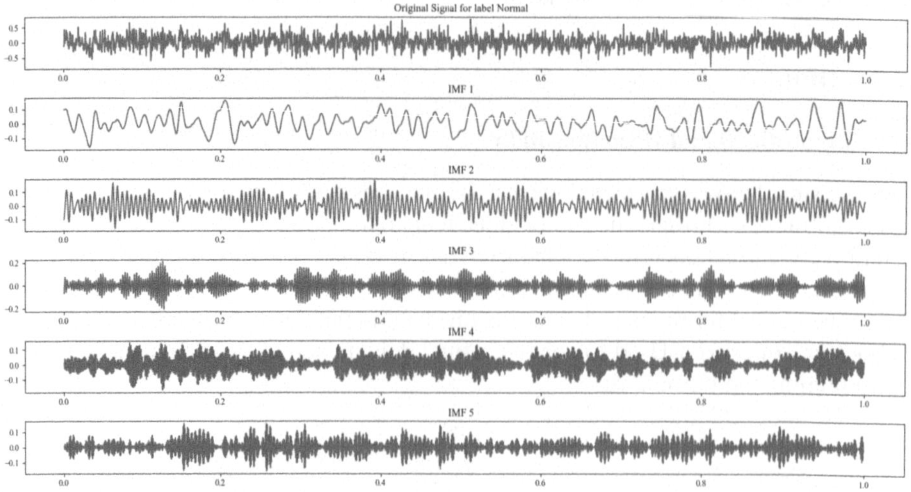

Fig. 1. Multi-Modal Noise Decomposition of Normal Operational Data.

2.2 Feature Extraction

CNN comprises convolutional layers, pooling layers, fully connected layers, activation function layers, batch normalization layers, and dropout layers. In the convolutional layer, the main operation is convolution, expressed mathematically as:

$$y_j^l = X * W = g(\sum W_{ij}^l \cdot X_i^{l-1} + b_j^l) \tag{5}$$

where, X represents the input of the convolution operation, $*$ represents the convolution operation itself, W describes the convolutional kernel, l is the non-linear activation function, and b_j^l is the bias term for the j-th convolution process in the l-th layer.

In the pooling layer, the selected formula is max pooling, and the mathematical expression for max pooling is as follows:

$$y_{ij} = max_{(p,q)\in P_{ij}} x_{pq} \tag{6}$$

Here, y_{ij} represents the output after pooling, P_{ij} denotes the pooling region, and x_{pq} is the input values within that region.

In the fully connected layer, the mathematical expression is as follows:

$$y_i = f(\sum\nolimits_{j=1}^{N} w_{ij} \cdot x_j + b_i) \tag{7}$$

Here, y_i is the value of the i-th neuron in the output of the fully connected layer. N is the number of neurons in the previous layer. x_j is the output of the j-th neuron in the b_i-th neuron previous layer. w_{ij} represents the weight connecting the j-th input neuron and the i-th output neuron. b_i is the bias term for the i-th output neuron. f is the activation function.

In the activation function layer, the chosen function is the ReLU (Rectified Linear Unit) function, with the mathematical formula as follows:

$$f(x) = max(0, x) \tag{8}$$

where $f(x)$ represents the output of the ReLU activation function, mapping the input x to non-negative values. Specifically, if x is greater than zero, the output is x; if x is less than or equal to 0, the output is 0.

In the batch normalization layer, input normalization is performed to center the mean around zero and standard deviation around one, aiding in accelerating the training process and enhancing the network's robustness. The fundamental concept of batch normalization involves normalizing each mini-batch and then learning appropriate offsets through scaling and shifting parameters. The batch normalization operation is as follows:

$$BN(x) = \gamma \frac{x - \mu}{\sqrt{\sigma^2 + \varepsilon}} + \beta \tag{9}$$

x represents the input mini-batch. μ denotes the mean of the mini-batch. σ^2 denotes the variance of the mini-batch. γ represents the learnable scaling parameter. β represents the learnable shifting parameter. ε is a small constant added to prevent division by zero errors.

In the dropout layer, dropout is employed during the training process by randomly deactivating some neurons in the neural network, thereby reducing overfitting. The dropout operation is as follows:

$$Dropout(x, p) = \begin{cases} x \ with\,probability\ 1 - p \\ 0 \quad with\,probability\,p \end{cases} \tag{10}$$

Here, x is the input, and p is the probability of deactivating neurons.

2.3 Anti-noise Diagnostic Framework

In this paper, we introduce the VCformer, a model specifically designed for anti-noise classification in fault diagnosis. The VCformer ingeniously utilizes VMD to segment vibration data into distinct modes, effectively reducing noise and enhancing signal clarity. This data is then processed through a dual-layer CNN where the first layer (Conv1) captures primary features, and the second layer (Conv2) delves into more complex feature extraction, establishing a nuanced understanding of the hierarchical structure of the data. The innovation extends to the integration of a Transformer architecture that excels in modeling time-series data, leveraging self-attention mechanisms to capture long-range dependencies within the sequence, surpassing the limitations of traditional fixed-length memory units like LSTMs. This synergy between the CNN layers and the Transformer facilitates a deep comprehension of both spatial and temporal data characteristics. The VCformer's decision-making process employs a voting system among six different outcomes, derived from the original data and the five VMD-based modal components, ensuring a robust diagnostic assessment. This unique approach enhances the fault diagnosis accuracy and generalizability of the model. In the subsequent feature fusion and classification phase, the outputs from the dual-layer CNN and Transformer are concatenated, then fed into fully connected layers for the final classification, adeptly mapping the fused features to specific fault categories.

The empirical results highlight the VCformer's superior diagnostic capabilities, marking it as a revolutionary advancement in enhancing machinery maintenance and reliability within the energy sector. This novel integration of VMD, CNN, and Transformer technologies in the VCformer establishes a significant breakthrough in the field of machinery fault diagnosis, as depicted in the network architecture diagram Fig. 2, providing a comprehensive and effective solution for diagnosing nuclear steam turbine faults.

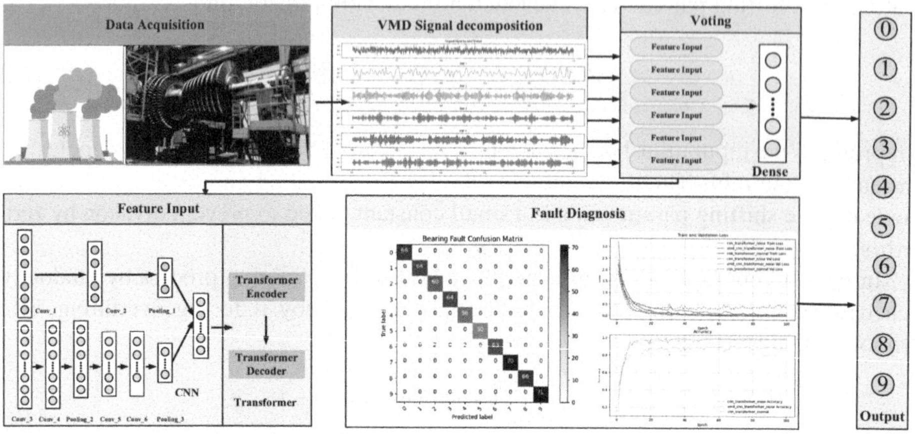

Fig. 2. The network architecture diagram for the VCformer.

3 Experiments and Data Analysis

3.1 Data Source

The experimental dataset utilized in this study is derived from the rolling bearing test rig at Case Western Reserve University (CWRU), illustrated in Fig. 3. The test rig comprises a motor, bearings, and a load motor. Failure modes include machining defects such as drilling or cutting, with acceleration sensors sampling vibration signals. Sampling frequencies are set at 12 kHz and 48 kHz. The dataset covers four standard categories of samples: inner race failure, roller failure, outer race failure.

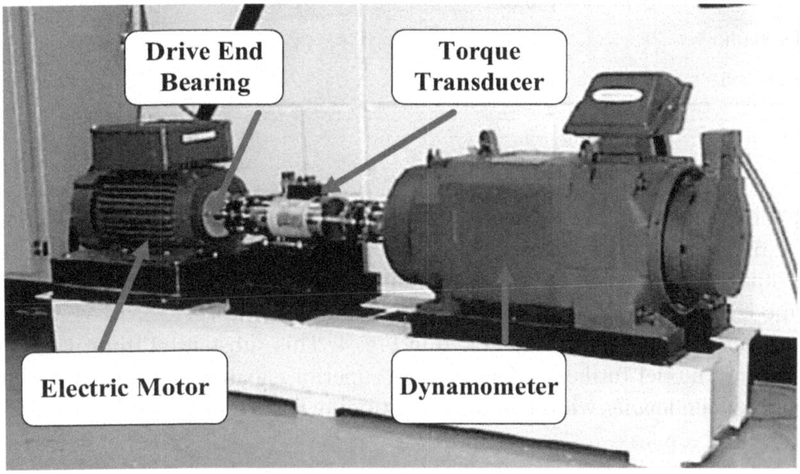

Fig. 3. Bearing experimental data workstation.

Data from the 48DriveEndFault category, operating at a rotational speed of 1730 RPM, were specifically chosen for experimentation. To ensure consistent data length extraction, a standardized collection of 2000 data points was employed. The data format is elaborated in Table 1 below.

3.2 Experiment Results and Analysis

During the training process, a learning rate of 0.0005 was employed. Initially, the dataset was partitioned into a training set, comprising 70% of the total dataset, and a validation set, constituting 30%. A batch size of 128 was configured to facilitate efficient training.

After completing 100 training epochs, the Transformer-based model demonstrated exceptional performance: achieving a training loss of 0.0125, a validation loss of 0.0400, and an impressive accuracy of 98.91%. The model also showcased superior statistical measures, with an MSE of 0.1172, RMSE of 0.3423, MAE of 0.0328, and an R2 Score of 0.9856, as illustrated in Fig. 4. These outcomes emphasize the model's excellent fit to the training data and its remarkable generalization ability on the validation set, affirming its effectiveness in managing the assigned task.

Table 1. Bearing experimental dataset details.

Bearing fault	Label	Defect size (inch)	No. Samples	Data length
Ball fault	0	0.007	240	2000
Inner race fault	1	0.007	240	2000
Outer race fault	2	0.007	240	2000
Ball fault	3	0.014	240	2000
Inner race fault	4	0.014	240	2000
Outer race fault	5	0.014	240	2000
Ball fault	6	0.021	240	2000
Inner race fault	7	0.021	240	2000
Outer race fault	8	0.021	240	2000
Normal	9	------	240	2000

Comparatively, when juxtaposed with the LSTM-based model, which yielded a Train Loss of 1.6904, Val Loss of 1.6753, and Accuracy of 78.59%, the transformer model clearly outperforms in terms of both loss minimization and accuracy enhancement. The LSTM model also recorded an MSE of 1.7297, RMSE of 1.3152, MAE of 0.5578, and an R2 Score of 0.7897, as illustrated in Fig. 4. This substantial improvement with the transformer model further underscores its superior capability in capturing intricate patterns and dependencies within the data, positioning it as a more effective and advanced choice for the given task.

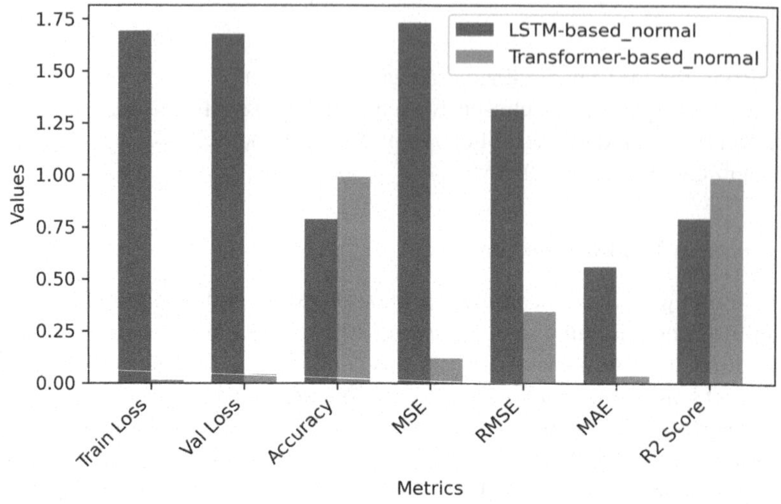

Fig. 4. Comparative Performance Metrics of Transformer-based vs LSTM-based Models.

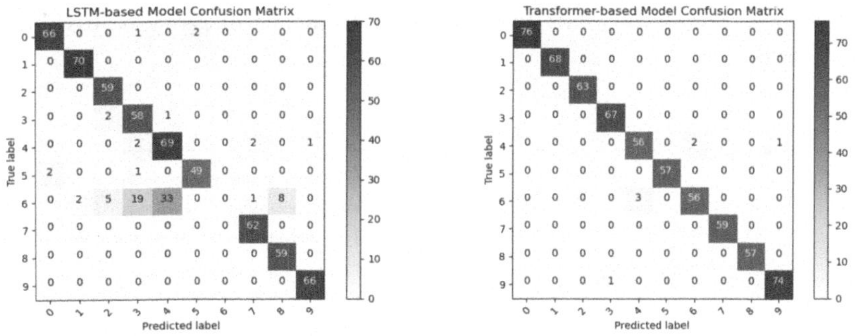

Fig. 5. Comparative Confusion Matrix Evaluation: LSTM (left) vs Transformer(right) Models.

In the confusion matrix for bearing classification, as illustrated in Fig. 5, right panel, it is evident that the transformer-based model showcases remarkable performance. Almost all categories are correctly classified, with only a negligible number of samples being misclassified. This exceptional performance is reflected in the sparse presence of confusion entries within the matrix, indicating the model's high precision and accuracy in bearing classification tasks. The model effectively distinguishes various categories of bearing states, underscoring its reliability and stability in practical applications. This proficiency is crucial for ensuring accurate monitoring and fault diagnosis of bearing conditions.

In contrast, upon examining the LSTM-based model, specific observations come to light, as depicted in Fig. 5, left panel. Notably, when the actual label is 6, the LSTM model tends to predominantly predict labels 2, 3, 4, and 8. This tendency is indicative of a certain limitation in the LSTM model's performance, particularly in accurately categorizing instances associated with label 6.

However, when transitioning to the transformer-based model, a significant improvement is discerned. The transformer model exhibits enhanced capabilities in correctly predicting instances with the actual label of 6, overcoming the limitations observed in the LSTM model. This notable advancement in performance underscores the transformative impact of the transformer architecture, particularly in scenarios where accurate classification of label 6 is imperative for comprehensive bearing condition monitoring and fault diagnosis.

3.3 Noise-Affected Data Experimentation and Analysis

To simulate real-world conditions, the original vibration signal data was deliberately corrupted by introducing 20% white noise, a common strategy to emulate environmental noise interference in actual operations. Following this augmentation, the performance metrics of the transformer-based network experienced a decline in accuracy when compared to the noise-free scenario. The specific indicators after noise addition were as follows: a Train Loss of 0.0583, Val Loss of 0.3607, an Accuracy of 90.47%, MSE of 0.2141, RMSE of 0.4627, MAE of 0.1266, and an R2 Score of 0.9743. When compared with the previous accuracy of 98.91%, there is an 8.6% decrease in model accuracy after the introduction of noise. These results provide a crucial evaluation of the model's robustness to noise and its ability to maintain performance integrity is less than ideal data conditions.

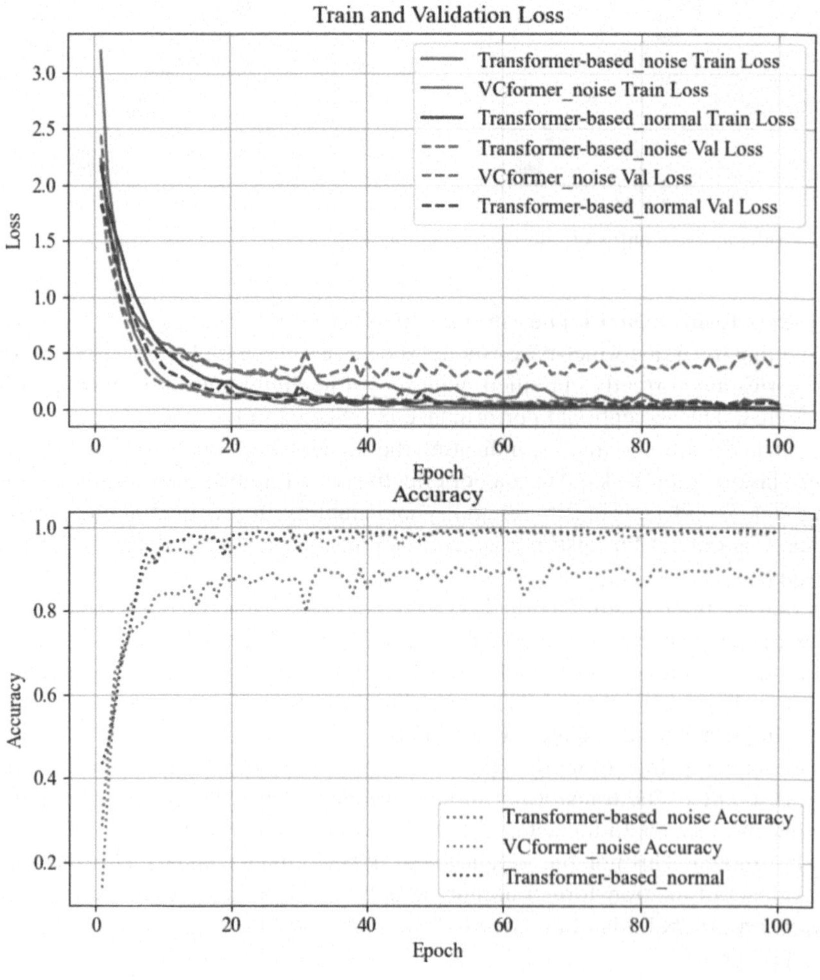

Fig. 6. Comparative Training Dynamics of VCformer vs Transformer Models.

In the experimental evaluation under noise-influenced conditions, the VCformer model effectively employs VMD to separate the signal into five modal components, with one specifically dedicated to capturing the noise component. This process ensures that noise is not merely isolated but is actively extracted and encapsulated within a distinct modal signal. The study meticulously inputs both this noise-specific modal signal and the remaining modal signals, alongside the original noisy signal, into the transformer-based model independently. This careful approach optimizes data utilization and retains essential information. During training, the model employs gradient optimization to dynamically adjust the voting weights of the six outputs, refining the final decision to reflect the most significant signals, including the extracted noise component, thus improving the diagnostic accuracy and reliability.

The performance metrics for the VCformer, with precision to four decimal places, are as follows: Train Loss is 0.0019, Val Loss is 0.0483, Accuracy is 98.75%, MSE is 0.0359, RMSE is 0.1896, MAE is 0.0172, and the R2 Score is 0.9955. In contrast, the transformer model under noisy conditions exhibited Train Loss of 0.0583, Val Loss of 0.3607, Accuracy of 90.4700%, MSE of 0.2141, RMSE of 0.4627, MAE of 0.1266, and R2 Score of 0.9743.

The comparative analysis clearly shows the VCformer's substantial advantage over the transformer model in all performance metrics, highlighting its refined capability to extract and utilize noise for enhancing fault diagnosis accuracy, and affirming its robustness and effectiveness in noise-pervasive operational environments.

When the VCformer is subjected to 20% white noise, its accuracy is only marginally reduced by 0.16% compared to the noise-free transformer-based model, as illustrated in Fig. 6. This minor discrepancy underscores the VCformer's efficacy in noise segregation and its substantial capability in addressing diagnostic challenges in noisy working conditions.

4 Conclusion

In conclusion, this paper introduces the VCformer, an innovative model that synergizes the strengths of VMD, a dual-layer CNN, and Transformer architecture, providing an in-depth analysis of vibration signals from nuclear steam turbines. A salient aspect of the VCformer is its inclusive approach to data input, which encompasses both the unfiltered noisy signal and the quintet of modal signals resultant from VMD. This is in stark contrast to conventional VMD techniques that are limited to utilizing a solitary or a constrained selection of decomposed signals. By harnessing the complete array of modal signals in concert with the raw data, the VCformer exploits the full gamut of available information, thereby maximizing the interpretive utility of the dataset. This holistic data assimilation, in concert with the dynamic ensemble learning voting weight adjustments, ensures that the VCformer navigates through noise with precision, significantly bolstering the accuracy of fault diagnostics. It boasts remarkable fortitude against noise, evidenced by the trifling 0.16% dip in accuracy even when confronted with up to a 20% infusion of noise. This robust performance marks a significant advancement in the development of intelligent diagnostic systems, setting a new standard for operational efficiency in power generation machinery and paving the way for future innovations in this field.

Acknowledgments. This research was supported by the Science and Technology Project of Shenzhen (Nos. GJHZ20210705141401005 and CJGJZD20220517141405012), the "Nanling Team Project" of Shaoguan City, and the Shenzhen-Hong Kong Joint Funding Project (A) (No. SGDX20230116092053005).

Disclosure of Interests. The authors have no competing interests to declare that are relevant to the content of this article.

References

1. Abulibdeh, A., Zaidan, E., Abulibdeh, R.: Navigating the confluence of artificial intelligence and education for sustainable development in the era of industry 4.0: Challenges, opportunities, and ethical dimensions. Journal of Cleaner Production **437**, 140527 (2024)
2. Zhou, J., An, Z., Yang, Z., et al.: PT-Informer: A Deep Learning Framework for Nuclear Steam Turbine Fault Diagnosis and Prediction. Machines **11**(8), 846 (2023)
3. Zhu, Z., Lei, Y., Qi, G., et al.: A review of the application of deep learning in intelligent fault diagnosis of rotating machinery. Measurement **206**, 112346 (2023)
4. Gao, Z., Cecati, C., Ding, S.X.: A survey of fault diagnosis and fault-tolerant techniques—Part I: Fault diagnosis with model-based and signal-based approaches. IEEE Trans. Industr. Electron. **62**(6), 3757–3767 (2015)
5. Lu, S., He, Q., Wang, J.: A review of stochastic resonance in rotating machine fault detection. Mech. Syst. Signal Process. **116**, 230–260 (2019)
6. Xiaofeng, L.I.U., Yingjie, C., Xiong, L., et al.: Intelligent fault diagnosis methods toward gas turbine: A review. Chinese Journal of Aeronautics **37**(4), 93–120 (2023)
7. An, Z., Cheng, L., Guo, Y., et al.: A novel principal component analysis-informer model for fault prediction of nuclear valves. Machines **10**(4), 240 (2022)
8. Yin, W., Xia, H., Huang, X., et al.: A fault diagnosis method for nuclear power plants rotating machinery based on deep learning under imbalanced samples. Ann. Nucl. Energy **199**, 110340 (2024)
9. Chen, H., Gao, P., Tan, S., et al.: Prediction of Automatic Scram during Abnormal Conditions of Nuclear Power Plants Based on Long Short-Term Memory (LSTM) and Dropout. Science and Technology of Nuclear Installations **2023**, 1–14 (2023)
10. Yi, S., Zheng, S., Yang, S., et al.: Robust transformer-based anomaly detection for nuclear power data using maximum correntropy criterion. Nuclear Engineering and Technology **56**(4), 1284–1295 (2023)
11. Yang, Y., Liu, H., Han, L., et al.: A feature extraction method using VMD and improved envelope spectrum entropy for rolling bearing fault diagnosis. IEEE Sens. J. **23**(4), 3848–3858 (2023)

Short-Term Photovoltaic Power Forecasting Using Multi-timescale Information Based FFTformer Model

Haohao Feng[1], Xiaoyun Niu[1], Mifeng Ren[1](✉), Junghui Chen[2], and Wenjie Zhang[1]

[1] College of Electrical Power and Engineering, Taiyuan University of Technology, Taiyuan, China
renmifeng@tyut.edu.cn

[2] Epartment of Chemical Engineering, Chung-Yuan Christian University, Taoyuan, Taiwan, Republic of China

Abstract. The addition of energy storage can enhance the dispatch capability of PV power plants and help them better adapt to grid demand. One of the keys to the stable control of a combined photovoltaic and storage power system is the accurate prediction of PV power generation. This paper proposes an FFTformer predictive model based on Fast Fourier Transform (FFT), 1-D CNN-LSTM and multi-timescale information. Firstly, the proposed model uses Fourier Transform to remove the noise of different timescale data. Secondly, the multi-attention mechanism is employed to strengthen the characterization ability of the data. At the same time, power features at different time scales are extracted using 1-D CNN networks and LSTM networks. Finally, the multi-timescale features extracted by the attention mechanism and 1D CNN-LSTM are used to predict the photovoltaic (PV) power. To validate the effectiveness of the model, a historical dataset from Hebei Province, China is used for various types of experiments. Among them, the effectiveness and advantages of the proposed model FFORMER are verified by comparison tests with other models.

Keywords: Photovoltaic Power Prediction · Multi-timescale information · Fast Fourier Transform · Attention Mechanism

1 Introduction

In recent years, global resource scarcity and environmental problems are increasingly apparent. Energy conservation and emission reduction have become a consensus in each country. In 2020, China launched the goal of "striving to peak carbon dioxide emissions by 2030 and striving to achieve carbon neutrality by 2060" at the United Nations General Assembly's 75th session [1, 2]. A PV storage hybrid system is an energy system that combines photovoltaic power and energy storage. PV power generation is intermittent, and energy storage systems can cope with intermittent energy fluctuations and improve energy utilization. Accurate PV power prediction can help energy storage respond faster to PV fluctuations and increase the efficiency and economy of a PV storage hybrid system, then help decision-makers coordinate the relationships between PV, energy storage, and the grid.

K. Li et al. (Eds.): LSMS/ICSEE 2024, CCIS 2218, pp. 79–90, 2025.
https://doi.org/10.1007/978-981-96-0232-2_7

PV power forecasting models can be divided into three aspects: The methods of physics, statistics, and artificial intelligence [3, 4]. The physical prediction method is used to achieve PV power prediction by simulating the process of converting solar energy into electricity by PV modules, but this method is less generalizable and robust, and needs to be modeled according to different parameters such as geographic location and installation information [5]. Statistical prediction uses historical data to analyze and learn characteristics such as trends, seasonal variations, and periodicity of the data. Then it predicts future PV power output based on the learned mapping laws. Common statistical prediction methods can be divided into time series prediction [6] and gray theory prediction [7]. However, the principal limitation of the statistical prediction method is that it requires a large number of historical data points.

The limitation of the above approaches is that the high need for datasets and the limitations of known models or hypotheses lead to weak generalization ability. With the development of modern information technology and artificial neural networks, artificial intelligence prediction is widely applied to predicting PV power. The literature [8] proposed a time-division long short-term memory network (time-division LSTM) that uses the Pearson correlation coefficient to select time series features and atmospheric features about PV power, and these features are inputted into LSTM networks to predict PV power. Literature [9] adds the Attention Mechanism to LSTM embedding to extract time series feature information and predict PV power. Literature [10] proposes an informer model that overcomes the gradient explosion or disappearance and the inability to parallelize the calculation of traditional recurrent neural networks (RNN) and LSTM. At the same time, the Informer model is based on Transformer, which adds ProbSpare Self-Attention Mechanism, parallel decoding, and dropout to the encoder section. These let the model reduce memory occupation and calculation complexity. Literature [11] proposes a FEDformer model that adds a frequency domain learning module, a frequency domain attention mechanism, and a cyclical-trend decomposition module, which reduces the calculation complexity further. This model also considers the impact of cyclical and trending data and high-frequency disturbances to increase the accuracy of the model. However, we learn from the above studies that the current intelligent forecasts mostly use a single-scale information. Multi-scale datasets can help model learning by embedding more information when training neural networks. In this paper, a PV power prediction model based on FFT is proposed. This model uses Multi-timescale datasets and the Fast Fourier Transform algorithm (FFT), transfers time domain data into frequency domain data to eliminate high-frequency disturbance, and then uses inverse FFT. Then the cross-attention mechanism is used to weight different time scales of data information, Simultaneously, 1-D CNN networks and LSTM networks are used to extract features from historical information at different time scales. Finally, PV power prediction is performed using feature information extracted from different time scales using cross-attention mechanism and 1D CNN-LSTM, respectively. In addition, the FFTformer model is vali-dated by using a dataset from a photovoltaic power station in Hebei and conducted comparative experiments to prove the validity and superiority of the FFTformer model.

2 Proposed Method

2.1 FFTformer Model Structure

PV power forecasting can be considered as a time series prediction problem, which can be solved by using the Transformer structure. However, the Informer model doesn't consider the impact of disturbances on the accuracy of models; the FEDformer model uses only single-timescale information to predict PV power [12]. The FFTformer model we proposed uses the FFT algorithm, attention mechanism, 1-D CNN and LSTM to deal with the high-frequency data of PV power and relevant Multi-timescale features. To enhance the accuracy of the model, we consider different timescales of dataset, which is expressed as $y_n = \{D_1, \ldots, D_n\}$, $D = \left(x_1^i, \ldots, x_L^i\right)$ where $i = 1, \ldots, M$. Here, D denotes the historical dataset of different timescale, L denotes the length of the sequence and M denotes dimension of each timestamp. In this study, the lookback window $(x_1, \ldots, x_l)$ is used to forecast T future values $(x_{l+1}, \ldots, x_{l+T})$. The FFTformer structure diagram proposed in this study is shown in Fig. 1.

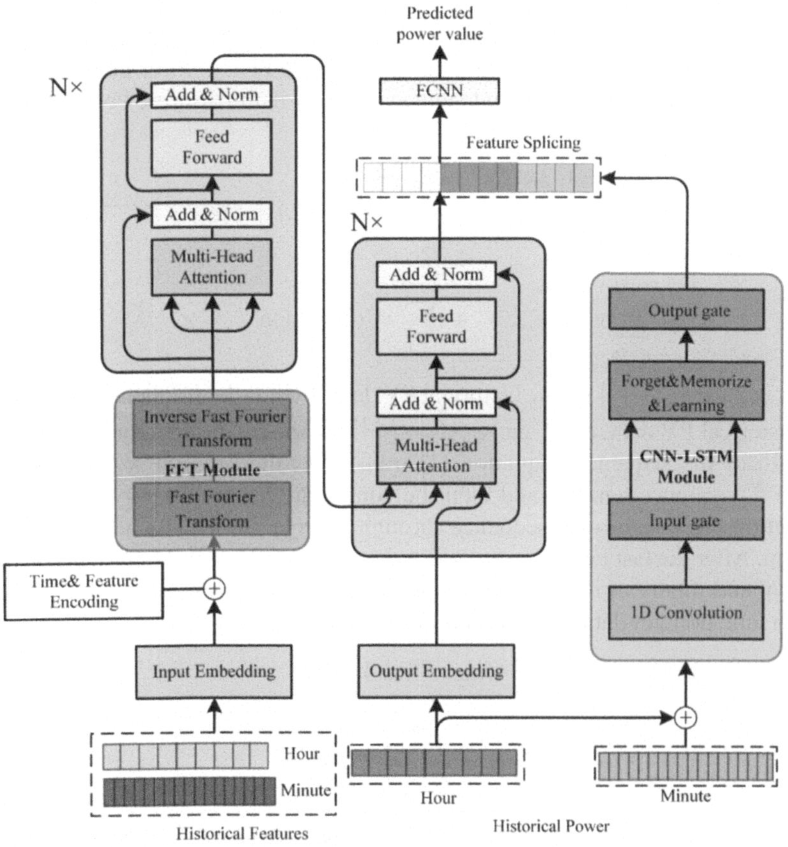

Fig. 1. FFTformer model structure.

The FFTformer model consists of three parts: encoder, decoder and 1D CNN-LSTM module. In the encoder, the FFT module is first used to remove the noise from the data, and after that, the self-attention mechanism is used to enhance the characterization of historical data of different time scales. In the decoder, the cross-attention mechanism was utilized to reorganize the output data of the encoder and establish the relationship between the encoder output and the historical PV power of different time scales. To further enhance the accuracy of the prediction results, a 1D CNN-LSTM module is introduced to extract the features of historical PV power. Finally the features extracted by the cross-attention mechanism and the 1D CNN-LSTM module respectively are used for PV power prediction.

2.2 Fast Fourier Transform Module

In this study, time domain data are transformed into frequency domain data, which can be observed some embedding information, such as the cyclic component and frequency component, that are beneficial to denoising filtering and remain effective sections of data. The FFTformer model this study proposed uses the FFT algorithm (an optimization algorithm of discrete Fourier transform) to decrease the time complexity of calculation from O(N^2)to O $NlgN$ and reduce the needed training time. Processed frequency domain data is transferred to time domain data through inverse FFT to filter noise and increase model accuracy. FFT is separating odd and even terms in the discrete Fourier transform. The formula is as follows:

$$\begin{aligned} Y(k) &= \sum_{n=0}^{\frac{N}{2}-1} y_n(2n)e^{-j\frac{4\pi nk}{N}} + \sum_{n=0}^{\frac{N}{2}-1} y_n(2n+1)e^{-j\frac{2\pi n(2k+1)}{N}} \\ &= \sum_{n=0}^{\frac{N}{2}-1} y'_{\frac{n}{2}}(n)e^{-j\frac{4\pi nk}{N}} + \sum_{n=0}^{\frac{N}{2}-1} y''_{\frac{n}{2}}(2n+1)e^{-j\frac{4\pi nk}{N}}e^{-j\frac{2\pi nk}{N}} \end{aligned} \tag{1}$$

where y(2n) and $y(2n+1)$ is the historical PV data, $Y(k)$ is the frequency domain output of the historical PV data, $e^{-j\frac{2\pi}{N}}$ and $e^{-j\frac{2\pi}{N}n(2k+1)}$ are the rotation factors, and N denotes the length of the sequence. Equation (1) transforms the original N sequence length into two $\frac{N}{2}$ sequence lengths, and then the computational complexity can be reduced by repeating the transformed sequence through Eq. (1), thus realizing the fast Fourier transform. After the fast Fourier transform to remove part of the high frequency for noise reduction, and finally through the fast Fourier inverse transform the frequency domain data into time domain data, the formula as follows:

$$Z(k) = sel(Y(K)) \tag{2}$$

$$Z(n) = \frac{1}{N}\sum_{n=0}^{N-1} Z(k)e^{\frac{2\pi nk}{N}} \tag{3}$$

where $Z(k)$ is the frequency domain data after noise reduction, $sel(.)$ denotes the noise reduction operation, $Z(n)$ is the time domain data after Fourier inversion, and N denotes the sequence length.

The Fast Fourier Transform Module removes the noise of the original dataset, allowing the FFTformer model to better learn the relationship between PV power and historical data of different time scales. Subsequently, the noise-removed data are reorganized and enhanced with features using the attention mechanism.

2.3 Attention Mechanism

A multi-head cross-attention mechanism and a multi-head self-attention mechanism are employed by the FFTformer model. Following the Fourier module's noise reduction process, the self-attention mechanism is used to the noise-reduced data to obtain new features containing global information, strengthening the data's characterisation. The multi-head attention is computed as follows:

$$\begin{cases} q = linear(Z(n)) \\ k = linear(Z(n)) \\ v = linear(Z(n)) \end{cases} \tag{4}$$

$$head_i = Attention_i(Q_i, K_i, V_i) = Soft\max(\frac{q_i k_i^T}{\sqrt{d_{m_i}}})v_i \tag{5}$$

where q, k and v are the vectors obtained by different linear transformations of the noise reduced time domain data $Z(n)$, and then the vectors q, k and v are segmented to obtain q_i, k_i and v_i, with i denoting the i th segment. d_{m_i} Denotes the dimension of the i th segment, and $Head_i$ denotes the vectors obtained by carrying out the i th head self-attention, and then the vectors obtained by splicing the self-attention of each head are spliced to obtain the new features. The new features, in order to prevent the gradient from disappearing by adding the residual structure, are calculated as follows:

$$\begin{cases} Attention(Q, K, V) = Concat(head_1, head_2, ...head_8) \\ Z_{out} = Attention(Q, K, V) + z(n) \end{cases} \tag{6}$$

where $Attention(Q, K, V)$ denotes the new features obtained by doing the splicing of the vectors obtained from the self-attention computation of each head, $z(n)$ denotes the data after noise reduction of the Fourier module, and z_{out} denotes the result obtained after the residual structure.

Using multi-head cross-attention, the correlation between the encoder's output data and the historical PV power is investigated. The data obtained after self-attention is weighted and summed to obtain new features related to the PV power. The vectors obtained after cross-attention at each head are spliced to obtain the final features of the decoder.

In the decoder module, the cross-attention mechanism only considers the relationship between the hourly-scale historical PV power information and the decoder output features. In order to make the prediction more accurate, the 1D-LSTM module was used to extract features from the hourly-scale historical power information and the minute-scale historical power information respectively, so that the features of the information in different time scales are complementary and the prediction accuracy of the PV power is greatly improved.

2.4 CNN-LSTM Module

LSTM is an improved recurrent neural network (RNN) model, although it can effectively solve the problems such as gradient explosion encountered in RNN backpropagation, it will still be difficult to retain the more distant contextual information for long sequence inputs. Therefore, in this paper, we first use 1D-CNN network to convolve the long sequence history power with different time scales into short sequence history power features with different time scales, and then extract the features by using the short sequence history power features with different time scales as the input of LSTM. The network structure of 1D CNN-LSTM is shown in Fig. 2.

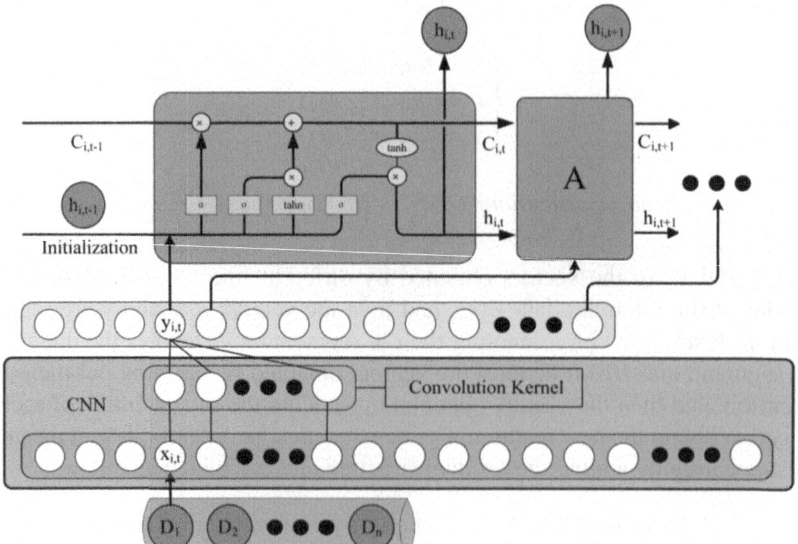

Fig. 2. 1D CNN-LSTM model structure.

The formulation of 1D-CNN network is as follows:

$$y_i = f(\sum_{i \in y_n} x_i k_i + b_i) \tag{7}$$

where y_n is a historical Multi-timescale PV dataset, k_i and b_i are the convolution kernel and bias for the ith time scale respectively. x_i is the historical power value at the ith time scale. y_i is the feature obtained after the convolution of the ith time scale.

The formula for the LSTM network is as follows:

$$\begin{cases} f_{i,t} = sigmoid(W_{i,f}[y_{i,t}, h_{i,t-1}] + b_{i,f}) \\ I_{i,t} = gigmoid(W_{i,j}[h_{i,t-1}, y_{i,t}] + b_{i,j}) \\ \widetilde{C}_{i,t} = \tanh(W_{i,c}[h_{i,t-1}, y_{i,t}] + b_{i,c}) \\ C_{i,t} = f_{i,t} C_{i,t-1} + I_{i,t} \widetilde{C}_{i,t} \\ O_{i,t} = sigmoid(W_{i,o}[h_{i,t-1}, y_{i,t}] + b_{i,t}) \\ h_{i,t} = O_{i,t} \tanh(C_{i,t-1}) \end{cases} \tag{8}$$

where $y_{i,t}$ is the input at moment t of the ith time scale. $f_{i,t}$, $I_{i,t}$ and $O_{i,t}$ and $h_{i,t}$ are the values of the forget gate, input gate, and output gate at the tth moment of the ith time scale, respectively. $W_{i,t}$, $W_{i,j}$, $W_{i,c}$ and $W_{i,o}$ represent the corresponding weight matrices, respectively. $b_{i,f}$, $b_{i,j}$, $b_{i,c}$ and $b_{i,t}$ represent the corresponding bias, respectively. $h_{i,t}$ is the output at moment t of the ith time scale.

Ultimately, the outputs of the decoder and the 1D CNN-LSTM of different time scales are used as inputs to the FCNN to pair the PV power values.

2.5 FFTformer Training Process

The FFTformer model proposed in this paper firstly encodes the historical data of PV power plants in different time scales with features and time, secondly, the encoded data is fast Fourier transformed to remove the high-frequency noise, and then converted to time-domain data for the self-attention mechanism to enhance the characterization of the data, and the residual network is added to prevent the gradient from disappearing, to reduce the risk of overfitting, and to improve the performance of the model, and then again, according to the correlation between the historical PV power values and the self-attention to the feature data to weight and sum the feature data. The correlation between the historical PV power values and the self-attention to the feature data to weighted sum of feature data, and then added to the residual network. Finally, the different time scale features obtained from the model decoder are feature spliced to predict the future PV power values. Table 1 illustrates the FFTformer model's training procedure.

Table 1. FFTformer modeling algorithm.

Algorithm: the training algorithm for the FFTformer model
Input: Historical characterization data of PV plants at different time scales $D_h = \left(x_1^i, \ldots, x_L^i\right)$, $D_m = \left(x_1^i, \ldots, x_L^i\right)$, PV power data $D_p = \left(y_1^i, \ldots, y_L^i\right)$
Output: Hourly PV power forecast for the next 2 days
Parameters: Maximum number of iterations:epoch; Training batch size:B; Learning rate: lr
Training process:
1: Random initialization: randomly initialize the weight parameter matrix and bias of the FFMORMER model θ_w and θ_b
2: Packed datasets: making small batches $\{X_b\}_{b=1}^B$, $\{Z_b\}_{b=1}^B$ and $\{y_b\}_{b=1}^B$ of data from D_h, D_m and D_p
3: repeat
4: for b = 1 to B do
5: Input X_b and Z_b into the FFTformer model to obtain the PV power prediction p_b, Then input p_b and y_b into the loss function to get the loss

(*continued*)

Table 1. (*continued*)

Algorithm: the training algorithm for the FFTformer model
6: Inverse update θ_w and θ_b by minimizing the loss function value
7: end for
8: until reach maximum number of iterations epoch
Testing process:
9: for b = 1 to B do
10: Input $testX_b$ and $testZ_b$ into the FFTformer model, obtain the PV power pre diction p_b

3 Experimental Simulation Result and Analysis

3.1 Data Preprocessing

This study uses a dataset from a photovoltaic plant in Hebei Province, China [13], which was collected between September 2018 and August 2019 using the Weather Research and Forecasting 3.9.1 model. The model is initialized with a global-scale numerical weather prediction of 0.125° × 0.125° every three hours, which is released by the European Weather Prediction Center (EWPC) at 12:00 Coordinated Time (CET). The dataset contains seven features: Global horizontal irradiance, direct normal irradiance, temperature, humidity, wind speed, wind direction, pressure, and PV output power, with a temporal sampling resolution of 15 min, as shown in Table 2.

Table 2. Collect data information.

Variable	Description	Ranges Value
Power (MW)	PV output of the station	0–16.0482
globalirrad (W/m^2)	Global irradiance	0–944.94
directirrad (W/m^2)	Direct irradiance	0–889.94
temperature (°C)	dry-bulb temperature	−11.18–38.96
humidity (%)	relative humidity	5.52–99.99
windspeed (m/s)	wind speed	0.08–15.71
wind direction (degree)	wind direction, zero north clockwise	0.05–359.97
pressure (hPa)	Atmospheric pressure	962.3–1010.28

This study uses historical data from the PV plant's first six days to forecast the PV for the next two days. Since the PV power output at night is almost zero, it is not favorable for model training. To address this issue, in the study, only data from 8:00 a.m. to 6:00 p.m. during the daytime were used as input data.

In order to speed up training, reduce gradient explosion or disappearance, improve the model's ability to generalize and help the model better understand the distributional characteristics of the data, data standardization is necessary because different factors such as temperature, humidity, wind speed and direction, pressure, photovoltaic output power, and horizontal and direct normal irradiance have different magnitudes and units of magnitude.

Normalization is a method of data standardization that usually involves scaling the data so that it falls into some small interval, transforming the data into dimensionless values. This article requires the normalization of seven measurements: Horizontal irradiance, direct normal irradiance, temperature, humidity, wind speed, wind direction and pressure. The seven measures in this research are standardized to fall between 0 and 1. Here's how to compute the measurement data normalization:

$$Y = \frac{x - x_{\min}}{x_{\max} - x_{\min}} \tag{9}$$

where x is the original input data, x_{min} are the minimum value of the original input data feature, and x_{max} are the maximum value of the original input data feature. Y is the normalized data. The year, month, day, and hour are normalized respectively, and the normalization range is from −0.5 to 0.5, and the formula for normalizing the timestamp information is:

$$\begin{cases} Z_y = \frac{1}{365}(X_y - 1) - 0.5 \\ Z_m = \frac{1}{30}(X_m - 1) - 0.5 \\ Z_d = \frac{1}{6}(X_d - 1) - 0.5 \\ Z_h = \frac{1}{23}(X_h - 1) - 0.5 \end{cases} \tag{10}$$

where X_y, X_m, X_d and X_h denote the original timestamp information year, month, day and hour, respectively, Z_y, Z_m, Z_d and Z_h denote the normalized timestamp information year, month and day, respectively.

3.2 Evaluation Indicators

In this study, various evaluation metrics such as mean absolute error (MAE), mean square error (MSE) and coefficient of determination R-squared (R^2) have been used to assess the performance of the model and are calculated as follows:

$$MAE = \frac{1}{n}\sum_{i=1}^{n} |\hat{y}_i - y_i| \tag{11}$$

$$MSE = \frac{1}{n}\sum_{i=1}^{n} (\hat{y}_i - y_i)^2 \tag{12}$$

$$R^2 = 1 - \frac{\sum_{i=1}^{n} (\hat{y}_i - y_i)^2}{\sum_{i=1}^{n} (\overline{y}_i - y_i)^2} \tag{13}$$

where n is the number of predicted samples, $\hat{y}_i$ is the predicted power output, y_i is the measured power output, and $\overline{y}_i$ is the measured average power output.

3.3 Results and Discussion

The five models are listed in Table 3 with their respective MSE, MAE, and R^2 values. M-FFTformer and S-FFTformer denote predictions using multi-timescale information and single-timescale information, respectively. The results show that the M-FFTformer model's MAE, MSE are minimum, which are 0.93258 and 1.39417 respectively. In addition, the M-FFTformer model has the maximum R^2 (0.92319) with others. It is clear that the M-FFTformer model fits the data the best, has a better prediction effect than other models.

The prediction results of the PV power values for the next 20 moments under mild and extreme weather conditions are shown in Fig. 3 and Fig. 4. The results show that M-FFTformer model has the closest predicted values to the actual values, while FEDformer model is the worst. Since the M-FFTformer model uses Fast Fourier Transform and Attention Mechanism and is trained with data from multiple time scales, it can better capture the temporal characteristics of the time series, and the predicted values of the M-FFTformer model are more accurate as compared to other models.

Table 3. Comparison of several models' prediction mistakes.

Model	MAE	MSE	R^2
M-FFTformer	**0.93258**	**1.39417**	**0.92319**
S-FFTformer	1.07196	1.96422	0.89269
Informer	1.34989	3.09921	0.83229
LSTM	2.2116	7.15347	0.61899
FEDformer	2.46401	9.10132	0.50856

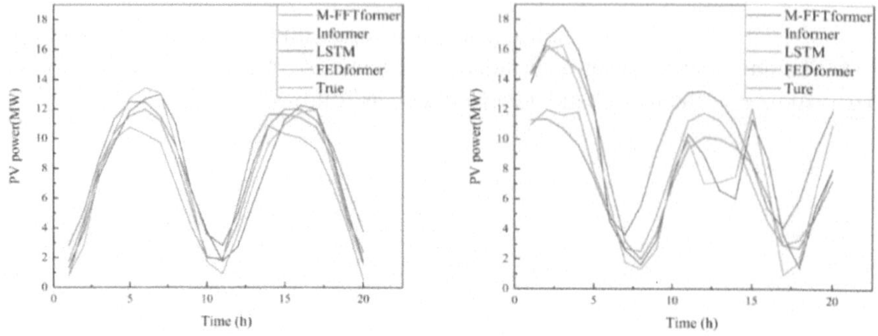

(a) moderate weather condition (b) harsh weather condition

Fig. 3. PV power forecasts under different weather conditions, (a) moderate weather condition, (b) harsh weather condition.

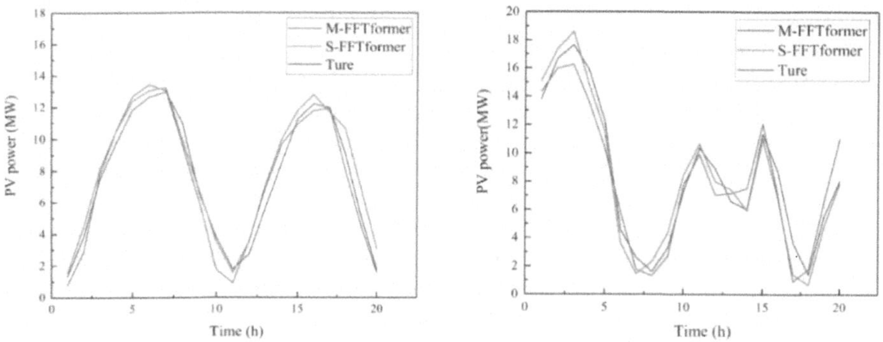

(a) moderate weather condition (b) harsh weather condition

Fig. 4. PV power forecasts with different time scale input data under different weather conditions, (a) moderate weather condition, b) harsh weather condition.

4 Conclusion

In this paper, a FFTformer model is proposed for short-term PV power prediction using multi-timescale information. To begin with, the FFT is employed to convert PV plant data from different time scales into frequency domain signals, then we eliminate the high-frequency signals to achieve the purpose of eliminating the data noise and retaining the effective data, this operation of data noise reduction establishes a solid foundation for the accurate predictions of FFTformer. To further enhance the prediction accuracy, 1D-CNN and LSTM are used to extract historical PV power features at different time scales. After that, feature extraction is performed on the data at different time scales through the attention mechanism. Finally, PV power prediction is performed using all the features at different time scales. This study employs testing data from a photovoltaic power plant located in Hebei Province, China. The prediction results of model are found to be superior to those of the LSTM, FEDformer and Informer models in both extreme and favorable weather scenarios. The proposed model can assist the storage system in more economically and sustainably planning the capacity size and configuration scheme of a combined photovoltaic and storage system. Further research can be conducted on how to reduce the computational complexity of the attention, reduce the training time of the model, and investigate how to accurately predict the PV power values for a longer period of time, so as to better provide the operation strategy and configuration scheme for the combined photovoltaic and storage system.

Acknowledgments. This research was funded by the Natural Science Foundation of China (52307247) and Shanxi (20210302123189, 202203021222124). These are gratefully acknowledged.

References

1. Wang, D., Liu, J., Huang, Y., et al.: Photovoltaic power prediction method combinating solar radiation calculation and CNN-LSTM. Acta Energiae Solaris Sinica **45**(02), 443–450 (2024)
2. Li, Z., Zhang, J., Xu, R.S., et al.: Short term photovoltaic power prediction based on similar day clustering and PCC-VMD-SSA-KELM model. Acta Energiae Solaris Sinica **45**(02), 460–468 (2024)
3. Dong, C., Wang, Z., Bai, J., et al.: Review of Ultra-short-term Forecasting Methods for Photovoltaic Power Generation. Gaodianya Jishu/High Voltage Eng. **49**(7), 2938–2951 (2023)
4. Zhan, H., Li, S.: A review of characterization and forecasting methods for distributed photovoltaic power generation. Sci. Technol. Revol. **03**, 182–183 (2020)
5. Leva, S., Dolara, A., Grimaccia, F., et al.: Analysis and validation of 24 hours ahead neural network forecasting of photovoltaic output power. Math. Comput. Simul. **131**, 88–100 (2017)
6. Jin, S., Xue, R.: Research on generation prediction of photovoltaic panel based on time series model. J. Chongqing Univ. Sci. Technol. (Nat. Sci. Edn.) **24**(03), 104–108 (2022)
7. Hou, W., Xiao, J., Niu, L.Y.: Analysis of power generation capacity of photovoltaic power generation system in electric vehicle charging station. Electr. Eng. **4**, 53–58 (2016)
8. Xu, K., Wang, J., Yao, T., et al.: Photovoltaic power forecast based on Time-Division LSTM. Front. Data Comput. **3**(05), 109–117 (2021)
9. Li, D., Qi, Y., Liu, L.: PV power prediction based on LSTM-ATTENTION fusion neural network. Journal of Inner Mongolia University of Technology Natural Science Edition **42**(04), 350–354+384 (2023)
10. Zhou, H., Zhang, S., Peng, J., et al.: Informer: Beyond efficient transformer for long sequence time-series forecasting. Proc. AAAI conf. Artif. Intell. **35**(12), 11106–11115 (2021)
11. Zhou, T., Ma, Z., Wen, Q., et al.: FEDformer: Frequency enhanced decomposed transformer for long-term series forecasting. In: International conference on machine learning, pp. 27268–27286. PMLR, Faridabad, India (2022)
12. Walczewski, M.J., Wöhrle, H.: Prediction of Electricity Generation Using Onshore Wind and Solar Energy in Germany. Energ. **17**(4), 844 (2024)
13. Yao, T., Wang, J., Wu, H., et al.: PVOD v1.0: A photovoltaic power output dataset. In: V4. Science Data Bank. Chin. Acad. Sci., Beijing, China (2021)

Wind Power Forecasting Model Based on Feature Processing and Deep Hybrid Kernel Extreme Learning Machine

Xiongfei Wei[1], Jing Wang[2](✉), Yi Ruan[2], and Yuanjie Fang[2]

[1] College of Electrical Engineering, Anhui Polytechnic University, Wuhu 241000, Anhui, China
[2] School of Electronic Engineering, Chaohu University, Chaohu 238024, Anhui, China
doudou140227@163.com

Abstract. To improve the forecasting accuracy of wind power, a wind power forecasting (WPF) model based on feature cleaning combined with deep hybrid kernel extreme learning machine (DHKELM) is proposed. The Pearson correlation coefficient (PCC) method is used to optimize the features, eliminate irrelevant features, and the isolated forest (IF) algorithm is used to clean the anomalous data in the retained features, which enhances the correlation between the selected features and wind power, and then combines the DHKELM to predict the wind power. Finally, the proposed model is applied to predict the wind power of a 1MW wind turbine, and the results of the progressive forecasting of the proposed method and the forecasting results of different models are compared. The results show that the PCC-IF-DHKELM forecasting model has a root-mean-square error of 0.01928 MW, an average absolute error of 0.015254 MW, and a correlation coefficient of 0.99435, which is significantly better than the other forecasting models, proving the effectiveness of the PCC-IF-DHKELM forecasting method.

Keywords: Pearson Correlation Coefficient · Isolated Forest Algorithm · Deep Hybrid Kernel Extreme Learning Machine · Wind Power Forecasting

1 Introduction

With the over-consumption of fossil energy, wind energy, as a clean and renewable clean energy source, has important research significance in the development and utilization of clean energy [1]. Wind energy has the advantage of being non-polluting and low-cost. However, because wind power generation uses wind energy as a direct energy source, the inherent stochastic, intermittent and uncertain characteristics of wind energy are also extended to wind power stability [2], which makes the accurate forecasting of wind power a difficult task, which brings great challenge for wind power grid integration and power dispatch and consumption [3]. Therefore, accurate and efficient WPF is essential to keep the grid running stable.

Most of the early WPF methods are based on the physical forecasting of numerical weather forecasting (NWP). The accuracy and reliability of the NWP data will directly determine the final forecasting accuracy of the model. NWP data contains many factors

K. Li et al. (Eds.): LSMS/ICSEE 2024, CCIS 2218, pp. 91–103, 2025.
https://doi.org/10.1007/978-981-96-0232-2_8

that have very low correlation with wind power, and their direct use in forecasting will greatly reduce the accuracy of the forecast. Wind power data can have abnormal data due to wind turbine part failures, sensor failures, meteorological fluctuations, etc. [4], which can also affect the accuracy of the forecasting.

At present, many scholars have combined physical and statistical methods in WPF to improve the accuracy of WPF [5]. With the wide application of artificial intelligence, statistical methods are not limited to gray forecasting, ARIMA and other time series methods, but also include ELM, support vector machine, deep learning and so on [6]. Since a single model cannot establish a good nonlinear relationship among variables in the case of dealing with a large amount of data, the forecasting accuracy needs to be improved. Literature [7] applies the exhaustive likelihood (ELM) model to time series forecasting, and its effectiveness is verified by case analysis, but the initial input weights and the bias of the implicit layer of the ELM are random, which can't fully reflect the performance of the model. The introduction of kernel function based on ELM can well solve above problems, and the kernel extreme learning machine is applied to power forecasting [8], which effectively solves the influence of random mapping on modeling and greatly improves the accuracy of power forecasting. DHKELM is a new type of intelligent forecasting model, which has a deeper structure based on ELM [9], and introduces the autoencoder and hybrid kernel function idea, which can effectively improve the problem of low stability of output due to the random generation of initial weights and thresholds.

In this paper, the PCC-IF-DHKELM wind power combination forecasting method is proposed. Firstly, PCC is used to remove the physical modes whose correlation coefficient with wind power is less than 0.2, and then the IF algorithm is used to clean the anomalies of the wind power data; the DHKELM model is chosen as the forecasting model; finally, the data processed by PCC-IF is used as the input of the DHKELM model for the forecasting of wind power, and the comparison is made between the different input models as well as the different combination of the forecasting models, to verify the effectiveness of the proposed method. Finally, the PCC-IF processed data are used as inputs to the DHKELM model for WPF, and compared with different input models and different combination forecasting models to verify the validity of the proposed method.

2 Methodology

2.1 Pearson Correlation Coefficient (PCC)

There are many modes in the data that affect turbine power, and some of these modes have less influence on the turbine power. Directly used in the model forecasting will make the forecasting accuracy lower. To address this issue, this paper introduces PCC to remove the modes with less correlation with wind turbine power.

The PCC is used to evaluate the correlation between two variables x and y. And it is mathematical expression formula is Formula (1).

$$P_{xy} = \frac{\left|\sum_{i=1}^{n} (x_i - \overline{x})(y_i - \overline{y})\right|}{\sqrt{\sum_{i=1}^{n} (x_i - \overline{x})^2}\sqrt{\sum_{i=1}^{n} (y_i - \overline{y})^2}} \tag{1}$$

where x_i and y_i denote the i-th observation of the modal characteristics and power, and $\overline{x}$ and $\overline{y}$ denote the average of the modal characteristics and power, respectively. The value range of P_{xy} is [0,1].

In this study, physical modes with correlation greater than 0.2 are retained for fan power forecasting.

2.2 Isolated Forest Algorithm (iF)

Through the modal characteristics selected by PCC, the measuring device of wind turbine sometimes fails due to the influence of many unfavorable factors. The measured wind power data will have abnormal data. These abnormal values will have a great impact on the forecasting results of wind power. To address this problem, this paper utilizes the IF algorithm, which is unaffected by dimensional catastrophe and maintains better performance even in high dimensional space, to clean the anomalies in wind power data.

IF construction can be divided into two stages. In the first stage, t isolated trees need to be trained to form an isolated tree (iTree). In the second stage, each sample is brought into each iTree in the forest to calculate the average height, and then the outlier score of each sample is calculated.

(1) First stage

Step1. $X = \{x_1, \cdots, x_n\}$ for the given dataset, $\forall x_i \in X, x_i = (x_{i1}, \cdots, x_{id})$, randomly select n sample points from X' to form a subset of X and put them into the root node.

Step2. Generate a random intersection point p in the current data by specifying a random dimension q from the d dimensions. $\min(x_{ij}, j = q, x_{ij} \in X') < p < \max(x_{ij}, j = q, x_{ij} \in X')$.

Step3. At the intersection point p, the current data space is divided into two subspaces: sample points with the specified dimension less than p will be placed in the left child node, and sample points with the specified dimension greater than p will be placed in the right child node.

Step4. Recursively Step2 and Step3 until all leaf nodes have only one sample or the iTree has reached the specified height.

Step5. Loop Step1 through Step4 until an iTree is generated as shown in Fig. 1.

(2) Second stage

For each data x_i, make it traverse each iTree, calculate the average height $h(x_i)$ in the forest, and normalize the average height of all points. The formula for calculating the outlier score is shown below:

$$s(x, n) = 2^{\frac{E(h(x))}{c(n)}} \tag{2}$$

where

$$c(n) = \begin{cases} 2H(n-1) - 2(n-1)/n, n > 2 \\ 1, n = 2 \\ 0, otherwise \end{cases}$$

$$H(n) = \ln(n) + \zeta, \zeta = 0.5772156649$$

2.3 Structure of NLECM Model

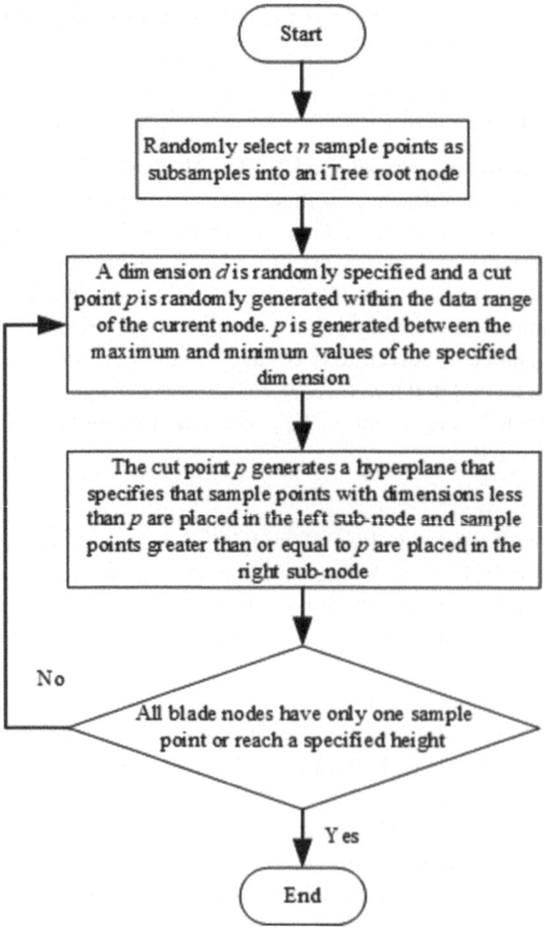

Fig. 1. Construction process of a single iTree.

2.4 Deep Hybrid Kernel Extreme Learning Machine (DHKELM)

Hybrid Kernel Extreme Learning Machine (HKELM)

Extreme learning machine (ELM) is essentially a single hidden layer feed-forward neural network structure. Compared with the traditional neural network, ELM structure is simple with less initial parameters, which has a faster convergence speed and higher learning efficiency [10].

The wind power dataset $\boldsymbol{X}_0 = \{(\boldsymbol{x}_i, \boldsymbol{y}_i), i = 1, 2, \cdots, N\}, \boldsymbol{x}_i = [x_{i1}, x_{i2}, \cdots, x_{in}] \in \boldsymbol{R}^n$ is the input feature vector, $\boldsymbol{t}_i = [t_{i1}, t_{i2}, \cdots, t_{im}] \in \boldsymbol{R}^m$ is the corresponding target vector, and $\boldsymbol{y}_i = [y_{i1}, y_{i2}, \cdots, y_{im}]$ is the output vector obtained through the ELM network. If the error between the output $\boldsymbol{y}$ and the target $\boldsymbol{t}$ can be approximated as 0, then $\sum_{i=1}^{N} \|\boldsymbol{t}_i - \boldsymbol{y}_i\| = 0$ is satisfied. For an ELM with the number of hidden layer nodes K and the activation function *g(x)*, the learning process can be characterized by solving the least squares solution, as shown in Eq. (3).

$$\hat{\boldsymbol{\beta}} = \boldsymbol{H}^{+}\boldsymbol{T} \tag{3}$$

where $\boldsymbol{H}^{+}$ is the Moore--Penrose generalized inverse transform of the Hidden layer output matrix $\boldsymbol{H}$. $\boldsymbol{T}$ is the matrix of true values of training samples.; $\hat{\boldsymbol{\beta}}$ is the optimal weight matrix of the output layer.

The idea of kernel function is introduced into ELM, and Kernel Extreme Learning Machine (KELM) is constructed. Because the kernel function has a good nonlinear mapping ability, which makes the KELM performance more stable and generalization ability is more excellent [8]. For the wind power dataset $\boldsymbol{T}_0$, the KELM output is:

$$f(x) = \begin{bmatrix} K(x, x_1) \\ \vdots \\ K(x, x_N) \end{bmatrix} \left(\frac{\boldsymbol{E}}{C} + \Omega_{ELM} \right)^{-1} \boldsymbol{T} \tag{4}$$

where $K(x, x_N)$ is the kernel function, C is the regularization coefficient, E is the unit matrix and Ω_{ELM} is the kernel matrix Ω_{ELM} is defined as follows:

$$\begin{cases} \Omega_{ELM} = \boldsymbol{H}\boldsymbol{H}^{T} \\ \Omega_{i,j} = h(x_i) \cdot h(x_j) = K(x_i, x_j) \end{cases} \tag{5}$$

To further improve the learning ability and generalization of KELM, The radial basis function (RBF) kernel function with strong learning ability and the polynomial (Poly) kernel function with strong generalization ability [11] are weighted and combined into a hybrid kernel function [12], which has the expression:

$$K_{mix} = \varphi \exp\left(-\frac{x_i - x_j^2}{2\sigma^2} \right) + (1 - \varphi)(x_i, x_j + c)^d \tag{6}$$

where σ is the width parameter of the RBF kernel function, c and d are the kernel parameters of the Poly kernel, and φ is the weight coefficient between the two kernel functions.

DKHELM Based on Auto-encoder

ELM Auto-Encoder can be regarded as a special ELM with equal inputs and outputs [13], which orthogonalizes the weights and thresholds of the randomly generated input layers in the ELM, which enables the input data to be mapped to different dimensional spaces, thus generating features that are easier to classify. The Deep Extreme Learning Machine (DELM) connects multiple ELM-AEs to form a feature learning layer, and finally the ELM is used as a classification layer [14], DELM effectively improves the ability to mine the hidden information of data samples.

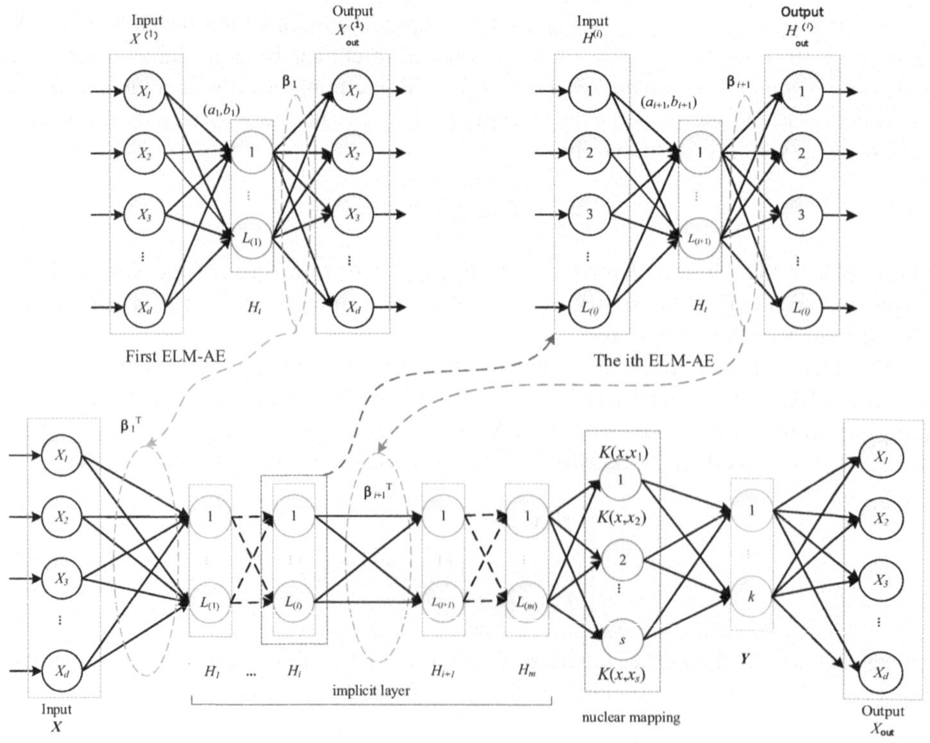

Fig. 2. DHKELM structure diagram.

The DHKELM based on ELM Self-Encoder combines DELM and HKELM and its structure is shown in Fig. 2. The flow of DHKELM model is as follows:

Step 1: Randomly generate the input weights a_i and biases b_i in ELM-AE;

Step 2: Calculate the output matrix $\boldsymbol{H}_i$ and output weights $\boldsymbol{\beta}_i$ for each hidden layer as in Eqs. (7) and (8).

$$\boldsymbol{H}_i = g(a_{in}\boldsymbol{X} + b_i) \tag{7}$$

$$\boldsymbol{\beta}_i = \left(\frac{\boldsymbol{E}}{C} + \boldsymbol{H}_i^T \boldsymbol{H}_i\right)^{-1} \boldsymbol{H}_i^T \boldsymbol{H}_{i-1} \tag{8}$$

In Eq. (7), $\sigma(x) = \frac{1}{1+e^{-x}}$ is taken as the nonlinear activation function *g(x)*.

Step 3: Update $\boldsymbol{H}_i$, according to Eq. (9) and use it as the input of one lower layer of ELM-AE, and so on, to compute the implied layer outputs of all ELM-AEs;

$$\boldsymbol{H}_i = g(\boldsymbol{\beta}_i^T \boldsymbol{H}_{i-1}) \tag{9}$$

Step 4: The output matrix of the last hidden layer is used as the decision input of the HKELM, and the output weights of the HKELM are calculated according to Eq. (10).

$$\boldsymbol{\beta} = \left(\frac{\boldsymbol{E}}{C} + \Omega_{ELM}\right)^{-1} \boldsymbol{H}_i \tag{10}$$

The output of DHKELM is:

$$f(x) = \begin{bmatrix} K_{mix}(x, x_1) \\ \vdots \\ K_{mix}(x, x_N) \end{bmatrix} \left(\frac{\boldsymbol{E}}{C} + \Omega_{ELM} \right)^{-1} \boldsymbol{T} \tag{11}$$

Wind power data has the characteristics of nonlinearity and low regularity. DHKELM uses ELM-AE to mine hidden information in data, and can also effectively learn nonlinear features. At the same time, it uses hybrid kernel function to map data to high-dimensional space, which effectively improves classification accuracy and speed.

3 Forecasting Model

The main work of this paper includes two parts: feature selection and data cleaning, and model testing and evaluation. The flow chart is shown in Fig. 3.

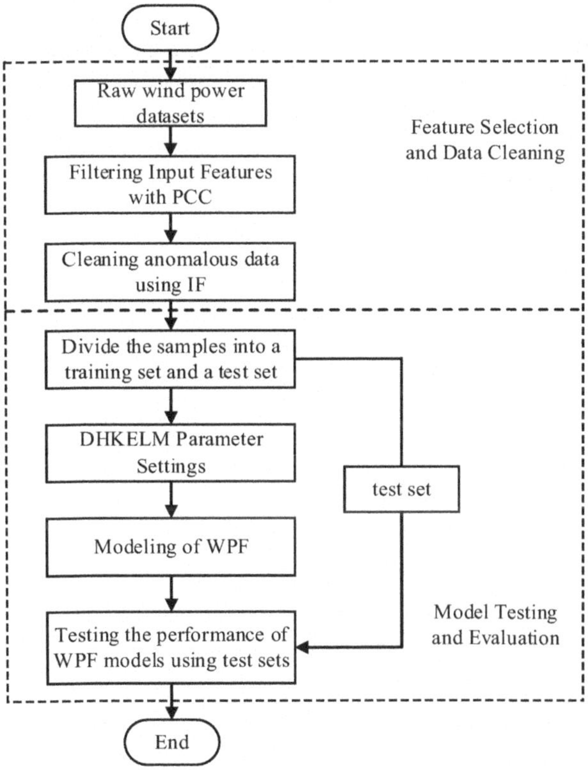

Fig. 3. Forecasting model flowchart.

4 Experimental Result Analysis

The data of this study are selected from public data sets. The data is the wind power data from April 1,2017 to May 31,2017 in eastern China, and the wind turbine model is GW _ 1.5 _ 70. The data is sampled every 1 h, which is divided into training set and test set in a ratio of 8: 2.

4.1 Data Normalization and Index Evaluation

In order to avoid the error caused by the order of magnitude between different variables, it is necessary to normalize the data and reduce the data to [0,1]. In order to prevent the leakage of future wind power data to the training set, the divided training set and test set are normalized respectively, and the normalization of the test set is based on the training set, and the test set finally adopts the same normalization parameters as the training set. The normalization formula (12) used in this paper is as follows:

$$x^* = \frac{x - x_{\min}}{x_{\max} - x_{\min}} \tag{12}$$

where x is the variable to be normalized; $x_{\min}$ and $x_{\max}$ are the minimum and maximum values in the x variable, respectively; and x^* is the normalized variable.

In this paper, the average absolute error e_{MAE}, the root mean square error e_{RMSE} and the coefficient of determination R^2 are selected as the evaluation indexes of forecasting accuracy, and the calculation formulas are as follows:

$$e_{MAE} = \frac{1}{\mathrm{n}} \sum_{i=1}^{n} |y_i - \hat{y}_i| \tag{13}$$

$$e_{RMSE} = \sqrt{\frac{1}{n} \sum_{i=1}^{n} (y_i - \hat{y}_i)^2} \tag{14}$$

$$R^2 = 1 - \frac{\sum_{i=1}^{n} (y_i - \hat{y}_i)^2}{\sum_{i=1}^{n} (y_i - \overline{y})^2} \tag{15}$$

where n denotes the number of samples, and y_i, $\hat{y}_i$ and $\overline{y}$ denote the actual power, predicted power and actual power values, respectively.

4.2 PCC Feature Selection

The wind data cosine function needs to be processed before correlation analysis. The above five features with wind power PCC are shown in Table1.

The characteristics of PCC greater than 0.2 are retained, that is, three characteristics of wind speed, humidity and wind direction are retained.

Table 1. The PCC of five features of wind power.

5 feature	6 wind speed	7 temperature	8 humidity	9 air pressure	10 wind direction
PCC	0.9306	0.1894	0.2847	0.1619	0.2648

4.3 Data Cleaning by iF Algorithm

The fault of wind power data acquisition equipment, data transmission error and other problems will lead to abnormal power data. In this paper, IF algorithm is used to clean abnormal data, and the proportion of abnormal points is set to 5%, which is relatively small and will not affect the authenticity of the original data. The features selected in Sect. 4.2 are cleaned by the IF algorithm, and the abnormal points of the cleaned data are reduced to the display of the three-dimensional data by t-SNE, as shown in Fig. 4.

After cleaning, the data was filled with linear interpolation. The formula of linear interpolation method is as follows (16). The data after filling and the PCC value of wind power are shown in Table 2.

$$x_i = \frac{x_{i+1} - x_{i-1}}{2} \tag{16}$$

Table 2. The PCC of the features and power processed by the IF.

11 feature	12 wind speed	13 humidity	14 wind direction
IF-PCC	0.9312	0.2883	0.3061

From Fig. 4, it can be seen that the abnormal points are distributed in the periphery of the space. In combination with Table 2, the PCC values of the selected three characteristics of wind speed, humidity and wind direction and power are increased, which verifies the effectiveness of IF algorithm in cleaning data.

4.4 Comparative Experiments and Analysis of Forecasting Models

To verify the validity of the proposed method, the model is progressively compared, and the DHKELM model parameters are in Table 3. Table 4 is the progressive forecasting error of the model, and Fig. 5 is the progressive forecasting curve of the model.

As can be seen from Table 4, among the three models of DKKELM, PCC-DHKELM and PCC-IF-DHKELM, the worst forecasting result is using the DHKELM model without any data processing. After PCC feature selection, the forecasting effect of DKKELM model is improved. On this basis, the IF algorithm is added, and the WPF effect is further improved.

To validate the DHKELM model, two comparison models of LSTM and BiLSTM are established respectively. Table 3 shows the parameter setting of different models,

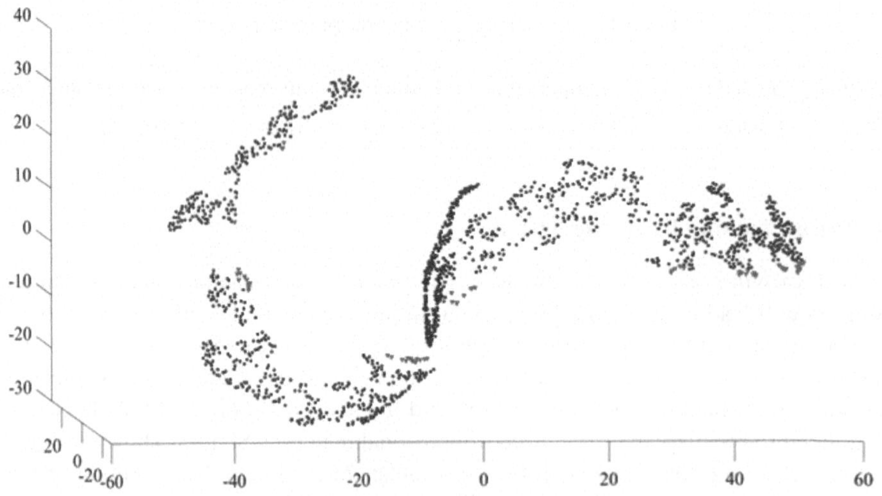

Fig. 4. The IF cleaned data display by t-SNE.

Table 3. Parameter settings of different models.

15. Model	16. P Arameter	17. V Alue
LSTM	Hidden layers	1
	Number of neurons	9
BiLSTM	Hidden layers	1
	Number of neurons	18
DHKELM	Hidden layers	3
	Number of neurons	75,52,7
	Kernel function	RBF,Poly
	RBF nuclear parameter	150
	Poly nuclear parameter	951,9
	Regularization coefficient	375
	Penalty coefficient	778
	Weight coefficient φ	0.06

Table 5 shows the forecasting error of different models, and Fig. 6 shows the forecasting curve of different models.

It can be seen from Table 5 that the data processed by PCC-IF are fed into three different forecasting models after filling, the forecasting result of LSTM model is the worst, and the forecasting effect of BiLSTM is better than that of LSTM. It's because BiLSTM considers both of past and future information, better maintains long-term memory,

Table 4. Progressive forecasting error of the models.

18 Model	19 e_{MAE}/MW	20 e_{RMSE}/MW	21 R^2
DHKELM	0.097707	0.11172	0.82778
PCC-DHKELM	0.019593	0.025784	0.99083
PCC-IF-DHKELM	0.015254	0.01928	0.99435

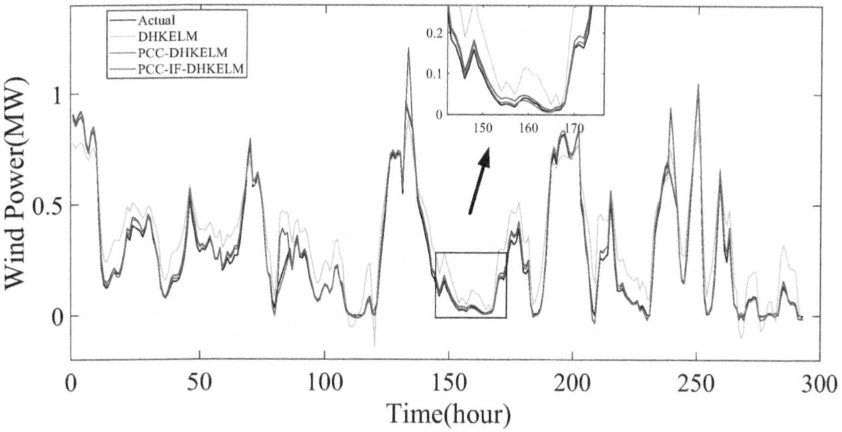

Fig. 5. Progressive forecasting curve of the models.

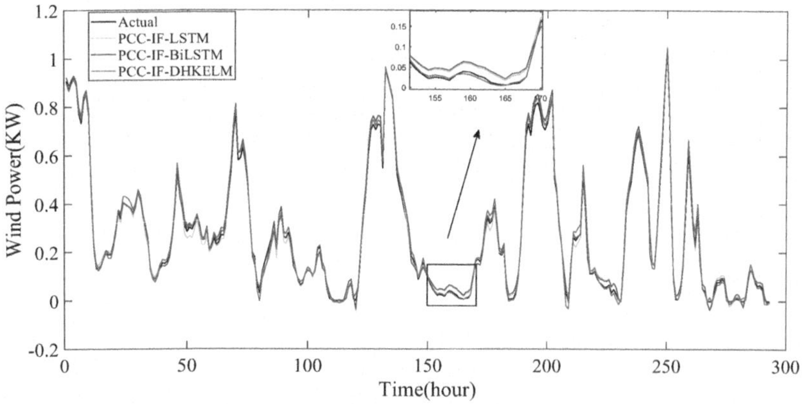

Fig. 6. Forecasting curves of different models.

extracts richer features and more accurately models sequence data, but its forecasting performance is seriously affected by key parameters. Among the three models, the DHKELM model has the best forecasting effect, and its kernel function is combined with the Gaussian kernel function and the polynomial kernel function, which makes up for the defects of the single kernel function. It can better mine the internal laws of

Table 5. Forecasting errors of different models.

Model	e_{MAE}/MW	e_{RMSE}/MW	R^2
PCC-IF-LSTM	0.01847	0.021726	0.99283
PCC-IF-BiLSTM	0.016526	0.019963	0.99394
PCC-IF-DHKELM	0.015254	0.01928	0.99435

wind power and has stronger forecasting performance. The results are more accurate and stable.

5 Conclusions

In this paper, a WPF model based on feature processing and DHKELM is proposed, which is verified with practical examples, and specific conclusions are drawn as follows:

(1) The PCC method is used to optimize the features, eliminate irrelevant features, and enhance the correlation between data features and wind power. The IF algorithm is used to remove abnormal data, which is conducive to the learning and training of the DHKELM model. At the same time, it also reduces the complexity of the model and improves the forecasting accuracy of wind power.

(2) The DHKELM model can better explore the internal law of wind power and has stronger prediction performance, which makes the results more accurate and stable.

Acknowledgments. This work is supported by National Natural Science Foundation of China (no.62303076),Natural Science Research Programme of Colleges and Universities of Anhui Province, China (nos.gxyq2022081, 2023AH052105, KJ2020ZD39, KJ2021A1025,2023AH052099,2023AH052109 and 2022AH020093), Natural Science Foundation of Anhui Province,China(no.2308085QF205),Subject Construction Promotion project of Chaohu University,China(nos.kj21gczx02,kj21bskc06 and hxkt20230149), Chaohu University Scientific Research Project(nos.KYQD-2023047, KYQD-202304).

References

1. Dong, Y., Zhang, H., Wang, C., et al.: A novel hybrid model based on Bernstein polynomial with mixture of Gaussians for wind power forecasting. Appl. Energy **286**, 116545 (2021)
2. Liu, Y., Ćetenović, D., Li, H., et al.: An optimized multi-objective reactive power dispatch strategy based on improved genetic algorithm for wind power integrated systems. Int. J. Electr. Power Energy Syst. **136**, 107764 (2022)
3. Wu, W., Zhu, J., Chen, Y., et al.: Modified Shapley Value-Based Profit Allocation Method for Wind Power Accommodation and Deep Peak Regulation of Thermal Power. IEEE Trans. Ind. Appl. **59**(1), 276–288 (2022)
4. Kaloop, M.R., Kumar, D., Zarzoura, F., et al.: A wavelet- particle swarm optimization-Extreme learning machine hybrid modeling for significant wave height prediction. Ocean Eng. **213**, 107777 (2020)

5. Cao, B., Chang, L.: Development of short-term wind power forecasting methods. In: 2022 IEEE 7th Southern Power Electronics Conference (SPEC). IEEE (2022)
6. Zhu, N., Dai, Z., Wang, Y., et al.: A contrastive learning-based framework for wind power forecast. Expert Syst. Appl. **230**, 120619 (2023)
7. Feng, Y., Hao, W., Li, H., et al.: Machine learning models to quantify and map daily global solar radiation and photovoltaic power. Renew. Sustain. Energy Rev. **118**, 109393 (2020)
8. Shang, L.Q., Li, H.B., Hou, Y.D.: Short-term photovoltaic power generation prediction based on VMD-ISSA-KELM. Power Syst. Prot. Cont. **50**(21), 138–148 (2022)
9. Chu, Y., Feng, C., Guo, C., et al.: Network embedding based on deep extreme learning machine. Int. J. Mach. Learn. Cybern. **10**(10), 2709–2724 (2019)
10. Xia, X., Wang, X.: A novelhybrid modelfor short-term wind speed forecasting base dontwice decomposition, PSR, and IMVO-ELM. Complex. **2022**, 4014048 (2022)
11. Guo, J., Cui, S., Guo, J., et al.: Ultra-short-term load forecasting based on VMD-SSA-HKELM. Foreign Electr. Measure. Technol. **41**(6), 105–111 (2022)
12. An, G., Jiang, Z., Chen, L., et al.: Ultra short-term wind power forecasting based on sparrow search algorithm optimization deep extreme learning machine. Sustain. **13**(18), 10453 (2021)
13. Zhang, Y., Ma, H., Wang, S., et al.: Indirect prediction of remaining useful life for lithium-ion batteries based on improved multiple kernel extreme learning machine. J. Energ. Storage **64**, 107181 (2023)
14. Liu, Y., Wang, J.: Transfer learning based multi-layer extreme learning machine for probabilistic wind power forecasting. Appl. Energ. **312**, 118729 (2022)

Optimal User-Side Energy Arbitrage Strategy in Electricity Market with Accurate Battery Model Using Benders Decomposition

Mengchen Zhu and Tao Chen(✉)

School of Electrical Engineering, Southeast University, Nanjing 210096, China
taoc@seu.edu.cn

Abstract. In this paper, a user-side battery energy storage system is modeled, using a linear programming approach to solve the problem of minimum cost and optimal operation strategy. In the accurate modeling of the battery energy storage system, segmented curves are used to subtly describe the charging performance of the battery at different charging levels. Additionally, the Rainflow algorithm is used to calculate the lifetime loss incurred during the charging and discharging of the chemical battery in the energy storage system. By using Benders Decomposition method, the lifetime loss cost of the energy storage system would be further incorporated into the overall energy arbitrage optimization process, in which the cyclic features of the energy arbitrage in electricity market with battery lifetime loss calculation is constructed. Finally, the performance of such an optimal user-side energy arbitrage strategy is analyzed through various simulations, meanwhile showcasing the advantages through comparison study.

Keywords: Battery energy storage system · Electricity market · Energy arbitrage · Battery degradation cost · Benders decomposition

1 Introduction

In recent years, with the integration of new energy sources and the increasing demand for electricity, the contradiction between electricity supply and demand has expanded, and electricity prices have become more volatile [1]. Electric energy storage has gradually become an important research direction. Battery energy storage technology has advantages such as high efficiency, application flexibility, and fast response speed. With the decrease of battery prices, the attractiveness of battery energy storage technology gradually increases [2], and it occupies an increasingly important position in the electricity energy storage market. Battery energy storage systems can store electricity when it is abundant and cheap, and release stored electricity when electricity is tight and electricity

K. Li et al. (Eds.): LSMS/ICSEE 2024, CCIS 2218, pp. 104–118, 2025.
https://doi.org/10.1007/978-981-96-0232-2_9

prices are high, reducing grid pressure and providing users with a certain degree of economy.

In the modeling of energy storage systems, the main constraints include power balance constraints, battery charging/discharging power and quantity constraints, or reliability index constraints [3], and different algorithm models are used according to requirements. Reference [4] adopts the numerical simulation method of optimal control, taking into account benefits and costs, the frequency fluctuation curve of the system, and the charging/discharging characteristics of the battery. With the goal of maximizing the annual revenue generated by the energy storage system, a numerical simulation model of the energy storage system for primary frequency regulation in the power grid is established. Some literature aims to plan for the best social benefits brought by energy storage systems for a specific purpose. For example, literature [5] focuses on the optical storage system in the distribution network, and studies the operation strategy and capacity allocation of the optical storage system in the distribution network from the perspective of planning and construction, with the objective of maximizing the system self-sufficiency rate and net present value.

Electricity market trading strategies for energy storage systems are important to ensure that the energy storage system is able to capture revenue [6,7]. An effective control strategy can optimize energy revenues and reduce maintenance costs by purchasing lower-priced power during low-peak periods and selling power during peak hours. Reference [8] proposes a stochastic planning formulation for energy storage systems that is able to perform storage arbitrage and maximize profits in the presence of price uncertainty in the day-ahead and real-time markets, which helps the energy storage system to make market bidding and operational decisions. Reference [9] proposes a dynamic planning approach to optimize the real-time operational decisions of the energy storage system using the nested structure of the problem to enable energy arbitrage and frequency regulation. Reference [10] proposed a deep reinforcement learning method based on noise networks to learn the optimal control strategy for storage charging and discharging strategies.

In terms of the precise description of the battery performance, a simple battery model for energy simulation and embedded intelligent control systems is presented in literature [12]. The drawback of this approach is that it only considers the constant current phase of the battery charging characteristics and ignores the current characteristics of the constant voltage phase that occurs when the battery is about to be fully charged. Literature [13] investigates the dependence of the battery charging capacity on its energy state and proposes an accurate linear battery charging model, which takes into account the constant-current phase and the constant-voltage phase of the battery, and describes the relationship between the current charging capacity of the battery and the battery capacity with a segmented function.

Some research works model the battery aging process by dividing battery aging into different stages. In literature [16] a multi-stage model for SoH (State of Health) estimation based on cyclic difference learning is proposed for estimating the degradation trend at different stages. Similarly, literature [17] proposes a deep learning based method for predicting the degradation trajectory of disposable batteries. In literature [18], aging mechanisms and SoH prediction over the entire life cycle were investigated based on three stages. All these works estimate the SoH of batteries at different stages, however they do not propose methods for its application in scheduling problems.

There have been several studies on battery life loss modeling using model-based or data-driven approaches [19,20], which can provide accurate life loss costs for different charging modes. Optimized operation of energy storage systems based on multifactor battery cycling models was proposed in the literature [21] using a rainflow counting algorithm (RCA, rainflow counting algorithm) for capacity loss estimation. The RCA algorithm is widely used in fatigue analysis and damage estimation. However, it is difficult to implement it explicitly in optimization problems [22] due to the lack of an analytical formulation for RCA. RCA is integrated as a nonlinear term into the scheduling model [23] using a problem-specific solver, while the scheduling model does not explicitly incorporate the battery life loss into the optimization problem, and only analyzes its impact after the fact. Literature [24] proposes a scheduling approach for battery storage systems that models the complex impact of participating markets on battery life using a dynamic procedure by modeling the regulated market signals through robust optimization.

This paper examines various types of energy constraints on energy storage systems, considering battery economics operating costs and market participation trading strategies. It helps assess the costs and benefits of energy storage systems and helps decision makers better understand the feasibility of their investments. By simulating the design scenarios, the optimal charging and discharging parameter configurations of the energy storage system can be determined, the charging and discharging arbitrage gains and battery life depletion costs can be coordinated, and the operation strategy of the energy storage system can be optimized to maximize the efficiency of the system.

2 Model Architecture

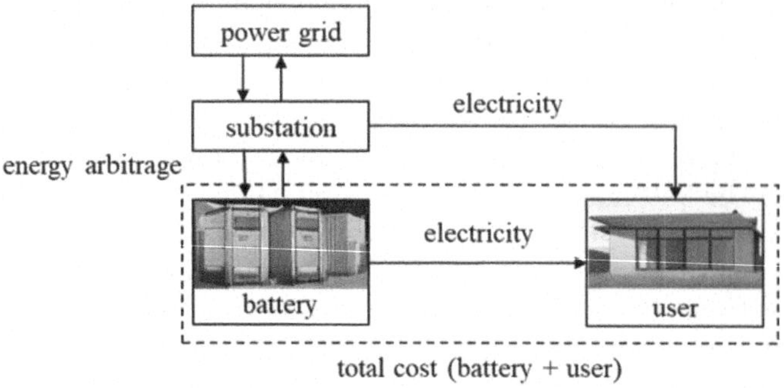

Fig. 1. The figure illustrates the energy storage system model considered in this paper, in which the energy storage battery and the user are considered as a whole for the cost calculation.

The structure of the energy storage system model used in this paper is shown in the following figure below and consists of four main components: the grid, the substation, the battery pack and the user. The grid is connected to the energy storage system and users through the substation. Through energy arbitrage, the battery pack can buy energy when the price of electricity is low and supply energy to the users when the price of electricity is high, and at the same time, it can sell electricity to the grid for energy arbitrage. The power demand from the user is satisfied by the energy released by the battery energy arbitrage and the energy provided by the grid through the substation. The total power balance in the system is always maintained during the operation of the model.

In this paper, the structure of benders decomposition is used to construct the cyclic algorithm model, and the specific structure is shown in Fig. In the main problem, the value of the limiting factor is corrected by the optimization model. In the subproblem, the energy storage system operation model considering the charging and discharging limiting factors is used to calculate the optimal scheduling strategy of the battery with the objective of minimizing the cost under the condition of satisfying the user's demand. The sensitivity of the limiting factor to the operating cost of the system is analyzed. In the battery loss model, the battery loss is calculated based on the battery scheduling data calculated from the subproblem. The relationship between loss and limiting factor is linearly approximated to evaluate the sensitivity of battery loss to the limiting factor. All the sensitivities will be used in the main problem for the evaluation of the limiting factor size.

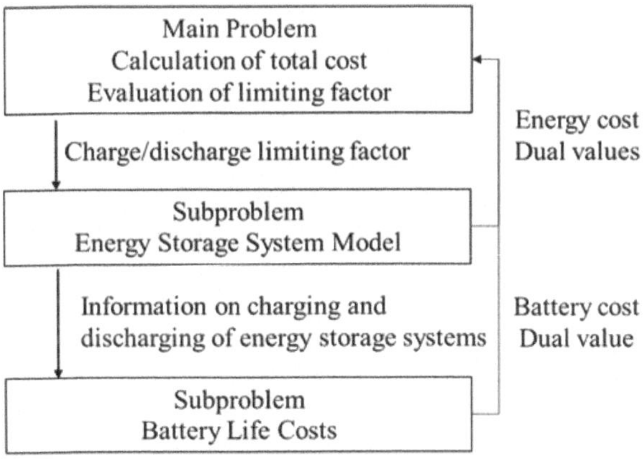

Fig. 2. Model Algorithm Architecture.

Table 1. Parameters of the Energy Storage System Model.

Parameters	Parameters name
H	Total duration.
Δt	Minimum time unit (set to one hour).
C_t^B	The electricity price at time t .
C^{SUB}	Unit power cost of substation (per day).
P_t^D	The electricity demand at time t .
$E_{\max}^B$	The maximum energy capacity of the battery.
$P_{\max}^B$	The maximum power capacity of the battery.
$SOC_{\min}, SOC_{\max}$	Battery energy state range.

3 Proposed Method

3.1 Energy Storage System Model

Considering the battery bank and the user as a whole, the optimization objective is to minimize the overall cost. In this case, the battery bank can make profits through energy arbitrage, such as buying power from the grid when the price of electricity is low, selling power to the grid when the price of electricity is high, or providing energy to users. The power demand of the users is met by either direct power from the grid or from the battery pack.

Table 2. Variables of the Energy Storage System Model.

Variables	Variables name
t	Indicates the current time. Range 0-H.
$energy$	Cost of purchasing electricity.
P_t^{SUB}	Power exchange with the substation.
P^{SUBMAX}	The maximum value of power exchange with the substation.
P_t^B	Energy arbitrage power (positive during charging, negative during discharging) at time t .
E_t^B	Battery capacity at time t .

Equation(1) describes the optimization objective as the lowest operating cost of the energy storage system, i.e., the minimum, where the cost of the energy storage system to meet the customer's electricity demand and to carry out energy arbitrage is considered by summing up the energy costs for each period of time and considering the cost of substation usage, which is related to the maximum exchange of electricity at the substation:

$$energy = \sum_{t \in T} (C_t^B P_t^{SUB}) + \frac{H}{24} * C^{SUB} * P^{SUBMAX} . \quad (1)$$

Equation(2) describes the substation power balance, where the power purchased from the grid is equal to the customer demand power plus the battery charging power.

$$P_t^{SUB} = P_t^D + P_t^B, \forall t \in T \; . \tag{2}$$

Equation(3) describes the change in battery energy due to charging or discharging associated with the energy arbitrage performed by the energy storage system:

$$E_{t+1}^B = E_t^B + P_t^B, \forall t \in T \; . \tag{3}$$

Equation(4) limits the battery energy range:

$$E_{\max}^B SOC_{\min} \leq E_t^B \leq E_{\max}^B SOC_{\max}, \forall t \in T \; . \tag{4}$$

Equation(5) limits the range of energy available for charging and discharging a substation:

$$-P^{SUBMAX} \leq P_t^{SUB} \leq P^{SUBMAX}, \forall t \in T \; . \tag{5}$$

Equation(6) limits the range of power that can be used by an energy storage system for energy arbitrage.

$$-P_{\max}^B \leq P_t^B \leq P_{\max}^B, \forall t \in T \; . \tag{6}$$

Equation(7) ensures that the start-end battery energy state is consistent and prevents misjudgment of the energy storage system economics due to the difference between the start and end energy:

$$E_1^B = E_T^B \; . \tag{7}$$

3.2 Segmented Battery Charging Curve

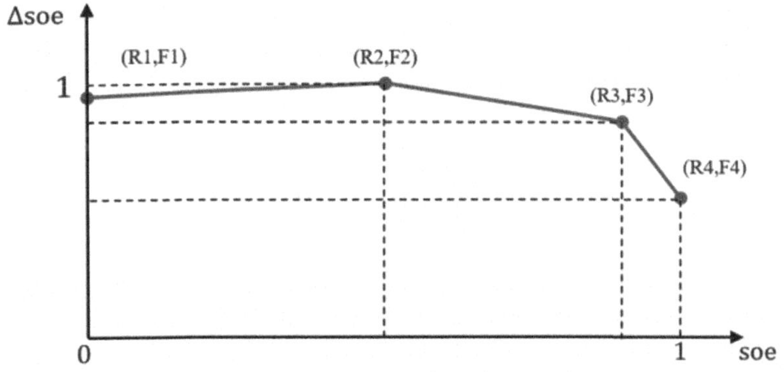

Fig. 3. Segmented Battery Charging Curve.

Due to the constant-current-voltage characteristic of lithium-ion battery charging [12] , its charging capacity is not the same under different battery capacities.

In this paper, segmented battery curves are used to describe the battery charging capacity under different battery energy states.

Constraints are placed on the battery's charging capacity in relation to the current charge to achieve accurate segmented battery charging modeling. Where I is the number of breakpoints of the segmented curve, the segmented curve in the above figure is divided into three segments with four breakpoints $(R_n, F_n), n \in \{1, 2, 3, 4\}$.

soe means state of energy, it indicates the battery's current charge percentage. Use four points $(R_n, F_n), n \in \{1, 2, 3, 4\}$ to describe the shape of the curve. The current battery charging capacity can be calculated by combining the slope of each segment of the curve with the current state of charge, and the formula is as follows:

Equation(8) indicates that Et at moment t, E_t is calculated by summing the $\Delta soe - soe$ of the three, corresponding to the seg function divided into three segments.

$$E_t = \sum_{i=1}^{I-1} E_{t,i}^B \ , t \in T. \tag{8}$$

Equation(9) indicates the maximum charge limit for each of the three curve segments in the calculation. Each segment $E_{t,i}^B$ is prevented from exceeding the transverse range of the segment.

$$E_{t,i}^B \leq R_{i+1} - R_i, t \in T, i \in I. \tag{9}$$

Equation(10) shows the maximum charging capacity of the battery at each time period. Since an increase in the charging capacity of the battery is beneficial to the system, charging will follow the $E_{t,i}^B$ order of precedence. As shown above, the $E_{t,i}^B$ will start to fill with electricity from the time when i is at its lowest, using segment 1 first, because the charging capacity of the battery is increased in this segment. Segment 2 will be used first, because the charging capacity of this segment is increasing. Then segment 2 will be used, because the charging capacity of this segment is decreasing, but not as much as the slope of the decline in segment 3, so segment 2 is preferred. Finally, use Sect. 3, which has the fastest decreasing charging capacity. The same applies when discharging.

$$\Delta soe_t = F_1 + \sum_{i=1}^{I-1} \frac{F_{i+1} - F_i}{R_{i+1} - R_i} E_{t-1,i}^B, t \in T. \tag{10}$$

Equation(11) is the charging power constraint. The charging power is the negative of the sum of the power released by the energy storage system for energy arbitrage and frequency regulation.

$$- P_t^B \leq \frac{\Delta soe_t}{\Delta t}. \tag{11}$$

3.3 Battery Life Cost

The depth of discharge of the battery over time and the corresponding number of discharges are calculated using the rainflow library.

$$R = rainflow.count\_cycles(E^B). \tag{12}$$

List $E^B = [E_1^B, E_2^B..., E_T^B]$ is the energy state of the battery at different moments.

Calculation result $R = [(D_1, n_1), (D_2, n_2)..., (D_l, n_l)]$, D_l is the calculated cycle depth, n_l is the number of cycles corresponding to the depth of the loop,l is the number of elements in R.

As an example, take a set of data $E^B = [0, -2, 1, -3, 5, -1, 3, -4, 4, -2, 0]$, the calculation result is $R = [(2, 1), (3, 0.5), (4, 1.5), (6, 0.5), (8, 0.5), (9, 0.5)]$.

Based on the calculated battery cycle depth and number of cycles, it is possible to calculate the cost of battery depletion for this period of time.

$$battery = \frac{C^{FB}}{C^{F100}} \sum_{i \in l} \left(\left(\frac{D_i}{E^B_{\max}} \right)^{kp} * n_i \right). \tag{13}$$

C^{FB} is the purchase cost of the battery.

C^{F100} is the maximum number of cycles of the battery at 100

kp is a constant, and different values are selected according to different batteries, generally in the range of 0.8–2.1.

The meaning of $\left(D_i/E^B_{\max}\right)^{kp}$ is the ratio of the life lost in one cycle to the life lost in one full cycle at cycle depth D_i.

Due to limitations in the pyomo framework, it is not possible to use functions from rainflow directly in linear programming. In the following section, the benders decomposition approach will be used to construct an overall loop architecture for power market arbitrage for energy storage systems that includes battery life calculations.

3.4 Main Problem Modeling

The objective of main optimization problem is to minimize the total cost *total_cost*:

$$\begin{aligned} total\_cost^{(m)} \geq energy^{(n)} + (L^{B(m)} - L^{B(n)})(\sum_{t \in T} \eta 1_t^{(n)} + \eta 2_t^{(n)}) \\ +battery^{(n)} + (L^{B(m)} - L^{B(n)})\alpha^{(n)}, \forall n \in m-1. \end{aligned} \tag{14}$$

$$L^{B\min} \leq L^{B(m)} \leq L^{B\max}. \tag{15}$$

where, *total_cost* indicates for the total cost taking into account the cost of purchasing electricity and the cost of battery life loss, L^B for limiting factor in the amount of limitation on the maximum charging, $\eta 1, \eta 2$ for dual values for the participation of sub-issue energy storage systems in electricity markets, α for

Dual values for the partial linearization approximation of battery life loss, and $L^{B\min}, L^{B\max}$ for minimum and maximum values of the limiting factors, taking into account the reality of the situation.

In this optimization, only the values of the variables with superscript m need to be adjusted, and the values with superscript n were determined in the previous iterations. That is, in the m th iteration, m-1 Eq.(14) are included, and these constraints work together to accomplish the determination of the maximum optimization objective and for the selection of the new constraints.

Dual values come from the constraints of the subproblem containing limiting factors to add formulas in the arbitrage model of energy storage system participation in the electricity market:

$$P_t^B \leq L^B : \eta 1_t, \forall t \in T. \tag{16}$$

$$-P_t^B \leq L^B : \eta 2_t, \forall t \in T. \tag{17}$$

Battery life depletion costs are calculated by linear approximation of the pairwise values.

(1) Calculation of Dual Values in Subproblems: In pyomo, when solving with a linear programming model, you can use the .dual attribute to get the dual values for each constraint or variable. Dual values are an important concept in linear programming, representing the Lagrange multiplier corresponding to each constraint or variable, and are used to indicate how much the objective function changes as the value of the constraint increases or decreases at the optimal solution.

For constraints: the Dual values (also known as the Lagrange multiplier) indicates the effect of a unit increase or decrease in each constraint on the value of the objective function under the optimal solution. If the dyadic value is positive, increasing the value of the right-hand side (RHS) of the constraint equation results in an increase in the objective function value; if the dyadic value is negative, increasing the value of the right-hand side of the constraint results in a decrease in the objective function value. The greater the absolute value of the dyadic value, the greater the effect of the constraints on the optimal solution.

(2) Calculation of Approximate Dual Values: Since the calculation of battery life loss cannot be described by a formula, it is not possible to calculate the pairwise values directly.

The sensitivity of battery life loss cost *battery* to the limiting factor L^B is approximated as linear. Then the approximate pairwise values can be calculated. ΔL can take smaller values, which reflect the approximate linear sensitivity of the battery life loss cost to the limiting factor.

$$\alpha \approx \frac{battery(L^B) - battery(L^B - \Delta L)}{\Delta L}. \tag{18}$$

4 Experiments

4.1 Simulation Parameters

The 24-hour tariff data and customer electricity demand used for the simulation are as follows:

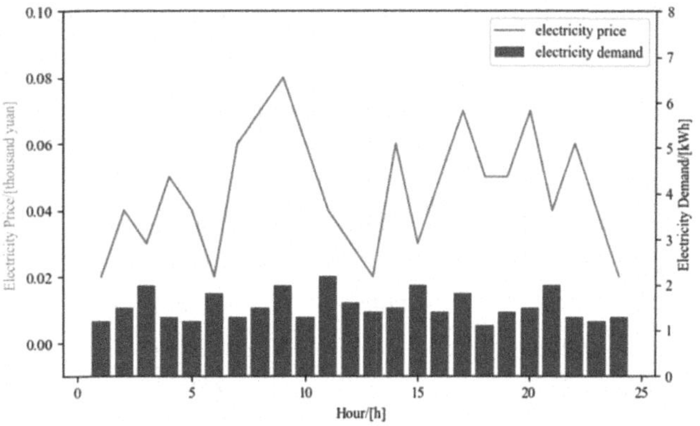

Fig. 4. 24-hour electricity price and electricity demand data.

The graph shows that the 24-hour tariff data is reasonable, with peaks in tariffs occurring during the day from 6:00 - 23:00 PM. This corresponds to the peaks in electricity consumption during the day, including those caused by production, living and entertainment. The peak valleys of the day's electricity demand are also positively correlated with the electricity price at the current moment.

The default parameters used for the simulation are as follows:

Table 3. Simulation parameters

Parameters name	Parameters	Parameter value
Limiting factor range/kW	$L^{B\min} - L^{B\max}$	1-50
Total time/h	H	24
Maximum battery energy capacity/kWh	$E^B_{\max}$	50
Maximum battery power capacity/kW	$P^B_{\max}$	L^B
Battery pack energy level range	$SOC_{\min} - SOC_{\max}$	0.1-0.9
Energy storage battery cost/k yuan	C^{FB}	32912
Substation cost/(k yuan/kw)	C^{SUB}	0.05
Maximum number of complete battery cycles	C^{F100}	10000
Battery cycle life constant	kp	2.0
Segmented Battery Curve Parameters	R	[0, 0.5, 0.9, 1.0]
Segmented Battery Curve Parameters	F	[0.95, 1.0, 0.85, 0.55]

4.2 Results and Analysis

The results of the iterative modeling of the participation of the energy storage system in the electricity market considering the battery losses are as follows:

The initial value of limiting factor L^B is set to 15.

Table 4. Iterative run results (default parameters).

Iterations	L^B	Main Prob. kCNY	Energy storage operating costs kCNY	Battery life depletion cost kCNY	Total cost kCNY
0	15.0000	/	−2.8250	5.3951	2.5701
1	1.0000	−0.7859	1.3600	0.0342	1.3942
2	5.4153	0.2725	−0.0529	1.0340	0.9811
3	3.1576	0.8460	0.6696	0.3496	1.0192
4	4.2365	0.9106	0.3243	0.6315	0.6315
5	4.7759	0.9429	0.1517	0.8034	0.3243
6	4.4562	0.9505	0.2540	0.6990	0.9551
7	4.5660	0.9521	0.2189	0.7341	0.9530
8	4.6209	0.9529	0.2013	0.7519	0.9529

It can be seen that at the beginning the value of L^B changes by a large margin, and with the increase in the number of iterative steps, the magnitude of the change gradually decreases, and finally the value of L^B gradually converges.

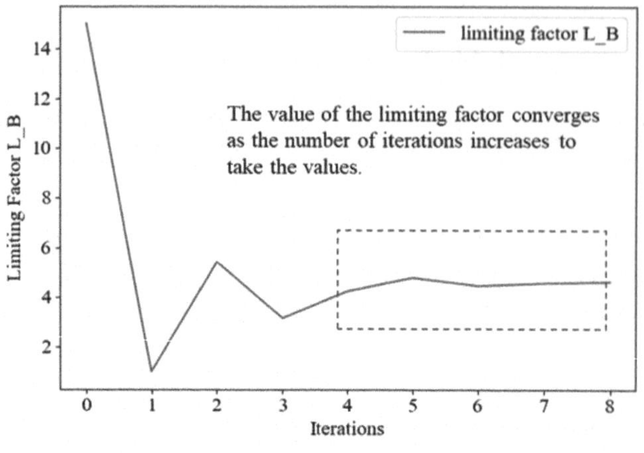

Fig. 5. Trends in limiting factors.

Energy cost and battery life cost stabilize after fluctuations, and the sum of the two curves maintains a downward trend. It means the total cost gradually decreases during the optimization process.

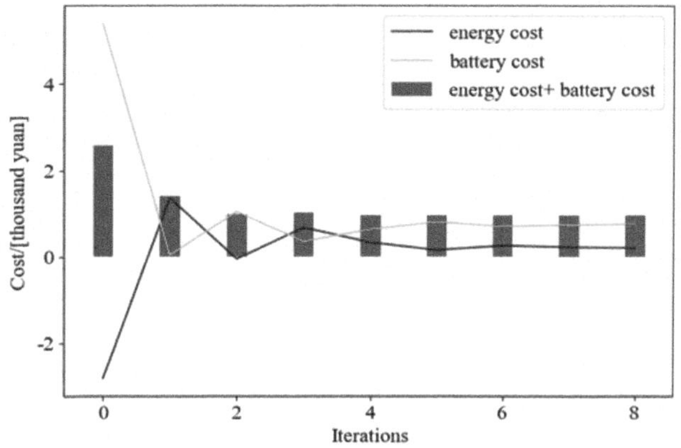

Fig. 6. Different types of costs with increasing number of iterations

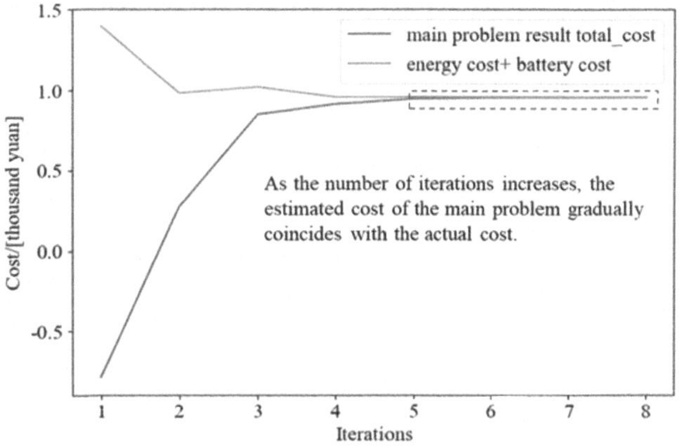

Fig. 7. Total cost with iterations in the main problem

As the number of iterations increases, the cost calculated by the main problem gradually matches the actual cost. From the main problem formula, we know that with the increasing number of iterations, the number of Eq.(14) is also increasing, that is, the main problem for the optimization objective is more and more accurate description, the coupling of subproblems and the main problem is more and more close, and finally make the main problem can fully reflect the nature of the subproblem, and the main problem as a linear programming problem, can easily find the optimal value (Figs. 1,2, 3, 4, 5, 6, 7, 8 and Tables 1, 2, 3, 4, 5).

4.3 Comparison of Pre-optimization and Post-optimization

The calculation results before and after optimization are shown in the table below:

Table 5. Iterative run results (pre-optimization and post-optimization).

Situation	L^B	Main Prob. kCNY	Energy storage operating costs kCNY	Battery life depletion cost kCNY	Total cost thousand yuan
pre-optimization	15.0000	/	-2.8250	5.3951	2.5701
post-optimization	4.6209	0.9529	0.2013	0.7519	0.9529

The energy arbitrage strategy of pre-optimization and post-optimization is presented as follows:

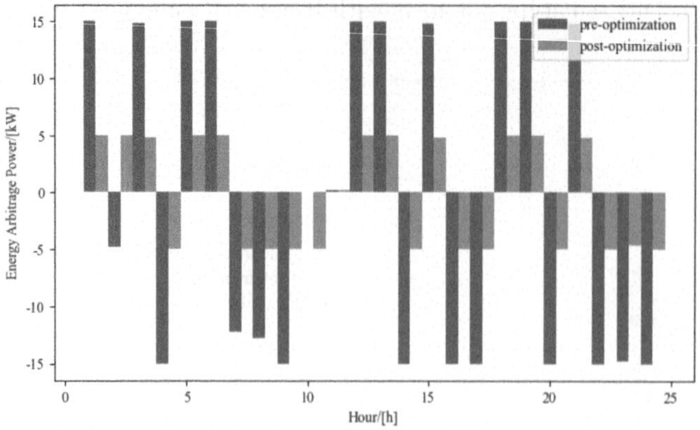

Fig. 8. Energy Arbitrage Strategy of pre-optimization and post-optimization.

It can be seen that the pre-optimization energy storage system is more aggressive in performing energy arbitrage. Although the arbitrage income is high, it also brings high battery life loss, leading to higher total cost. After optimization, the arbitrage power size of the energy storage system is limited. So that it can obtain as large arbitrage income as possible while ensuring lower battery life loss.

5 Conclusion

This paper presents a research methodology for the accurate modelling of user-side energy storage systems and power market trading strategies considering battery life. The characteristics are that the energy storage system meets the user's electricity demand and carries out charging and discharging arbitrage on the basis of the use of segmented curves to describe the charging performance of the battery in different energy states, the use of the rainfall cycle to calculate the battery's loss in the operation process, and the use of the idea of benders decomposition to incorporate the power market arbitrage model of the energy storage system and the battery's life loss during operation into the overall cyclic structure. A balanced result is achieved between the charging and discharging gains and the battery life loss to form an optimized arbitrage strategy.

Acknowledgments. This work is supported by National Natural Science Foundation of China under Grant No.52107079, Natural Science Foundation of Jiangsu Province under Grant No.BK20210243, Young Elite Scientists Sponsorship Program by Jiangsu Association for Science and Technology (JSTJ-2023-XH050) and "Zhishan Youth Scholar" Program of Southeast University under Grant No.2242023R40048.

References

1. Chen, T., Su, W.: Indirect customer-to-customer energy trading with reinforcement learning. IEEE Trans. Smart Grid **10**(4), 4338–4348 (2019)
2. Peng, Q., Li, W., Fowler, M., Chen, T., Jiang, W., Liu, K.: Battery calendar degradation trajectory prediction: data-driven implementation and knowledge inspiration. Energy **294**, 130849 (2024)
3. Wang, C., Sun, W., Yi, T., Yan, Z., Zhang, Y.: Review on energy storage application planning and benefit evaluation methods in smart grid. Proc. CSEE **33**(7), 33–41 (2013)
4. Oudalov, A., Chartouni, D., Ohler, C.: Optimizing a battery energy storage system for primary frequency control. IEEE Trans. Pow. Syst. **22**(3), 1259–1266 (2007)
5. Zhang, Y., Lundblad, A., Campana, P.E., Benavente, F., Yan, J.: Battery sizing and rule-based operation of grid-connected photovoltaic-battery system: a case study in Sweden. Energy Convers. Manage. **133**, 249–263 (2017)
6. Hassan, A., et al.: Second-life batteries: a review on power grid applications, degradation mechanisms, and power electronics interface architectures. Batteries **9**(12) (2023)
7. Sang, L., Xu, Y., Long, H., Hu, Q., Sun, H.: Electricity price prediction for energy storage system arbitrage: a decision-focused approach. IEEE Trans. Smart Grid **13**(4), 2822–2832 (2022)
8. Krishnamurthy, D., Uckun, C., Zhou, Z., Thimmapuram, P.R., Botterud, A.: Energy storage arbitrage under day-ahead and real-time price uncertainty. IEEE Trans. Pow. Syst. **33**(1), 84–93 (2018)
9. Cheng, B., Powell, W.B.: Co-optimizing battery storage for the frequency regulation and energy arbitrage using multi-scale dynamic programming. IEEE Trans. Smart Grid **9**(3), 1997–2005 (2018)

10. Cao, J., Harrold, D., Fan, Z., Morstyn, T., Healey, D., Li, K.: Deep reinforcement learning-based energy storage arbitrage with accurate lithium-ion battery degradation model. IEEE Trans. Smart Grid **11**(5), 4513–4521 (2020)
11. Wu, X., Zhao, J., Conejo, A.: Optimal battery sizing for frequency regulation and energy arbitrage. IEEE Trans. Pow. Deliv. **37**
12. Homan, B., Smit, G.J., van Leeuwen, R.P., ten Kortenaar, M.V., Ten, B.V.: A comprehensive model for battery state of charge prediction. In: 2017 IEEE Manchester PowerTech, pp. 1–6 (2017)
13. Pandžić, H., Bobanac, V.: An accurate charging model of battery energy storage. IEEE Trans. Pow. Syst. **34**(2), 1416–1426 (2019)
14. Wankmüller, F., Thimmapuram, P.R., Gallagher, K.G., Botterud, A.: Impact of battery degradation on energy arbitrage revenue of grid-level energy storage. J. Energy Storage **10**, 56–66 (2017)
15. Brivio, C., Mandelli, S., Merlo, M.: Battery energy storage system for primary control reserve and energy arbitrage. Sustain. Energy Grids Netw. **6**, 152–165 (2016)
16. Qin, Y., Yuen, C., Yin, X., Huang, B.: A transferable multistage model with cycling discrepancy learning for lithium-ion battery state of health estimation. IEEE Trans. Ind. Inf. **19**(2), 1933–1946 (2023)
17. Li, W., Sengupta, N., Dechent, P., Howey, D., Annaswamy, A., Sauer, D.U.: One-shot battery degradation trajectory prediction with deep learning. J. Pow. Sources **506**, 230024 (2021)
18. Liu, J., et al.: Capacity fading mechanisms and state of health prediction of commercial lithium-ion battery in total lifespan. J. Energy Storage **46**, 103910 (2022)
19. Ashwin, T., Chung, Y.M., Wang, J.: Capacity fade modelling of lithium-ion battery under cyclic loading conditions. J. Pow. Sources **328**, 586–598 (2016)
20. Xu, B., Oudalov, A., Ulbig, A., Andersson, G., Kirschen, D.S.: Modeling of lithium-ion battery degradation for cell life assessment. IEEE Trans. Smart Grid **9**(2), 1131–1140 (2018)
21. Abdulla, K., et al.: Optimal operation of energy storage systems considering forecasts and battery degradation. IEEE Trans. Smart Grid **9**(3), 2086–2096 (2018)
22. Xu, B., Zhao, J., Zheng, T., Litvinov, E., Kirschen, D.S.: Factoring the cycle aging cost of batteries participating in electricity markets. IEEE Trans. Pow. Syst. **33**(2), 2248–2259 (2018)
23. Schneider, S.F., Novák, P., Kober, T.: Rechargeable batteries for simultaneous demand peak shaving and price arbitrage business. IEEE Trans. Sustain. Energy **12**(1), 148–157 (2021)
24. Kazemi, M., Zareipour, H.: Long-term scheduling of battery storage systems in energy and regulation markets considering battery's lifespan. IEEE Trans. Smart Grid **9**(6), 6840–6849 (2018)

A Novel Distributed Economic Dispatch Method for Smart Grid with Time-Varying Feedback Gain

Hui Guo[1(✉)], Xue Li[1], Junlin Yang[1], and Runze Tian[2]

[1] Shanghai University, Shanghai 200444, China
guohuibingbing@163.com, lixue@i.shu.edu.cn, yjlll@shu.edu.cn
[2] Shanghai University of Finance and Economics, Shanghai 200433, China
trz@163.sufe.edu.cn

Abstract. The feedback gain in the traditional economic dispatch consensus algorithm is a fixed value, which leads to a slow convergence speed. To solve the problem, a novel distributed economic dispatch method for smart grid with time-varying feedback gain is proposed in this paper. First, by adding the direct cost, overestimation cost and underestimation cost of wind power to the objective function, an economic dispatch model considering wind power uncertainly is established, which can be flexibly scheduled under random wind speeds. Second, an improved consensus algorithm with time-varying feedback gain is proposed to solve the economic dispatch problem, which provides a design method of dynamic feedback gain instead of a fixed value and improves the convergence speed. Finally, several case studies are carried out in the IEEE 14-bus system to illustrate the effectiveness and superiority of the proposed method.

Keywords: distributed economic dispatch · smart grid · wind power uncertainty · improved consensus algorithm

1 Introduction

Smart grid is the next-generation digital power grid that integrates advanced information, communication and intelligent control technologies. It supports two-way flow of power flow and communication [1]. As one of the basic problems of energy management, economic dispatch plays an important role in the operation and control of smart grid [2]. With the continuous development of smart grid technology, the access of renewable energy and flexible load in smart grid is gradually increasing. Most of the units are plug-and-play devices, which means that the stability requirements of the smart grid are getting higher and higher. The traditional energy management mode is to meet the load demand by dispatching power generation. However, the increase in user-side participation makes it impossible to achieve optimal control of the smart grid in this mode. Therefore, in order to ensure the economic and stable operation of smart grid, it is particularly important to study the economic dispatch problem.

K. Li et al. (Eds.): LSMS/ICSEE 2024, CCIS 2218, pp. 119–133, 2025.
https://doi.org/10.1007/978-981-96-0232-2_10

Centralized optimization algorithms are often used to solve economic dispatch problems, such as quadratic programming [3], particle swarm optimization [4], and so on. However, the centralized economic dispatch mode imposes high communication and computation burden on dispatch center. It is vulnerable to single-point failures and cannot meet the plug-and-play requirements of distributed units [5].

Due to the above shortcomings of centralized optimization algorithms, distributed optimization algorithms have received extensive attention. In the related research of distributed optimization algorithms, consensus algorithm [6–8], distributed gradient method [9] and alternating direction method of multipliers [10] are the three most important methods. Among the three algorithms, the consensus algorithm has the best performance and is the most widely used. However, its convergence has always been a bottleneck in practical applications. The convergence rate is an important indicator for evaluating the scheduling performance of a system. Designing more effective and more practical distributed economic dispatch algorithms to improve convergence speed and other performance of the system has become an important research direction and has received attention from many scholars.

The feedback gain of the traditional consensus algorithm is a fixed value determined by artificial experience [11]. The feedback gain of the consensus algorithm affects the convergence speed [12]. Therefore, it is imperative to improve the convergence performance of the consensus algorithm by determining the feedback gain through optimization.

In recent years, a large number of distributed units have been integrated into the smart grid, including distributed generation units such as wind power and photovoltaic, distributed energy storage units and flexible load units on the user side. The traditional distributed economic dispatch model needs to be improved and optimized. These distributed units should be added to the model and the impact of their own characteristics on the system scheduling should be considered. Due to the free availability, environmental friendliness of wind energy and the maturity of turbine technology, wind power has been widely used [13]. However, the randomness and intermittence of wind energy make wind power uncertain, which makes it difficult to model the distributed economic dispatch problem.

In most existing studies, wind power is not considered in the distributed economic dispatch model of the smart grid. The model established in [14] only considers thermal generators and non-dispatchable loads. In [15], the cost function of thermal generator, the profit function of load and the transmission loss are considered at the same time to maximize social welfare. In the articles considering wind power, wind power output is mostly placed as a constant in the power balance constraint [16–18]. However, the wind speed fluctuates randomly during the actual scheduling process. Wind power should be flexibly scheduled under random wind speed. Therefore, the impact of wind power uncertainty should be fully taken into account in the actual scheduling.

Under the above background, a novel distributed economic dispatch method for smart grid with time-varying feedback gain is proposed in this paper. The main contributions are summarized as follows:

1) An economic dispatch model considering wind power uncertainty is established by adding the direct cost, overestimation cost and underestimation cost of wind power to the objective function. The established model can be flexibly scheduled under random wind speeds.
2) An improved consensus algorithm with time-varying feedback gain is proposed to solve the economic dispatch problem. The algorithm provides a design method of dynamic feedback gain instead of a fixed value, which can improve the convergence speed.

2 Preliminaries And Problem Formulation

2.1 System Architecture

Assuming that there are m thermal generators, n wind turbines and k flexible loads in the smart grid. The smart grid architecture considered in this paper is shown in Fig. 1. The multi-agent theory is used to construct the communication network between distributed units. Each distributed unit is regarded as an agent node. Each agent node adjusts its output to obtain the optimal power output by communicating and exchanging information with its neighbor agent nodes. The control center is not required to participate in the calculation during the above distributed economic dispatch process.

2.2 Graph Theory

An undirected graph $G = (V, E, A)$ is used to model the communication topology. $V = V_G \cup V_W \cup V_D$ represents the set of nodes. $V_G = (g_1, g_2, g_3, \ldots, g_m)$ represents the set of thermal generators. $V_W = (w_1, w_2, w_3, \ldots, w_n)$ represents the set of wind turbines. $V_D = (d_1, d_2, d_3, \ldots, d_k)$ represents the set of flexible loads. $E = \{(i, j) \mid i, j \in V\} \subseteq V \times V$ represents the set of communication links. A represents the adjacency matrix of the communication topology. The diagonal elements a_{ii} of the adjacency matrix A are all 0. The non-diagonal element a_{ij} represents the connection between node i and node j. If the two nodes are connected, $a_{ij} = 1$; otherwise, $a_{ij} = 0$. The diagonal element of the Laplacian matrix L is denoted by $l_{ii} = \sum a_{ij}$. The non-diagonal element is denoted by $l_{ij} = -a_{ij}$. Furthermore, the element in the state transition matrix H is represented as $h_{ij} = \frac{|l_{ij}|}{\sum_{j=1}^{n}|l_{ij}|}$.

2.3 Economic Dispatch Model

The economic dispatch model established in this paper includes two parts: objective function and constraint conditions. The objective function is to minimize

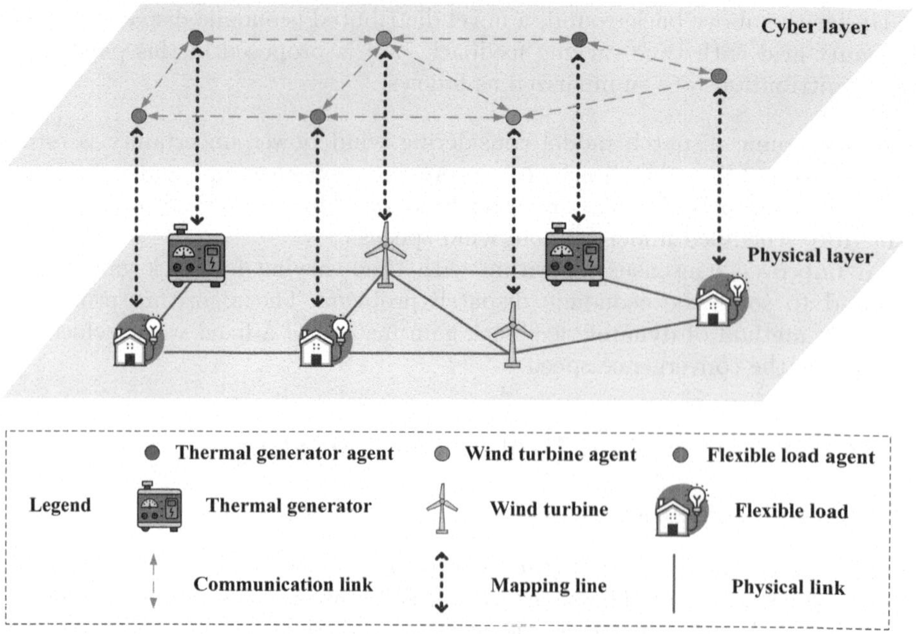

Fig. 1. System architecture

the system operation cost. The constraints include power balance constraint and upper and lower limit constraint. The decision variable is the active output of each distributed unit.

The cost function of thermal generator i is expressed as

$$C_i\left(P_{Gi}\right) = a_{Gi}P_{Gi}^2 + b_{Gi}P_{Gi} + c_{Gi} \tag{1}$$

where P_{Gi} denotes the active power output of thermal generator i; a_{Gi}, b_{Gi} and c_{Gi} are the fuel cost coefficients of thermal generator i; $C_i\left(P_{Gi}\right)$ represents the fuel cost of thermal generator i.

Due to the randomness of wind speed, the actual available power of wind power is a random variable, which may be different from the scheduling power. The energy storage system is integrated into the wind turbine to compensate the deviation between the scheduling power and the actual available power, so the total output of wind power is equal to the scheduling power. If the scheduling power is greater than the actual available power, it is called overestimation. The energy storage system will generate electricity to fill the shortage of the actual available power of wind power, resulting in overestimation penalty cost. If the scheduling power is less than the actual available power, it is called underestimation. The energy storage system will use the surplus wind power to charge, resulting in underestimation penalty cost. Overestimation penalty cost and Underestimation penalty cost are used to measure the impact of wind power uncertainty on system scheduling. The cost of wind power is expressed as the sum of direct cost, overestimation cost and underestimation cost. The cost function of wind turbine j is expressed as

$$C_j(W_j) = d_j W_j + C_{rwj} E(Y_{oe,j}) + C_{pwj} E(Y_{ue,j}) \tag{2}$$

where the first term represents the direct cost, the second represents the overestimation cost and the third represents the underestimation cost; d_j, c_{rwj} and c_{pwj} denote the cost coefficients of wind turbine j; $E(Y_{oe,j})$ is the mean value of random variable $Y_{oe,j}$, $Y_{oe,j}$ denotes the difference between the scheduling power and the actual available power; $E(Y_{ue,j})$ is the mean value of random variable $Y_{ue,j}$, $Y_{ue,j}$ denotes the difference between the actual available power and the scheduling power.

$E(Y_{oe,j})$ and $E(Y_{ue,j})$ are expressed as [19]

$$\begin{aligned} E(Y_{oe,j}) = & W_j \left[1 - exp\left(-\frac{v_{in}^k}{c^k}\right) + exp\left(-\frac{v_{out}^k}{c^k}\right)\right] \\ & + \left(\frac{W_r v_{in}}{v_r - v_{in}}\right)\left[exp\left(-\frac{v_{in}^k}{c^k}\right) - exp\left(-\frac{v_j^k}{c^k}\right)\right] \\ & + \frac{W_r c}{v_r - v_{in}}\left\{\Gamma\left[1 + \frac{1}{k}, \left(\frac{v_j}{c}\right)^k\right] - \Gamma\left[1 + \frac{1}{k}, \left(\frac{v_{in}}{c}\right)^k\right]\right\} \end{aligned} \tag{3}$$

$$\begin{aligned} E(Y_{ue,j}) = & (W_r - W_j)\left[exp\left(-\frac{v_r^k}{c^k}\right) - exp\left(-\frac{v_{out}^k}{c^k}\right)\right] \\ & + \left(\frac{W_r v_{in}}{v_r - v_{in}}\right)\left[exp\left(-\frac{v_r^k}{c^k}\right) - exp\left(-\frac{v_j^k}{c^k}\right)\right] \\ & + \frac{W_r c}{v_r - v_{in}}\left\{\Gamma\left[1 + \frac{1}{k}, \left(\frac{v_j}{c}\right)^k\right] - \Gamma\left[1 + \frac{1}{k}, \left(\frac{v_r}{c}\right)^k\right]\right\} \end{aligned} \tag{4}$$

where v_{in}, v_{out} and v_r are the cut-in wind speed, the cut-out wind speed and the rated wind speed, respectively; k and c are the shape factor and the scale factor of Weibull distribution of wind speed, respectively; W_r is the rated wind power; $\Gamma(a, x)$ is the form of standard incomplete gamma function; v_j is the intermediate variable.

v_j is expressed as

$$v_j = v_{in} + \frac{(v_r - v_{in}) W_j}{W_r} \tag{5}$$

The electricity benefit function of flexible load x is expressed as

$$C_x(P_{Lx}) = a_{Lx} P_{Lx}^2 + b_{Lx} P_{Lx} + c_{Lx} \tag{6}$$

where a_{Lx}, b_{Lx} and c_{Lx} denote the cost coefficients of flexible load x; P_{Lx} is the required power of flexible load x, which is negative in this paper.

Then, considering the power balance constraint and the upper and lower limit constraint, the optimization problem can be formulated as

$$\begin{aligned} &\min_{P_{Gi},W_j,P_{Lx}} \quad C_i\left(P_{Gi}\right)+C_j\left(W_j\right)+C_x\left(P_{Lx}\right) \\ &s.t. \sum_{i=1}^{m} P_{Gi}+\sum_{j=1}^{n} W_j+\sum_{x=1}^{k} P_{Lx}=0 \\ &P_{Gi}^{\min} \leq P_{Gi} \leq P_{Gi}^{\max} \\ &0 \leq W_j \leq W_r \\ &P_{Lx}^{\min} \leq P_{Lx} \leq P_{Lx}^{\max} \end{aligned} \tag{7}$$

where $P_{Gi}^{\min}$ and $P_{Gi}^{\max}$ represent the lower and upper limits of the thermal generator i, respectively; $P_{Lx}^{\min}$ and $P_{Lx}^{\max}$ represent the lower and upper limits of the flexible load x, respectively.

The convexity of (3) and (4) is proved in [19]. Because the cost functions of thermal generator, wind turbine and flexible load are all convex functions, the objective function is convex. In addition, the constraints are convex. Therefore, the above distributed economic dispatch problem is a convex optimization problem.

3 Improved Consensus Algorithm Design

3.1 Feedback Gain Design

The convergence speed of the consensus algorithm is determined by the second largest eigenvalue $|\mu_2|$ of the system matrix M. The optimal convergence rate will be obtained when $|\mu_2|$ is minimum [12].

The topology G of the system and the feedback gain η of the algorithm affect the second largest eigenvalue $|\mu_2|$ of the system matrix M and further affect the convergence speed. Next, we will explore the change of the second largest eigenvalue $|\mu_2|$ of the system matrix M under different topologies G and different feedback gains η.

Because the cost function of wind power is not a quadratic function, the system matrix is not fixed and will change with the change of wind power during the convergence process. The system matrix is expressed as

$$M\left(k\right)=\begin{bmatrix} H & \eta \\ \delta\left(k\right)\left(I-H\right)Z-\eta\delta\left(k\right) \end{bmatrix} \tag{8}$$

In the above formula, Z is the transpose matrix of the state transition matrix H. The topologies of the three five-node systems are shown in Fig. 2. Node 1 is a thermal generator, node 2 is a wind turbine and nodes 3, 4 and 5 are flexible loads. The parameters of thermal generator and flexible loads are shown in Table 1. Taking $\delta\left(k\right)$ of the wind turbine as 24.15984151, 24.30509293 and 24.76345089, respectively. The changes in the second largest eigenvalue $|\mu_2|$ of

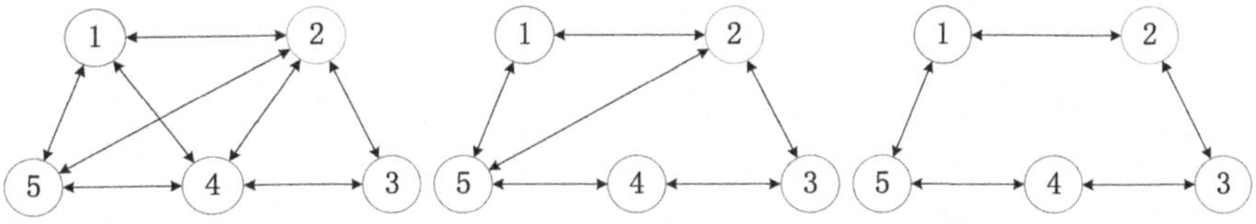

(a) Topology 1 (b) Topology 2 (c) Topology 3

Fig. 2. Topologies of three five-node systems

Table 1. Parameters of Thermal Generator and Flexible Loads

Node	a($/kW^{2}h)	b($/kWh)	c($/h)	$P_{Gi}^{\max}$(kW)	$P_{Gi}^{\min}$(kW)
1	0.040	2.25	300	80	30
3	0.036	8.25	0	−10	−30
4	0.033	7.2	0	−5	−20
5	0.0375	7.8	0	−5	−30

the system matrix M under different topologies G and feedback gains η are shown in Tables 2, 3 and 4, respectively.

The following conclusions can be drawn from Tables Tables 2, 3 and 4:

1) In three cases, the second largest eigenvalue $|\mu_2|$ of the system matrix M is slightly different, but the change trend is the same.
2) For a specific topology G, the second largest eigenvalue $|\mu_2|$ of the system matrix M decreases first and then increases as the feedback gain η increases. When the feedback gain η exceeds a certain threshold, the first largest eigenvalue $|\mu_1|$ of the system matrix M is greater than 1. Then, the system will diverge.
3) Under different topologies G, the second largest eigenvalue $|\mu_2|$ of the system matrix M changes at different speeds as the feedback gain η increases.

For a specific distributed economic dispatch problem, the topology G is fixed. Therefore, the improvement of feedback gain η is considered. For a fixed system matrix M, there exists an optimal feedback gain η^* such that the second largest eigenvalue $|\mu_2|$ is the minimized and the convergence rate is the fastest. The system matrix M of the distributed economic dispatch problem in this paper changes with the change of wind power in the iterative process. The system

Table 2. Change of $|\mu_2|$ When $\delta(k)$ of Wind Turbine is 24.15984151

Topology	$\eta = 0.0005$	$\eta = 0.005$	$\eta = 0.01$	$\eta = 0.02$	$\eta = 0.03$	$\eta = 0.035$	$\eta = 0.04$
1	0.9917	0.9160	0.8263	0.8012	0.8332	0.9081	1($\|\mu_1\|$=1.0323)
2	0.9919	0.9188	0.8325	0.8594	0.9480	1($\|\mu_1\|$=1.0702)	1($\|\mu_1\|$=1.1905)
3	0.9921	0.9207	0.8367	0.8687	0.9487	1($\|\mu_1\|$=1.0713)	1($\|\mu_1\|$=1.1920)

Table 3. Change of $|\mu_2|$ When $\delta(k)$ of Wind Turbine is 24.30509293

Topology	$\eta = 0.0005$	$\eta = 0.005$	$\eta = 0.01$	$\eta = 0.02$	$\eta = 0.03$	$\eta = 0.035$	$\eta = 0.04$
1	0.9916	0.9158	0.8260	0.8012	0.8332	0.9124	$1(\|\mu_1\|=1.0373)$
2	0.9919	0.9187	0.8323	0.8595	0.9510	$1(\|\mu_1\|=1.0737)$	$1(\|\mu_1\|=1.1947)$
3	0.9921	0.9206	0.8367	0.8688	0.9519	$1(\|\mu_1\|=1.0751)$	$1(\|\mu_1\|=1.1963)$

Table 4. Change of $|\mu_2|$ When $\delta(k)$ of Wind Turbine is 24.76345089

Topology	$\eta = 0.0005$	$\eta = 0.005$	$\eta = 0.01$	$\eta = 0.02$	$\eta = 0.03$	$\eta = 0.035$	$\eta = 0.04$
1	0.9916	0.9154	0.8248	0.8012	0.8332	0.9160	$1(\|\mu_1\|=1.0531)$
2	0.9918	0.9183	0.8316	0.8597	0.9605	$1(\|\mu_1\|=1.0851)$	$1(\|\mu_1\|=1.2079)$
3	0.9920	0.9201	0.8366	0.8692	0.9619	$1(\|\mu_1\|=1.0869)$	$1(\|\mu_1\|=1.2100)$

matrix M of each iteration corresponds to an optimal feedback gain η^*. However, the optimal feedback gain η^* is difficult to obtain. In this case, the form of fixed feedback gain determined by artificial experience is not conducive to the acceleration of convergence speed. Therefore, it is considered to design the feedback gain as a function that changes dynamically with the number of iterations within a certain range. The feedback gain of the consensus algorithm is improved from the fixed value η to the designed dynamic function $\eta(k)$.

3.2 Improved Consensus Algorithm with Time-Varying Feedback Gain

The Lagrange multiplier method is used to transform the power balance constraint into the objective function. The solution of the distributed economic dispatch problem is transformed into the Lagrange extremum solution problem, which is expressed as

$$\begin{aligned}\min L = &\sum_{i=1}^{m} C_i(P_{Gi}) + \sum_{j=1}^{n} C_j(W_j) + \sum_{x=1}^{k} C_x(P_{Lx}) \\ &- \lambda\left(\sum_{i=1}^{m} P_{Gi} + \sum_{j=1}^{n} W_j + \sum_{x=1}^{k} P_{Lx}\right)\end{aligned} \tag{9}$$

where λ represents the Lagrangian multiplier.

According to the necessary condition for finding the extremum of the Lagrangian function: $\frac{\partial L}{\partial P_{Gi}} = \frac{\partial L}{\partial W_j} = \frac{\partial L}{\partial P_{Lx}} = 0$, the following equations can be obtained

$$\frac{\partial L}{\partial P_{Gi}} = \frac{\partial C_i(P_{Gi})}{\partial P_{Gi}} - \lambda = 0 \tag{10}$$

$$\frac{\partial L}{\partial W_j} = \frac{\partial C_j(W_j)}{\partial W_j} - \lambda = 0 \tag{11}$$

$$\frac{\partial L}{\partial P_{Lx}}=\frac{\partial C_x\left(P_{Lx}\right)}{\partial P_{Lx}}-\lambda=0 \tag{12}$$

It can be inferred that when the incremental cost of each distributed unit in the system is equal, the distributed economic scheduling problem obtains the optimal solution.

The incremental cost of thermal generator i is expressed as

$$\lambda_i=\frac{\partial C_i\left(P_{Gi}\right)}{\partial P_{Gi}}=2a_{Gi}P_{Gi}+b_{Gi} \tag{13}$$

The incremental cost of wind turbine j is expressed as

$$\begin{aligned}\lambda_j&=\frac{\partial C_j\left(W_j\right)}{\partial W_j}\\&=d_j+C_{rwj}\frac{\partial E\left(Y_{oe,j}\right)}{\partial W_j}+C_{pwj}\frac{\partial E\left(Y_{ue,j}\right)}{\partial W_j}\\&=d_j+C_{rwj}\left[1+exp\left(-\frac{v_{out}^k}{c^k}\right)-exp\left(-\frac{v_j^k}{c^k}\right)\right]\\&+C_{pwj}\left[exp\left(-\frac{v_{out}^k}{c^k}\right)-exp\left(-\frac{v_j^k}{c^k}\right)\right]\end{aligned} \tag{14}$$

The incremental cost of flexible load x is expressed as

$$\lambda_x=\frac{\partial C_x\left(P_{Lx}\right)}{\partial P_{Lx}}=2a_{Lx}P_{Lx}+b_{Lx} \tag{15}$$

The incremental cost of each distributed unit is selected as the consensus variable. The incremental cost iteration formula based on the proposed fully distributed consensus algorithm with time-varying feedback gain is expressed as

$$\lambda_i\left(k+1\right)=\sum_{j\in N_i}h_{ij}\lambda_j\left(k\right)+\eta\left(k\right)y_i\left(k\right) \tag{16}$$

where $\lambda_i\left(k+1\right)$ denotes the incremental cost of the distributed unit i in the $k+1$th iteration; N_i denotes the set of neighbor units of node i; h_{ij} is the weight element in the state transition matrix H; $\eta\left(k\right)$ is the feedback gain or convergence coefficient, which is a constant small enough determined by artificial experience for traditional consensus algorithm; $y_i\left(k\right)$ represents the feedback variable or the mismatch variable of node i in the kth iteration. The function of the feedback variable is to make the solution meet the power balance constraint.

The iterative formula of the mismatch variable is expressed as

$$y_i\left(k+1\right)=\sum_{j\in N_i}z_{ij}y_j\left(k\right)-\left[P_i\left(k+1\right)-P_i\left(k\right)\right] \tag{17}$$

where z_{ij} is the element in the transpose matrix Z of the state transition matrix H; $P_i\left(k+1\right)$ is the active power output of the distributed unit i in the $k+1$th iteration.

Since each thermal generator, wind turbine and flexible load in the system must meet the upper and lower limit constraint, the active power output of each distributed unit should be limited in the solution process, which is expressed as

$$P_{Gi}(k) = \begin{cases} P_{Gi}^{\min}, & \frac{\lambda_{Gi}(k)-b_{Gi}}{2a_{Gi}} \le P_{Gi}^{\min} \\ \frac{\lambda_{Gi}(k)-b_{Gi}}{2a_{Gi}}, & P_{Gi}^{\min} \le \frac{\lambda_{Gi}(k)-b_{Gi}}{2a_{Gi}} \le P_{Gi}^{\max} \\ P_{Gi}^{\max}, & \frac{\lambda_{Gi}(k)-b_{Gi}}{2a_{Gi}} \ge P_{Gi}^{\max} \end{cases} \tag{18}$$

$$W_j(k) = \begin{cases} 0, & C_j'^{-1}(\lambda_j(k)) \le 0 \\ C_j'^{-1}(\lambda_j(k)), & 0 \le C_j'^{-1}(\lambda_j(k)) \le W_r \\ W_r, & C_j'^{-1}(\lambda_j(k)) \ge W_r \end{cases} \tag{19}$$

$$P_{Lx}(k) = \begin{cases} P_{Lx}^{\min}, & \frac{\lambda_{Lx}(k)-b_{Lx}}{2a_{Lx}} \le P_{Lx}^{\min} \\ \frac{\lambda_{Lx}(k)-b_{Lx}}{2a_{Lx}}, & P_{Lx}^{\min} \le \frac{\lambda_{Lx}(k)-b_{Lx}}{2a_{Lx}} \le P_{Lx}^{\max} \\ P_{Lx}^{\max}, & \frac{\lambda_{Lx}(k)-b_{Lx}}{2a_{Lx}} \ge P_{Lx}^{\max} \end{cases} \tag{20}$$

The proposed improved consensus algorithm is summarized as Algorithm 1.

Algorithm 1: An Improved Consensus Algorithm

1: Set $P_i(0)$ to arbitrary value, $y_i(0)$ to $-P_i(0)$ and $\lambda_i(0)$ to $C_i'(P_i(0))$.
2: Set iteration index $k = 0$.
3: while system mismatch power error variable $\varepsilon > 0.01$.
4: Update $\lambda_i(k+1)$ according to formula (16).
5: Update $P_i(k+1)$ according to formula (18), (19) and (20).
6: Update $y_i(k+1)$ according to formula (17).
7: Set $k = k+1$.
8: end while.

4 Simulation Results

In this section, the effectiveness and superiority of the proposed method are verified in the IEEE 14-bus system. The structure and communication topology of the IEEE 14-bus system are shown in Fig. 3. Nodes 1, 2, 3 and 6 are thermal generators. Node 7 is a wind turbine. Nodes 4, 5, 8, 9, 10, 11, 12, 13 and 14 are flexible loads. The parameters of thermal generators and flexible loads are shown in Table 5. The parameters of the wind turbine are shown in Table 6. The feedback gain is a function of the number of iterations k, which is $0.03 \times \tan^{-1}\left(\frac{t}{1000}\right)$. The iteration termination condition is set as the system mismatch power error variable $\varepsilon \le 0.01$.

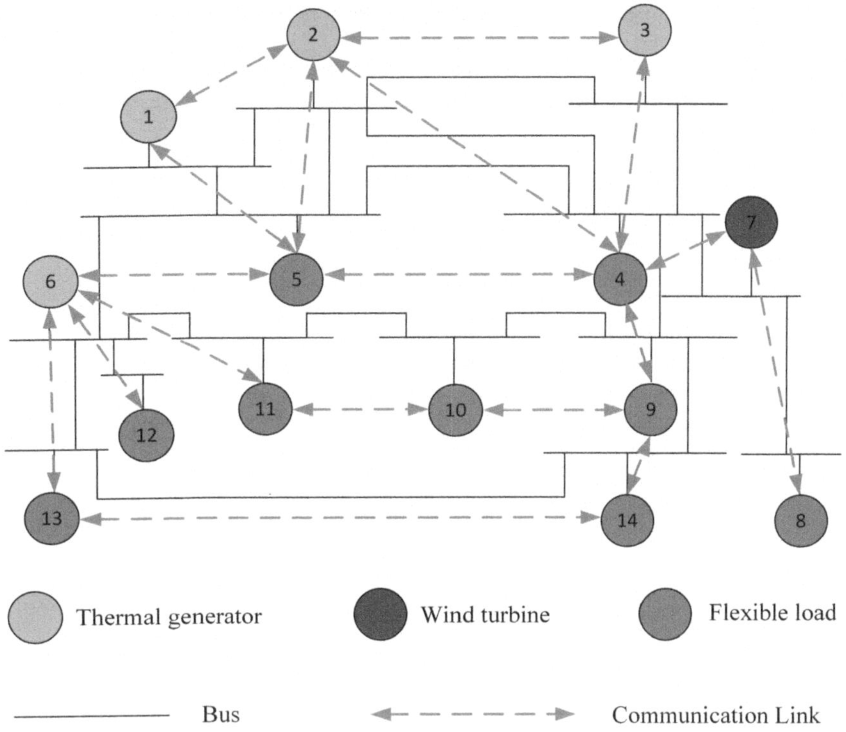

Fig. 3. Structure and communication topology of IEEE 14-bus system

4.1 Scheduling Results of Improved Consensus Algorithm

The improved consensus algorithm is used to solve the distributed economic dispatch problem of the IEEE 14-bus system. The simulation results are shown in Fig. 4. It can be seen from Fig. 4(a) that the improved consensus algorithm converges after 143 iterations. At this time, the system mismatch power is 0.00967709, which is less than 0.01, indicating that the iteration termination condition is reached. It can be seen from Fig. 4(b) that the mismatch power of each distributed unit eventually tends to 0, indicating that the system satisfies the power balance constraint. Figure 4(c) shows the incremental cost $\lambda^* = 6.41376\$/\text{kW}$. Figure 4(d) shows that the output of each distributed unit changes continuously at the beginning and tends to be stable after a certain number of iterations.

4.2 Comparison with Traditional Consensus Algorithm

The traditional consensus algorithm is used to solve the distributed economic dispatch problem of the IEEE 14-bus system. In the traditional consensus algorithm, the fastest convergence occurs after 178 iterations with the fixed feedback gain of 0.003. The simulation results when the feedback gain is 0.003 are

Table 5. Parameters of Thermal Generators and Flexible Loads

Node	a(\$/kW^{2}h)	b(\$/kWh)	c(\$/h)	$P_{Gi}^{\max}$(kW)	$P_{Gi}^{\min}$(kW)	P_0(kW)
1	0.040	2.25	300	80	30	35
2	0.031	4.2	210	70	20	30
3	0.0375	3.25	290	70	30	40
4	0.036	8.25	0	−10	−30	−10
5	0.033	7.2	0	−5	−20	−5
6	0.035	4.05	320	70	20	20
8	0.0375	7.8	0	−5	−30	−15
9	0.030	8.05	0	−10	−35	−20
10	0.039	8.45	0	−15	−35	−20
11	0.040	8.75	0	−15	−40	−15
12	0.0425	9	0	−15	−40	−20
13	0.0345	7.05	0	−5	−25	−20
14	0.0385	8.15	0	−10	−35	−25

Table 6. Parameters of Wind Turbine

v_{in}	v_{out}	v_r	(c,k)	d_j	C_{pwj}	C_{rwj}	W_r	W_0
5	45	15	(8,2)	6	3.1	3.1	160	5

shown in Fig. 5. It can be seen from Fig. 5(a) that the traditional consensus algorithm needs 178 iterations to converge at the fastest, which is slower than the proposed improved consensus algorithm. From Fig. 5(b), it can be seen that the optimal incremental cost obtained by the traditional consensus algorithm is 6.41374\$/kW, which is consistent with the optimal incremental cost obtained by the proposed improved consensus algorithm. The simulation results show that the proposed improved consensus algorithm improves the convergence speed compared with the traditional consensus algorithm. The effectiveness of the proposed improved consensus algorithm is verified by comparing the simulation results of the two algorithms.

4.3 Scheduling Results Under Random Wind Speed

Because the wind speed is random in the actual scheduling process, the values of Weibull distribution factors c and k will change with the change of wind speed. Next, the scheduling results under random wind speed are explored. Under the condition of keeping the shape factor k constant, change the values of the scale factor c to 12, 18, and 24, respectively. The experimental results are shown in Table 7. Then, under the condition of keeping the scale factor c unchanged, the values of the shape factor k are changed to 1, 3 and 5, respectively. The experimental results are shown in Table 8. The experimental results show that

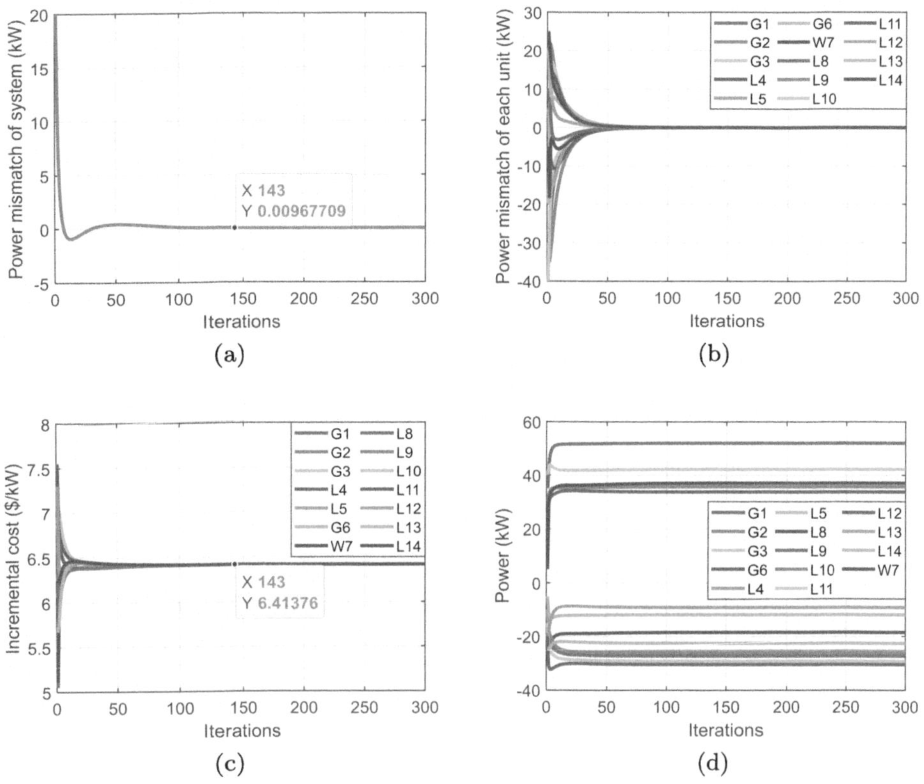

Fig. 4. Simulation results of improved consensus algorithm

when the value of c or k becomes larger, the number of iterations required for convergence increases and the output of wind turbine becomes larger. This is because the proportion of wind energy with higher wind speeds is higher and there is more available wind energy when the value of c or k is larger. At this time, increasing the scheduling of wind power is a more economical behavior. The experimental results verify the flexibility and effectiveness of the established model under random wind speed.

Table 7. Experimental Results of Different Scale Factors

c	8	12	18	24
Number of iterations	143	152	201	207
Wind power output(kW)	37.033	85.2748	145.617	160

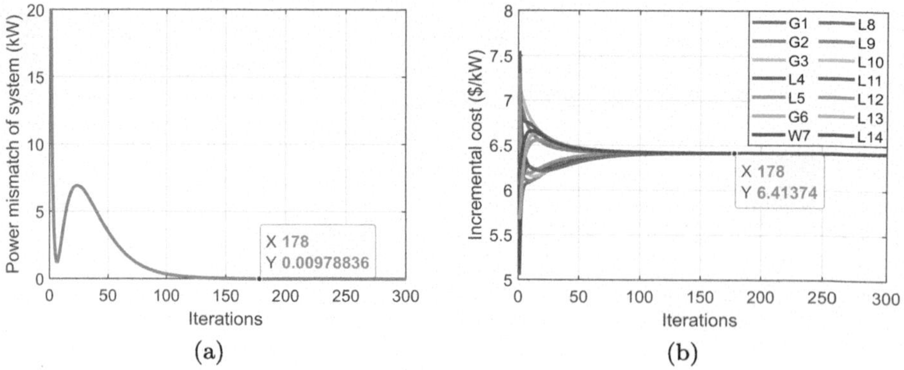

Fig. 5. Simulation results of traditional consensus algorithm

Table 8. Experimental Results of Different Shape Factors

k	1	2	3	5
Number of iterations	123	143	158	167
Wind power output(kW)	28.2487	37.033	40.2639	43.1276

5 Conclusion

A novel distributed economic dispatch method for smart grid with time-varying feedback gain is proposed in this paper. Firstly, a distributed economic dispatch model including thermal generators, wind turbines and flexible loads is established. The overestimation cost and the underestimation cost of wind power are used to measure the impact of wind power uncertainty on system scheduling. Furthermore, an improved consensus algorithm with time-varying feedback gain is proposed to accelerate the convergence speed. The feedback gain of the consensus algorithm is improved from a fixed value to a function that changes dynamically with the number of iterations. Finally, experiments confirm the effectiveness and superiority of the proposed model and algorithm. Moreover, the research on distributed economic dispatch for smart grid considering the non-ideal communication environment and security is future work.

References

1. Kumar, P., Lin, Y., Bai, G., Paverd, A., Dong, J.S., Martin, A.: Smart grid metering networks: a survey on security, privacy and open research issues. IEEE Commun. Surv. Tutor. **21**(3), 2886–2927 (2019)
2. Chatterjee, A., Paul, S., Ganguly, B.: Multi-objective energy management of a smart home in real time environment. IEEE Trans. Ind. Appl. **59**(1), 138–147 (2023)

3. Subathra, M.S.P., Selvan, S.E., Victoire, T.A.A., Christinal, A.H., Amato, U.: A hybrid with cross-entropy method and sequential quadratic programming to solve economic load dispatch problem. IEEE Syst. J. **9**(3), 1031–1044 (2015)
4. Wang, B., Zhang, P., He, Y., Wang, X., Zhang, X.: Scenario-oriented hybrid particle swarm optimization algorithm for robust economic dispatch of power system with wind power. J. Syst. Eng. Electron. **33**(5), 1143–1150 (2022)
5. Firouzbahrami, M., Nobakhti, A.: Finite-time distributed economic dispatch over network systems with coupled local costs. IEEE Control Syst. Lett. **7**, 325–330 (2023)
6. Chen, W., Liu, L., Liu, G.P.: Privacy-preserving distributed economic dispatch of microgrids: a dynamic quantization-based consensus scheme with homomorphic encryption. IEEE Trans. Smart Grid **14**(1), 701–713 (2023)
7. Zhao, C., Duan, X., Shi, Y.: Analysis of consensus-based economic dispatch algorithm under time delays. IEEE Trans. Syst. Man Cybern. Syst. **50**(8), 2978–2988 (2020)
8. Pourbabak, H., Luo, J., Chen, T., Su, W.: A novel consensus-based distributed algorithm for economic dispatch based on local estimation of power mismatch. IEEE Trans. Smart Grid **9**(6), 5930–5942 (2018)
9. Guo, F., Li, G., Wen, C., Wang, L., Meng, Z.: An accelerated distributed gradient-based algorithm for constrained optimization with application to economic dispatch in a large-scale power system. IEEE Trans. Syst. Man Cybern. Syst. **51**(4), 2041–2053 (2021)
10. Chen, G., Yang, Q.: An ADMM-based distributed algorithm for economic dispatch in islanded microgrids. IEEE Trans. Industr. Inf. **14**(9), 3892–3903 (2018)
11. Yan, X., Zhong, H., Wang, J., Tan, Z.: Consensus-based distributed economic dispatch with optimized transition matrix. In: 2019 IEEE Power and Energy Society General Meeting (PESGM), pp. 1–5 (2019)
12. Wang, R., Li, Q., Zhang, B., Wang, L.: Distributed consensus based algorithm for economic dispatch in a microgrid. IEEE Trans. Smart Grid **10**(4), 3630–3640 (2019)
13. Mu, G., Wu, F., Zhang, X., Fu, Z., Xiao, B.: Analytic mechanism for the cumulative effect of wind power fluctuations from single wind farm to wind farm cluster. CSEE J. Power Energy Syst. **8**(5), 1290–1301 (2022)
14. Wen, G., Yu, X., Liu, Z.W., Yu, W.: Adaptive consensus-based robust strategy for economic dispatch of smart grids subject to communication uncertainties. IEEE Trans. Industr. Inf. **14**(6), 2484–2496 (2018)
15. Zhao, C., He, J., Cheng, P., Chen, J.: Consensus-based energy management in smart grid with transmission losses and directed communication. IEEE Trans. Smart Grid **8**(5), 2049–2061 (2017)
16. Yan, L., Chen, X., Liu, D., Zhou, J., Chen, Y.: Fully distributed energy management in smart grids based on diffusion strategy. Int. J. Electr. Power Energy Syst. **129**, 106759 (2021)
17. He, Y., Wang, W., Wu, X.: Economic dispatch strategy for distributed generators considering communication delays. Power Syst. Technol. **45**(6), 3015–3025 (2021)
18. Ruan, B., Yu, D., Li, S.: Consensus algorithm based distributed energy management strategy of microgrids. Power Syst. Protect. Control **46**(15), 23–28 (2018)
19. Liu, X., Xu, W.: Minimum emission dispatch constrained by stochastic wind power availability and cost. IEEE Trans. Power Syst. **25**(3), 1705–1713 (2010)

Benefits, Barriers, and Human Factors of V2X Implementation – A Brief Review

Oghenewvogaga Oghorada(✉), Shihao Zhao, Yifei Zhang, and Kang Li

University of Leeds, Leeds LS2 9JT, UK
o.oghorada@leeds.ac.uk

Abstract. To achieve the decarbonisation of the transport sector, a mass roll-out of electric vehicles (EVs) is crucial. Transport decarbonisation, however, poses significant pressure on the existing electrical grid to meet the peak demand of charging these electric vehicles. The concept of vehicle-to-everything (V2X) will aid the entire power grid through various applications, including vehicle-to-home (V2H), vehicle-to-building (V2B), vehicle-to-load (V2L), vehicle-to-grid (V2G), and vehicle-to-vehicle (V2V). V2X technology allows electric vehicles to interact with the power grid and other devices, providing flexibility and support to the energy infrastructure. This paper presents a brief review of the benefits and barriers of V2X deployment, as well as the challenges and potential solutions related to human factors in its implementation. In addition, the benefits of V2G to the UK electricity market are discussed. By examining these aspects, this paper aims to provide insights into how V2X can be effectively integrated into the energy and transport sectors to support sustainable development and reduce carbon emissions.

Keywords: Electric Vehicle (EV) · Vehicle to Everything (V2X) · Benefits · and Human Factors

1 Introduction

Electric vehicles (EVs) are becoming increasingly popular since the last decade because of climate change problems and environmental regulations, hike in energy costs and the supply challenge with fossil fuels. In response to increasing demands on EVs, automobile manufacturers have added more models to their lineups. During the 2020 outbreak, the world vehicle sales declined by 16%, but the registration of EVs raised by 41% [1]. Across Europe in 2020, the EVs registered had an increment of 141% against the previous year [2].

EVs are conventionally viewed as electrical loads connected to electrical network. This implies the charging of these vehicles from the grid is the predominant part of the power flow. It was illustrated in [3] that EV charging through a grid powered home will substantially increase homeowner energy tariffs. EVs' large batteries and their high charging power, on the other hand, have made it more difficult for the electric grid to adequately meet their energy demand. An EV power flow in both directions is a key method for not only mitigating the impact of charging EVs, but also supporting the grid.

K. Li et al. (Eds.): LSMS/ICSEE 2024, CCIS 2218, pp. 134–146, 2025.
https://doi.org/10.1007/978-981-96-0232-2_11

An EV can be charged or discharged at home or at a charging station through bidirectional power flow control. Vehicle to everything (V2X) is an approach to electrifying residential and commercial buildings, other electrified cars, and the grid by using the energy from EV batteries.

V2X technology can help EVs reduce electricity costs, regulate frequency, shift loads, control voltage, and arbitrage energy. The International Energy Agency (IEA) regularly share information about the latest progress in V2X [4]. Stakeholders should engage in market, social, and technological activities, according to the report. The approaches, processes, and technologies of V2X were reviewed in [5]. [6] examined the performance of V2X energy services and found that deferred payments and fee administration are more beneficial than spinning reserves. Authors in [7, 8] discussed some difficulties associated with the deployment of V2X. An implementation of V2X was proposed in [9], examined the driving distance anxiety, electric market data, and charging station site to EV drivers. A study was conducted in [10] to determine the financial impact of battery degradation for EVs with lithium-ion batteries participating in V2X. The paper also investigated how several battery energy management strategies for various V2X applications. It was not discussed in the literature what measures might be taken to overcome the barriers to integrating V2X.

The emphasis of this paper is to review the potential benefits and barriers of V2X, and the potential measures to address the human factors affecting the deployment of EVs for V2X. The contributions of this paper include a brief review of the following:

1. Benefits of V2X, with a focus on the UK markets.
2. Barriers posed by the human factors and strategies for overcoming these barriers.

The advantages of deploying EVs for V2X as well as V2G in the UK markets are presented in Sects. 2 and 3. Section 4 discusses the challenges and remedies of human factors in the use of EVs for V2X, followed by the conclusion in Sect. 5.

2 Services and Benefits of V2X

The concept of V2X helps to harness extra benefits of electric vehicle batteries during their non-use periods. By providing benefits to the electricity network, reducing the energy consumption of residential and commercial buildings, or providing energy support to loads, V2X services produces income from the battery facility by engaging bi-directional power flow. To offer System Operators and key stakeholders with much needed flexibility for the practical functionality of the electric network, energy services are defined as providing dynamic charge control as accumulated flexible size in the bulk and secondary Services outlets.

2.1 Spinning Reserves

In case of an unforeseen event owing to the loss of power supply, to meet load demand, an additional available power generating unit is required. This additional generating unit to keep demand and supply balanced is known as spinning reserves [10]. This service is not a regular but rather occur seldomly. This results in the providers remunerated for having supply capacity in hand and as well as responding to issued demand signal. According to the record of the PJM system in the US, they occur at intervals of 20 to 50 [11, 12] times annually. The spinning reserve is a service that EVs could participate in conditions when battery degradation is a major issue for EVs [13].

2.2 Energy Arbitrage and Related Services

In energy arbitrage, energy is purchased at an affordable price and sold at an expensive price. Price variation can be utilized to offset power losses and running costs inherent in energy storage, and it is possible to profit from energy storage. Using a retail 'time-of-day' or real-time energy price model, this can be employed at the retail point. Residential customers have access to real-time retail pricing in some jurisdictions, while large industries could access more frequently. While it is one of the most straightforward operational schemes to execute, it also has the tendency to be very complex in terms of results [14].

2.3 Renewable Energy Integration

Electrification of the transportation sector presents one of the most compelling opportunities, as well as the highest means of actualizing carbon emission reduction in the global energy systems. Due to the networked nature of electricity, it is difficult to determine which load receives green electricity, but several investigations have highlighted the tendency of intelligently coordinating electric vehicles to maximize the usage of intermittent renewable energy sources (RES) [15].

2.4 Load Management

An array of several goals can be accomplished by a load management system, which is a very broad term. Several studies, including [16], examine the feasibility of integrating photovoltaic plant with V2H energy storage and V2H with a hybrid EV. Essentially, they want a house to operate without grid power for the longest amount of time possible. By using a variable operational horizon of up to 10 years, [17] attempts to reduce electricity tariffs in a more holistic manner.

2.5 Frequency Control

Whenever there is a power imbalance between power generated and demand, particularly for ac power systems, the frequency component of the system deviates from its rated value resulting in the market demanding a swift supplier response to restore the frequency to its nominal value. This phenomenon of quickly restoring the system frequency is known as 'frequency control'. Battery energy storage system can participate in providing this service, since their response time is within the response requirement of this service [18, 19], which is usually issued in two- to five-second time steps, with response times generally not exceeding a few minutes. In addition, frequency regulation markets are increasingly focusing on quicker response, that respond to schedules and with commitment to control signals not viable with synchronous generators. In light of the highly competitive transaction prices for the Frequency Regulation (FR), early V2X work assumed the preferred service for vehicle batteries would be Frequency Regulation [20].

2.6 Voltage Regulation and Power Factor Improvement

Certain types of equipment can be affected by deviations from a perfect sinusoidal voltage and current profile over a certain period. There are a variety of types of grid-connected loads that can affect Power Quality, including those that introduce harmonics. Therefore, power quality throughout the grid can vary based on distribution systems' spatial scales. A power enhancement service is one that monitors the voltage and current quantities locally and corrects alterations by accurately timing power inflows or outflows within a fraction of the time it takes for voltage to cycle. The authors in [21] developed a sensitivity matrix to examine the possibility of EVs supporting voltage levels. In addition to examining cases where users consider only local impacts, as well as cases where users consider all impacts.

2.7 Summary of Potential Services

In Table 1, all the services that V2X can participate in are listed. It is important to note, however, that these benefits are only available if aggregators can meet the requirements for market participation in terms of power rating, response time, and service duration.

3 Benefits of V2G from UK Markets

EVs can participate in a variation of balancing services in the existing UK market, including demand flexibility services, frequency response services, and balancing reserve services. Aside from this, electric vehicles can also help the power sector reduce wind curtailment expenses through their participation in the wind curtailment market. The following content will elaborate on detailed market information and potential benefits.

3.1 Demand Flexibility Service Market

The ESO introduced the Demand Flexibility Service (DFS) for the winter period in 2022. With the emergence of the Demand Flexibility Service (DFS), the ESO will be able to utilize extra flexibility when national consumption is peak particularly during winter – a period that is not presently available to it in real time [22]. In this novel service, users will be rewarded for actively altering the period of electricity usage. During the winter of 2023/24, the ESO will conduct 12 tests between November and March 2024.

According to [23], the National Grid has offered up to £5/kWh or £5000/MWh as incentives for decreasing demand. Although the pricing for the actual implementation phase of this service is uncertain, we have considered a price range of £1/kWh to £5/kWh. Based on the data released by the ESO [24], demand flexibility service (DFS) primarily occurs between 9:00am and 10:00pm and 17:00pm to 19:00pm. Therefore, EV users have significant potential to arbitrage this market by flexibly adjusting charging and discharging times.

Table 1. Potential benefits from V2X.

Service	Definition	V2G	V2B	V2H	V2L
Frequency Regulation	Maintaining the grid frequency within the designated range through charging and discharging	√			
Energy Arbitrage	At cheap tariff periods, EV battery is charged, while at high tariff it is discharged	√			
Spinning Reserves	Maintain electricity balance by discharging or charging in response to grid contingencies	√			
Peak Shaving	Discharge at peak hours to reduce demand	√			
Renewable Generation Integration	Charge to absorb excess renewable generations	√	√		
Reactive Power Support	Adjusting reactive power output to assist the grid in maintaining voltage balance	√	√		
Network Deferral	Reduce consumption or increase generation at peak to meet projected load growth in capacity-constrained areas	√	√		
Backup Power	Supply emergency power in event of outage		√	√	

3.2 Wind Curtailment Market

Wind farms supported by government green subsidies might be paid more for switching off their turbines than generating power. Current constraints in the transmission system and a lack of long-duration grid storage often mean that the UK is producing more electricity from wind power sources than it can use. The only way to manage the grid imbalance, which happens when renewable power generation outstrips demand, is to curtail wind power [25]. The report by financial think tank Carbon Tracker [26] warns that grid enhancements are being surpassed by wind projects in the pipeline. The problem might cause the cost of inaction more than triple by 2030, amounting to £3.5bn annually and losing higher than 20% of wind generation in Scotland resulting in an extra £200 per household. Due to the flexible charging demand of EVs, they have the ability to have a substantial impact in reducing wind curtailment. UK Wind Curtailment Monitor [27] provides daily wind power curtailment information in the UK, along with the associated curtailment costs. In this regard, the f statistics on wind power curtailment during the period from September 2022 to September 2023 is illustrated in Fig. 1.

3.3 Environment Benefits

The environmental benefits brought by electric vehicles mainly manifest in their utilization of renewable energy sources to reduce carbon emissions originating from the electricity grid. According to [28], the current average carbon emissions from the UK's electricity grid are 157.33 g of CO_2 per kilowatt-hour consumed. If electric vehicle users can adjust their charging times to coincide with periods of ample renewable energy availability, it will be beneficial in reducing carbon emissions originating from the power sector. The carbon factor and carbon price from 2024 to 2050 can be shown in the Fig. 2.

3.4 Frequency Response Service Market

According to recent Dynamic Response Services Provider Guide of the ESO, the provision of dynamic frequency response services for the power grid needs to meet the following requirements [29]:

- For each response unit, a minimum of 1 MW is required.
- Automated triggering is necessary from supply or demand, or energy limited re source like the battery.
- Energy restricted resources must meet with the "level of energy" governance rules.
- Service execution is conducted within four-hour EFA Blocks.
- Purchase requests may only be made in MWs.
- Submitted quotations need to have a minimum period of validity of 4 h in accordance with EFA blocks.
- Tenders must therefore only commence, and finish, at the given periods: 23:00, 03:00, 07:00, 11:00, 15:00 and 19:00.

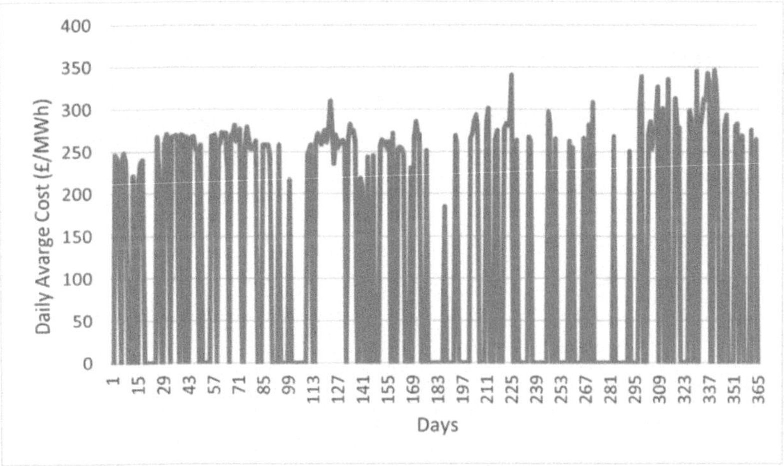

Fig. 1. Wind curtailment price from September 2022 to September 2023.

Considering the flexibility of scheduling charging and discharging during idle times of electric vehicles, EVs have the capability to participate in services such as Dynamic

Containment High (DCH), Dynamic Moderation High (DMH), Dynamic Regulation High (DRH), Dynamic Containment Low (DCL), Dynamic Moderation Low (DML) and Dynamic Regulation Low (DRL) frequency response services to the grid.

According to the dynamic containment, regulation, and moderation auction results of the ESO [30], from Sep 2022 until now, the average prices for DCH, DMH, and DRH are £3.68, £4.3, and £2.41, respectively. The average prices for DCL, DML, and DRL are £2.82, £2.01, and £8.79, respectively. The price information and cost benefits on frequency response services are highlighted in Table 2.

The summary of the environmental, social, and economic benefits of V2G to the different stakeholders is presented in Table 3.

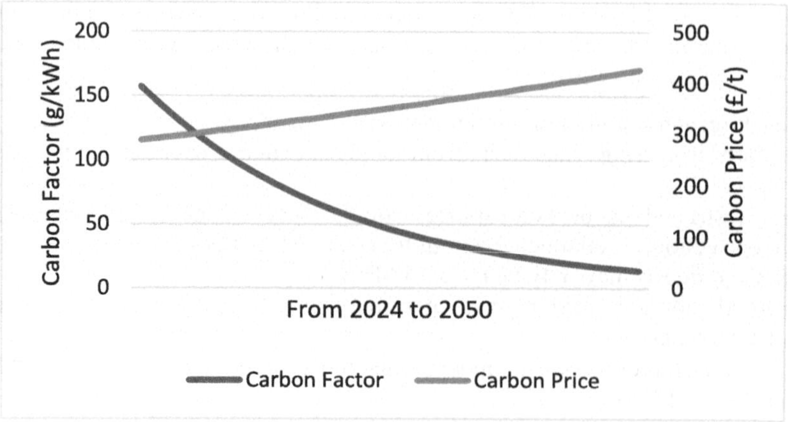

Fig. 2. UK carbon factor and price from 2024 to 2050.

Table 2. Frequency response services clearing price information.

Services	Maximum Price (£)	Minimum Price (£)
DCH	35	0.01
DMH	12.5	0
DRH	52.9	0
DCL	45.5	0.01
DML	7.04	0
DRL	55.99	0

4 Human Factors

There are many technical challenges that need to be overcome before wide-scale V2G applications can be used. However, many studies in this field concentrate only on the technical side. The field sampled by Sovacool et al. [31] only accounted for 2.1% of

Table 3. V2G benefits to the UK market.

Service	Beneficiaries	Environment Benefit	Social Benefit	Economic Benefit
Demand Flexibility Service	EV Users; Power sector		√	√
Frequency Response Service	EV Users; Power sector		√	√
Wind Curtailment Market	Power sector	√	√	√

papers addressing user attitudes and activities, and 1.1% of papers addressed range anxiety. In spite of a general lack of interest, user behaviours and routines play a crucial part in how V2G will be successful in its implementation.V2G applications can face social and practical challenges (availability of vehicles for V2G events, predictability of users, behavioural incentives). Evidence for V2G applications can be gathered by examining user routines and vehicle use patterns. V2G services can significantly affect load management in different building types by vehicle usage patterns, battery state on arrival, and preferred charging status on leaving [31]. Usage patterns can also inform public charging infrastructure planning [32].

The "human factor" is a valuable component of V2X system of electric vehicles. Among these factors are the driver's preferences for vehicle charging, acceptance of new technologies, and concerns about data security and privacy. To ensure the effectiveness, acceptability, and safety of V2X technology when designing and implementing it, these factors must be taken into consideration when designing and implementing it.

4.1 Aspects of Human Factors

In the context of V2X regulation, the various aspects of human factors with their challenges are discussed below.

Charging Behavior. V2X is affected by plug-in rate, or the percentage of time an EV is plugged into a charging point. EV users in [33] charge several times, whereas in [34], 30% charge when their batteries run low. This difference demonstrates that incentive initiatives encourage users to charge. Commercial fleets are better suited to V2X because they have established routines and are parked at one location, which minimizes infrastructural costs and supports smart charging solutions [35]. In contrast, domestic vehicles do not have a predefined schedule, resulting in variations in mobility needs. Charging behavior has a significant impact on electric vehicle dispatch. Depending on user preferences, charging stations may be overloaded during certain periods and underutilized during others [36]. With the expectation for smart charging facilities, V2X applications can improve efficiency and user experience, optimize electricity resources through predictive scheduling, and V2G and V2H applications can be made more practical and attractive.

Acceptance of New Technologies. Several factors influence user acceptance, including awareness and understanding, perceived benefits, ease of use, and privacy [37]. The user-friendliness of the technology, including the intuitive design and accessibility of V2X services, will result in improved user satisfaction and usage rates. Concerns regarding the collection and use of personal information, as well as whether the technology is secure, may cause hesitation and slow down the adoption of the technology if it is not adequately addressed.

Concern for Data Confidentiality and Security. The security of data is one of the major significant determinants that influence people's decisions. How their information is stored, protected, and used also influences how they participate in V2X [38]. A lack of clear privacy policies and practices for handling personal data may discourage user adoption in V2X. Hacking, unauthorized access to vehicle systems, and data breaches are possible risks to V2X adaptation. To maintain user trust and integrity of the system, V2X communication requires robust security measures. Data sharing, access control, and anonymity of vehicle and driver information are also important considerations to be addressed to avoid hindrance to the adoption of V2X.

4.2 Measures to Overcome Human Factor Barriers

To overcome the human factor barriers, several strategies are recommended as discussed below [38–40].

User Education and Engagement. Educate users about the importance of efficient charging behaviors and motivate them to engage in demand-response initiatives or use mobile apps that provide information on optimal charging times. Engage users through incentives such as discounts for off-peak charging or rewards for energy-efficient driving. In addition, effective communication and public relations are essential for fostering long-term acceptance of V2G. Accurate information about the benefits and risks associated with V2G needs to be communicated clearly to enhance general acceptance.

Data Analysis and Predictive Modeling. Utilize data analytics and predictive modelling to understand user behavior patterns, such as charging habits and driving preferences. This information can help optimize dispatch strategies and anticipate peak demand periods.

Dynamic Pricing and Incentives. Implement dynamic pricing schemes at charging stations to encourage users to charge during lower demand hours. Offer incentives such as discounts or loyalty rewards for users who comply with dispatch recommendations or exhibit energy-efficient driving behavior.

Driver Training and Feedback. Provide training programs for drivers on energy-efficient driving techniques and the effects of driving habits on vehicle range and battery life. Offer real-time feedback to drivers through in-vehicle displays or mobile apps, highlighting areas for improvement and incentivizing eco-friendly driving practices.

Optimization Algorithms and Decision Support Systems. Develop advanced optimization algorithms and decision support systems that consider human factors alongside technical constraints when dispatching electric vehicles. These systems can dynamically adjust dispatch plans based on user behavior, traffic conditions, and charging station availability.

User-Centric Design. Design user interfaces for electric vehicle dispatch systems that are intuitive and user-friendly, providing relevant information such as charging station locations, availability, and real-time energy prices. Solicit user feedback to continually improve the system's usability and effectiveness.

Strategy for Addressing Range Limitations. Because V2G applications primarily involve the discharging of electric vehicles, range anxiety may affect the acceptance of these technologies. To address the fear of losing control of their electric vehicle batteries' charging procedure, EV users are provided with control flexibility via a user interface when participating in V2G. In this user interface, users have multiple options to select their preferences rather than being confined to accepting V2G usage or opting out entirely. By selecting any desired charging mode along with certain targets such as state of charge (SoC), date and time, and charging/discharging rates based on electricity prices, the user can customize their charging preferences.

Communication System Interface. All EV preferences can be integrated into the user interface. It should be easy to operate and have touchscreen capabilities. This interface's home screen should provide the most relevant information and the different modes available. Additionally, the interface should provide a brief description of the various modes available. The interface should also be designed with the user in mind as well as the aggregator in mind. The SoC target will need to be converted into a range target that may be more useful to the user. Since the range target depends on the driving style and terrain covered by the driver, this estimate can be difficult to achieve. Depending on the mode selected, an additional window for SoC, date/time targets will appear. As soon as the user selects the appropriate option, it sends the desired information to the charging unit's microcontroller board.

5 Conclusion

This paper presents a brief review of the benefits of V2X and V2G in the world and UK markets respectively. Additionally, it discusses human factors related to V2X implementation, along with barriers and remedies. This review will be of benefit to manufacturers of electric vehicles, electric vehicle users, aggregators, and power system operators. V2X can benefit from several transmission and distribution network services like frequency regulation, spinning reserves, voltage support, energy arbitrage and renewable energy integration. V2G can enhance UK energy markets by providing demand flexibility services, frequency response services, and wind curtailment. These various services benefit the EV user and the power system operator. Human factors barriers have been discussed, along with measures recommended for addressing challenges associated with charging behaviour, acceptance of new technologies, and privacy and security concerns.

Acknowledgements. This paper is partially supported by the Electric Vehicle Fleet Bi-directional Charging (FLEXET) project funded by DESNZ and Innovate UK.

Disclosure of Interests. The authors have no competing interests to declare that are relevant to the content of this article.

References

1. International Energy Agency, Global EV Outlook 2021, https://www.iea.org/reports/global-ev-outlook-2021, last accessed 2024/02/24
2. European Environment Agency, New registrations of electric vehicles in Europe, https://www.eea.europa.eu/ims/new-registrations-of-electric-vehicles, last accessed 2024/02/24
3. Merrington, S., Khezri, R., Mahmoudi, A.: Optimal planning of solar photovoltaic and battery storage for electric vehicle owner households with time-of-use tariff. IET Gener. Transm. Distrib. **16**(2), 535–547 (2022)
4. International Energy Agency, Final Report (2014–2018), Task 28: Home Grids and V2X Technologies, https://ieahev.org/wpcontent/themes/newTheme/assets/pdfs/Task28_Final_Report.pdf, last accessed 2024/02/25
5. Pearre, N.S., Ribberink, H.: Review of research on V2X technologies, strategies, and operations. Renew. Sustain. Energy Rev. **105**, 61–70 (2019)
6. Thompson, W., Perez, Y.: Vehicle-to-Everything (V2X) energy services, value streams, and regulatory policy implications. Energy Policy **137**, 111136 (2020)
7. Gschwendtner, C., Sinsel, S.R., Stephan, A.: Vehicle-to-X (V2X) implementation: An overview of predominate trial configurations and technical, social and regulatory challenges. Renew. Sustain. Energy Rev. **145**, 110977 (2021)
8. Corchero, C., and Sanmarti, M.: Vehicle- to- Everything (V2X): Benefits and Barriers. In: 15th International Conference on the European Energy Market (EEM), pp. 1–4. IEEE, Lodz, Poland (2018)
9. Ferreira, J.C., Monteiro, V., Afonso, J.L.: Vehicle-to-Anything Application (V2Anything App) for Electric Vehicles. IEEE Trans. Industr. Inf. **10**(3), 1927–1937 (2014)
10. US Federal Energy Regulatory Commission, Order 888: Promoting wholesale competition through open access non-discriminatory transmission services by public utilities; recovery of stranded costs by public utilities and transmitting utilities. p. 24 (1996)
11. Kempton, W., Tomić, J.: Vehicle-to-grid power fundamentals: calculating capacity and net revenue. J. Power Sources **144**(1), 268–279 (2005)
12. Kempton, W., Udo, V., Huber, K., Komara, K., Letendre, S., Baker, S., et al.: A test of Vehicle-to-Grid (V2G) for energy storage and frequency regulation in the PJM system. Results from an Industry-University Research Partnership **32**(2008), 1–32 (2008)
13. Mullan, J., Harries, D., Bräunl, T., Whitely, S.: The technical, economic and commercial viability of the vehicle-to-grid concept. Energy Policy **48**, 394–406 (2012)
14. Dallinger, D., Wietschel, M.: Grid integration of intermittent renewable energy sources using price-responsive plug-in electric vehicles. Renew. Sustain. Energy Rev. **16**(5), 3370–3382 (2012)
15. Mwasilu, F., Justo, J., Kim, E., Do, T., Jung, J.: Electric vehicles and smart grid interaction: a review on vehicle to grid and renewable energy sources integration. Renew. Sustain. Energy Rev. **34**, 501–516 (2014)
16. Tuttle, D., Fares, R., Baldick, R., Webber, M.: Plug-In Vehicle to Home (V2H) duration and power output capability. In: Proceedings of IEEE transportation electrification conference and expo (ITEC), pp. 1–6. IEEE, Detroit, MI, USA (2013)
17. Wu, X., Hu, X., Teng, Y., et al.: Optimal integration of a hybrid solar-battery power source into smart home nanogrid with plug-in electric vehicle. J. Power Sources **363**, 277–283 (2017)

18. Lachs, W., Sutanto, D.: Application of battery energy storage in power systems. In: Proceedings of 1995 international conference on power electronics and drive systems, pp. 700–705. IEEE, Singapore (1995)
19. Leadbetter, J., Swan, L.: Battery storage system for residential electricity peak demand shaving. Energy and Buildings **55**, 685–692 (2012)
20. Kempton, W., Tomić, J.: Vehicle-to-grid power implementation: from stabilizing the grid to supporting large-scale renewable energy. J. Power Sources **144**(1), 280–294 (2005)
21. Beaude, O., He, Y., Hennebel, M.: Introducing decentralized EV charging coordination for the voltage regulation. In: Proceedings of IEEE PES ISGT Europe 2013, pp. 1–5. IEEE, Lyngby, Denmark (2013)
22. National Grid ESO. Demand Flexibility Service (DFS), https://www.nationalgrideso.com/industry-information/balancing-services/demand-flexibility-service-dfs, last accessed 2024/04/16
23. Catalyst Digital Energy. Demand Flexibility Service, https://www.catalyst-commercial.co.uk/demand-flexibility-service/, last accessed 2024/04/16
24. National Grid ESO. Demand Flexibility Service Utilisation Report Summary – Test, https://www.nationalgrideso.com/data-portal/demand-flexibility-service-test-events, last accessed 2024/04/10
25. Energy Live News. Are Switched-off Wind Turbines Generating more Cash, https://www.energylivenews.com/2023/01/30/are-switched-off-wind-turbines-generating-more-cash/, last accessed 2024/04/08
26. Carbon Tracker. Wasted Wind Power Costing Britain £1.5bn, https://www.edie.net/report-wasted-wind-power-costing-britain-1-5bn/, last accessed 2024/04/06
27. UK Wind Curtailment Monitor, https://wind-curtailment-app-ahq7fucdyq-lz.a.run.app/, last accessed 2023/11/24
28. 2023 Government Greenhouse Gas Conversion Factors for Company Reporting, https://assets.publishing.service.gov.uk/media/647f50dd103ca6001303 9a8a/2023-ghg-cf-methodology-paper.pdf, last accessed 2024/03/17
29. National Grid ESO. New Dynamic Response Service Provider Guidance, https://www.nationalgrideso.com/document/276606/download, last accessed 2024/03/23
30. National Grid ESO. Dynamic Containment, Regulation and Moderation Auction Results., https://www.nationalgrideso.com/data-portal/dynamic-containment-data, last accessed 2024/03/09
31. Sovacool, B., Noel, L., Axsen, J., Kempton, W.: The neglected social dimensions to a vehicle-to-grid (V2G) transition: A critical and systematic review. Environ. Res. Lett. **13**(1), 013001 (2018)
32. Kuang, Y., Chen, Y., Hu, M., Yang, D.: Influence analysis of driver behaviour and building category on economic performance of electric vehicle to grid and building integration. Appl. Energy **207**, 427–437 (2017)
33. Quiros-Tortos, J., Ochoa, L., Butler, T.: How electric vehicles and the grid work together: lessons learned from one of the largest electric vehicle trials in the world. IEEE Power Energ. Mag. **16**(6), 64–76 (2018)
34. Franke, T., Krems, J.: Interacting with limited mobility resources: psychological range levels in electric vehicle use. Transportation Research Part A: Policy and Practice **48**, 109–122 (2013)
35. Will, C., Schuller, A.: Understanding user acceptance factors of electric vehicle smart charging. Transp. Res. Part C: Emerging Technol. **71**, 198–214 (2016)
36. Waldron, J., Rodrigues, L., Gillott, M., et al.: Towards an electric revolution: a review on vehicle- to-grid, smart charging and user behaviour. In: Proceedings of the 18th International Conference on Sustainable Energy Technologies, pp. 1–9. Kuala Lumpur, Malaysia (2019)

37. Schmidt, T., Philipsen, R., Ziefle, M.: From V2X to control2trust. In: Proceedings of the 3rd International Conference on Human Aspects of Information Security, Privacy, and Trust, pp. 570–581. Springer, Berlin, Heidelberg, Germany (2015)
38. Yoshizawa, T., Singelée, D., Muehlberg, J., et al.: A survey of security and privacy issues in v2x communication systems. ACM Comput. Surv. **55**(9), 1–36 (2023)
39. Mehdizadeh, M., Nayum, A., Nordfjærn, T., Klöckner, A.: Are Norwegian car users ready for a transition to vehicle-to-grid technology? Transp. Policy **146**, 126–136 (2024)
40. Mehdizadeh, M., Nordfjaern, T., Klöckner, A.: Estimating financial compensation and minimum guaranteed charge for vehicle-to-grid technology. Energy Policy **180**, 113649 (2023)

A General Pattern of Permanent Magnet Arrangement for Design Optimization of a Reluctance Magnetic Gear with Enhanced Torque Density for Electric Vehicles

Yanding Bi[1], Weinong Fu[2(✉)], Shuangxia Niu[1], and Jiahui Huang[1]

[1] Department of Electrical and Electronic Engineering, The Hong Kong Polytechnic University, Hong Kong, China
shuangxia.niu@polyu.edu.hk
[2] Shenzhen University of Advanced Technology, Shenzhen 518055, China
wn.fu@siat.ac.cn

Abstract. A novel general pattern of permanent magnet (PM) arrangement for the optimal design of magnetic gears (MGs) is presented in this work. The proposed general pattern can generate six different types of PM arrangements. By using this optimization pattern, the PM arrangement with high torque and low torque ripple in MGs can be obtained within a short computing time. To demonstrate the effectiveness of this optimization approach, the proposed pattern is applied to a reluctance magnetic gear (RMG), a promising type of MG suitable for power transmission applications in lightweight electric vehicles (EVs). The transmission torque and torque ripple characteristics of the RMG model are evaluated using finite element analysis (FEA), while its design optimization is performed utilizing a non-dominated sorting genetic algorithm II (NSGA-II) method. The superiority of this general pattern is demonstrated through a comparative analysis between the proposed RMG and a conventional RMG model.

Keywords: Finite element analysis (FEA) · magnetic flux density · magnetic gear · optimization · permanent magnet (PM)

1 Introduction

Permanent magnets (PMs) are a class of materials that possess intrinsic properties to generate magnetic flux and magnetic force without consuming any external operating energy. Due to these exceptional advantages, PMs have found extensive applications in various electromagnetic devices [1, 2]. PM materials play a crucial role in the electric machines employed in the propulsion system of electric vehicles (EVs) [2]. The development of PM synchronous machines (PMSMs), PM-assisted synchronous reluctance machines (PMa-SynRMs), and flux modulation PM machines has witnessed rapid progress in recent years [3]. Apart from their application in electric machines, PMs also hold significant importance in magnetic gears (MGs), which serve as innovative power transmission

K. Li et al. (Eds.): LSMS/ICSEE 2024, CCIS 2218, pp. 147–159, 2025.
https://doi.org/10.1007/978-981-96-0232-2_12

devices [4]. MGs enable non-contact power transmission for potential EV applications, offering benefits such as lubrication-free operation, maintenance-free usage, low noise and vibration levels, and inherent overload protection [5].

A conventional radial-flux coaxial MG is shown in Fig. 1(a). PMs are mounted on both high-speed rotor (HSR) and low-speed rotor (LSR), and a stationary component of ferromagnetic segments is placed between the two rotors, which is named a flux modulator. Many studies have been made to investigate different PM arrangements in coaxial MGs [6–11]. MGs with various consequent-pole and multi-layer PM arrangements were investigated in [6], and a comprehensive comparison of 15 MGs with different topologies was presented, focusing on torque performance and cost estimation. The usage of a Halbach PM array in MGs was explored in [7] and [8]. A comparison between an MG with the Halbach array and a conventional MG is presented in [7], where both simulation and experiments demonstrate that the proposed MG with the Halbach array demonstrated higher torque, higher efficiency, and lower torque ripple. Furthermore, spoke-type PM arrangements were studied for their potential to enhance torque density significantly. This includes an MG with spoke-type PM solely on the high-speed rotor [9] as well as those equipped with spoke-type PM on both rotors [10, 11]. Due to the flux-focusing effect of the spoke-type PM, the torque density can be significantly enhanced in this type of MGs, which further improves the efficiency of the gear compared to conventional types of MGs [9]. In addition, a high active region torque density of 239 Nm/L is achieved in an MG with spoke-type PM arrangements on both high-speed and low-speed rotors [11]. Nonetheless, applications of conventional coaxial MGs are mainly focused on low-speed, high-torque situations. In high-speed power transmission applications, coaxial MGs face problems with low mechanical strength and low efficiency due to eddy-current loss in PMs [5].

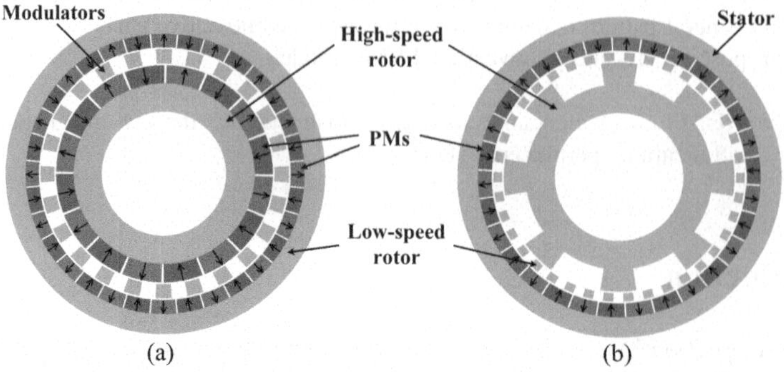

Fig. 1. (a) Surface-mounted coaxial magnetic gear. (b) Conventional surface-mounted reluctance magnetic gear (Gear I).

Recently, reluctance magnetic gears (RMGs) have been proposed, which have been considered as a promising choice in high-speed, low-cost applications [12–15]. As shown in Fig. 1(b), the RMG, which is referred to as Gear I, consists of a stator with surface-mounted PMs, an LSR with a flux modulator, and an HSR with a salient pole structure

[12]. The salient pole rotor is purely made of an iron core, which is robust in high-speed operation or harsh operation environment [13]. The performance of an RMG was compared to that of a conventional coaxial MG in [14], considering the entire driving system, including the motor and gears. Consequently, due to the capability of the RMG to transmit power from a high-speed machine, the overall size of the driving system is significantly reduced. Simultaneously, efficiency improves as PM eddy-current losses are greatly diminished by eliminating PMs in one rotor. In addition, flux-switching MG (FSMG) has been proposed based on RMGs [15]. PMs are inserted between the flux modulator segments in the FSMG, leading to an improvement in torque density. However, the efficiency is inferior compared to the RMG due to the extra PM eddy current loss generated in the additional PMs. Various studies investigated the influence of the HSR structures on the performance of RMGs [16–18]. The high torque ripple is a severe problem that influences the potential applications of RMGs. Four different types of skewed HSR were employed in the RMG to mitigate the torque ripple [16]. The RMG with one-side skew HSR exhibits a 55.4% decrease in torque ripple compared to the conventional RMG with a salient-pole HSR. However, the torque density is also reduced by 14.8%. Subsequently, a flux barrier type HSR is employed in the RMG [17]. A three-layer flux barrier rotor is adopted and optimized in this study, achieving substantial improvement in torque density and a notable reduction in torque ripple. Furthermore, a composite high-speed rotor (HSR) was implemented in the RMG [18], incorporating ferromagnetic materials in the rotor teeth and non-ferromagnetic materials for the remaining rotor core. Consequently, this design significantly enhances the gravimetric torque density of the gear while simultaneously reducing losses attributed to iron and permanent magnets.

However, the low torque density and high torque ripple of RMGs hinder the application potential in the industry [12]. All reported RMGs employ a surface-mounted PM arrangement in academic literature to the best of the authors' knowledge, and relevant studies focus on the structures of the HSR. The general pattern of PM arrangement is a powerful method for optimizing devices that utilize PM as excitation sources [19]. In this study, a novel general pattern of PM arrangement is proposed, which is able to produce six types of PM arrangements automatically. This general pattern of PM arrangement is applied to an RMG model. Magnetic torque and torque ripple are analyzed using finite-element analysis (FEA), and the model is optimized using a non-dominated sorting genetic algorithm II (NSGA-II) method. The optimization result presents a novel topology of RMG, which has the advantages of high torque density and low torque ripple compared to a conventional RMG with a surface-mounted PM arrangement.

2 Proposed Configuration and Optimal Design

2.1 General Pattern of PM Arrangement

The schematic of the proposed novel general pattern of PM arrangement and its design variables are shown in Fig. 2(a). By adopting this general pattern, five common PM topologies can be generated, including surface-mounted, consequent pole, Halbach array, simplified Halbach array, and spoke-type array PM arrangements as shown in Fig. 2(b).

Besides, Fig. 2(b) also presents a novel PM topology, which can be regarded as a combination of spoke-type and Halbach array. The proposed RMG model shown in Fig. 2(c) is constructed by using this general pattern, which is referred as Gear II. The optimization process and performance analysis of this topology are illustrated in subsequent sections.

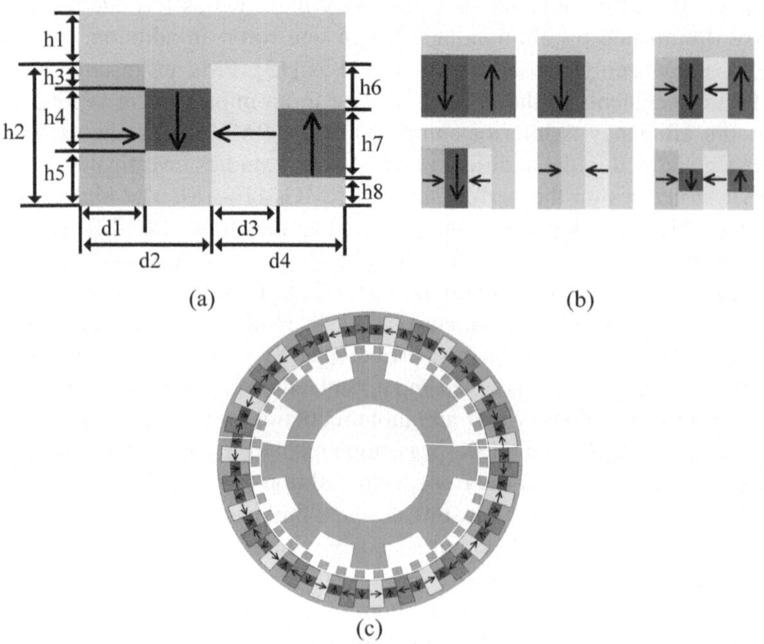

Fig. 2. (a) The general optimization pattern of PM arrangement. (b) The PM arrangements generated by the general pattern. (c) The proposed RMG model with PM arrangement using the general pattern (Gear II).

2.2 Operating Principle and Proposed Configuration

An RMG typically consists of three layers of components, as shown in Fig. 1(b). The general design parameters of the conventional and proposed RMGs are shown in Table 1.

The PMs on the stator generate magnetomotive force (MMF), which is the sole source of the magnetic field. The LSR rotor is made of ferromagnetic pole pieces, and the HSR is a salient pole rotor. Both the LSR and HSR modulate the MMF of stator PMs. Due to the modulation effect, the magnetic flux harmonics with different pole-pair numbers and speeds are generated in the air-gap.

The MMF of the stator PMs is described in (1):

$$F_s(\theta) = \sum_{i=1,3,5...}^{+\infty} F_i \cos(iP_s(\theta - \theta_{s0})) \tag{1}$$

where F_i is the amplitude of the MMF spatial harmonic, P_s is the pole-pair number of stator PMs, θ is the mechanical angle of the stator, and θ_{s0} is the initial position of the stator.

Table 1. General Design Parameters of the RMG

Parameter	Values
Axial length, L_a (mm)	160
Outer radius of stator, R_{out} (mm)	80
Length of air gap, T_a (mm)	1
Duty ratio of stator PMs, R_{pm}	0.95
Thickness of modulators, T_m (mm)	5
Duty ratio of modulators, R_m	0.5
Thickness of rotor teeth, T_t (mm)	8
Duty ratio of rotor teeth, R_t	0.5
Pole pair number of stator PM, P_s	15
Number of modulators, N_L	38
Number of teeth, N_H	8
PMs material	NdFeB35
Steel material	DW310-35

Due to the uneven distribution of ferromagnetic materials, the permeance distribution of the LSR and HSR can be stated as:

$$\Lambda_L(\theta, t) = \Lambda_{L0} + \sum_{j=1}^{+\infty} \Lambda_j \cos(jN_L(\theta - \omega_L t - \theta_{L0})) \tag{2}$$

$$\Lambda_H(\theta, t) = \Lambda_{H0} + \sum_{k=1}^{+\infty} \Lambda_k \cos(jN_H(\theta - \omega_H t - \theta_{H0})) \tag{3}$$

where Λ_{L0} and Λ_{H0} are the average permeance components of LSR and HSR, respectively. Λ_j and Λ_k are the amplitude of the permeance spatial harmonic of LSR and HSR. N_L and N_H are the number of modulator pole pieces on LSR and the number of teeth on HSR. ω_L and ω_H are the rotational speeds of LSR and HSR. θ_{L0} and θ_{H0} are the initial positions of LSR and HSR.

The magnetic flux density due to the modulation effect of LSR and HSR can be stated as:

$$B_{SL}(\theta, t) = F_s(\theta, t)\Lambda_L(\theta, t) \tag{4}$$

$$B_{SH}(\theta, t) = F_s(\theta, t)\Lambda_H(\theta, t) \tag{5}$$

The effective harmonics with the same pole-pair numbers and rotation speeds are coupled in the inner air-gap between LSR and HSR to transmit torque [12]. The relationship between pole-pair number and rotational speed is presented as:

$$|i_a P_s + jN_L| = |i_b P_s + kN_H| \tag{6}$$

$$\begin{aligned} &\frac{jN_L\omega_L}{i_a P_s + jN_L} = \frac{kN_H\omega_H}{i_b P_s + kN_H} \\ &i_a, i_b = 1, 3, ..., +\infty; \quad j, k = 0, \pm 1, ... \pm \infty \end{aligned} \tag{7}$$

When the high-order harmonics are ignored, the relationship of pole-pair number and rotational speed in (6) and (7) can be derived as follows:

$$N_L = 2P_s \pm N_H \tag{8}$$

$$N_L\omega_L = \pm N_H\omega_H \tag{9}$$

The gear ratio of the RMG is the ratio of the rotational speed of HSR and LSR, which can be stated as:

$$G_r = \frac{\omega_H}{\omega_L} = \pm\frac{N_L}{N_H} \tag{10}$$

where the minus sign indicates that the two rotors rotate in opposite direction.

As shown in Table 1, the RMG model in this model adopts a combination of $P_s = 15$, $N_L = 38$, and $N_H = 8$, which satisfies the relationship in (8) when taking the plus sign. By using (10), the gear ratio can be derived, which is 4.75 in this study. In the meantime, the designed rotational speeds of LSR and HSR are 2000 rpm and 9500 rpm, respectively.

2.3 Optimization Process

To perform a fair comparison between the conventional RMG and the proposed RMG, the structural design parameters of both gears are optimized. Due to the large number of design parameters in the proposed general pattern of PM arrangement, an intelligent optimization algorithm is adopted, which is named as NSGA-II. The main steps of the optimization process are shown in Fig. 3. The NSGA-II algorithm imitates the process of natural selection [20]. The objective functions of the elements are evaluated by the facilitation of FEA. The elements with a relatively high ranking are reproduced in the next generation. The remaining elements in the new generation are produced by mutation and crossover.

In this study, the objective functions are maximizing the average torque of LSR and minimizing the torque ripple of HSR. The optimization generation is selected as 50, and the population in each generation is 100. In addition, the crossover factor and the mutation factor are selected as 0.9 and 0.1, respectively. The Pareto front of selected generations of Gear II is shown in Fig. 4. As it can been seen, along with the increasing optimization generation, the average torque of LSR increases, and the torque ripple of HSR decreases. The optimal design of Gear II is marked in the plot, and the values of the optimal design parameters are listed in Table 2.

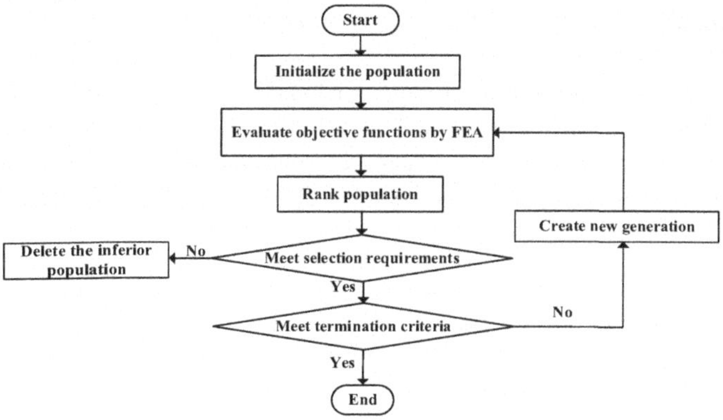

Fig. 3. The flowchart of the global optimization process by using NSGA-II.

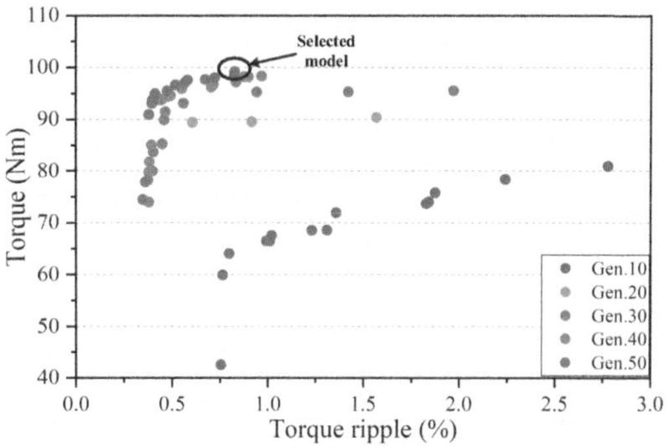

Fig. 4. The value of optimization elements in different generations.

Table 2. Optimal Design Parameters of Gear I and Gear II

Parameter	Gear I	Gear II	Parameter	Gear I	Gear II
h_1 (mm)	10.25	2.01	d_2 (deg)	/	4.92
h_2 (mm)	6.42	14.38	d_3 (deg)	/	7.08
h_3 (mm)	/	0.58	d_4 (deg)	/	4.92
h_4 (mm)	/	6.63	R_{pm}	0.97	/
h_5 (mm)	/	7.18	T_m (mm)	4.02	4.02
h_6 (mm)	/	0.14	R_m	0.48	0.48
h_7 (mm)	/	7.55	T_t (mm)	10.5	9.9
h_8 (mm)	/	6.69	R_t	0.49	0.47
d_1 (deg)	/	7.08			

3 Performance Analysis and Comparison

The performance of the optimal design of Gear I and Gear II is analyzed by using FEA. The optimal design of Gear II is shown in Fig. 5. In this chapter, the flux distribution, output torque, loss, and efficiency of both gears are simulated and compared. The simulation results reveal the merits of the proposed RMG with a novel PM arrangement.

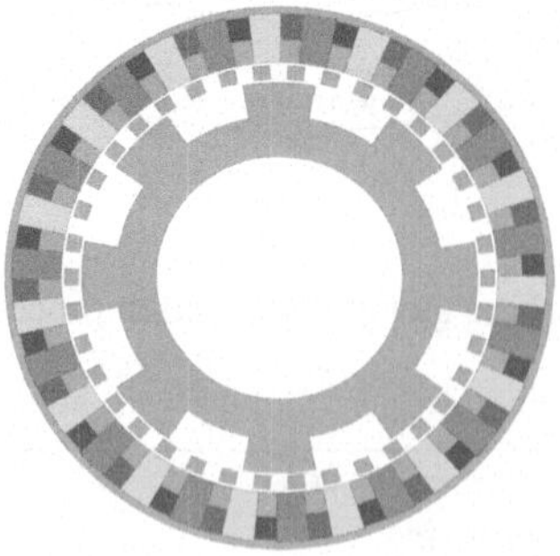

Fig. 5. Optimal design of Gear II.

3.1 Flux Distribution

The flux lines and flux density distribution of the optimized gears are shown in Fig. 6. As shown in Fig. 6(a), the surface-mounted PM arrangement requires a thick stator yoke for constructing the magnetic flux path. In contrast, the stator yoke of Gear II is largely reduced due to the novel PM arrangement of a combination of Halbach array and spoke-type, as shown in Fig. 6(b), which increases the inner radius of the stator and enhances the output torque of the RMG. In the meantime, due to the spoke-type PM arrangement, the flux concentration effect is obvious in Gear II. The magnetic flux density in the iron core of Gear II is notably higher than that of Gear I, which reveals that Gear II has a higher utilization ratio of iron core than Gear I.

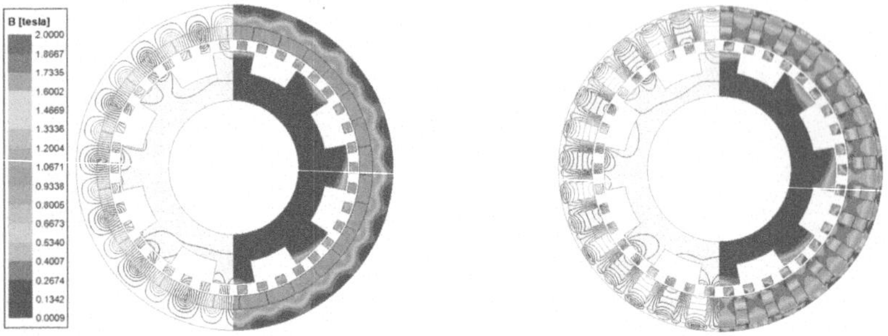

(a) (b)

Fig. 6. FEA results of flux distribution and flux density of (a) Gear I. (b) Gear II.

The magnetic flux density in the inner air-gap is shown in Fig. 7(a). The magnitude of the air-gap magnetic flux density in Gear II is significantly higher than in Gear I. To have a clear visualization of the magnitude of certain harmonics due to the flux modulation effect, the normalized fast Fourier transform (FFT) of the radial airgap flux density is shown in Fig. 7(b). The magnitudes of the 23rd effective harmonics and 15th main flux of Gear II are considerably higher than Gear I, which exhibits the merit of this novel RMG model.

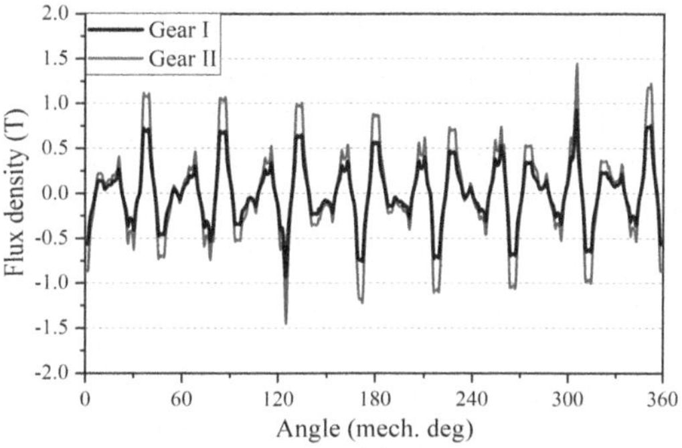

(a)

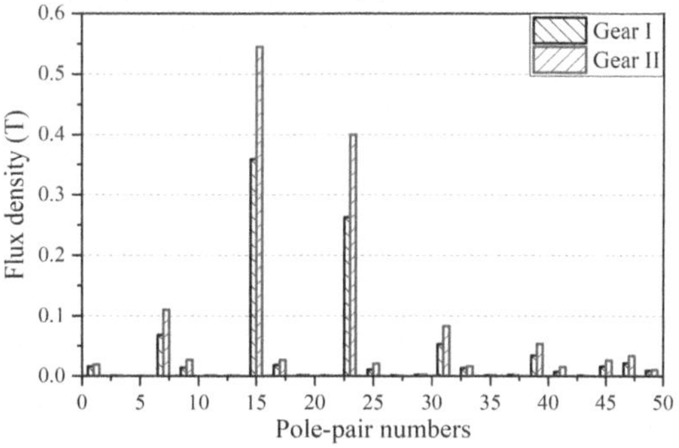

(b)

Fig. 7. (a) Radial flux density distribution. (b) Normalized FFT of the radial flux density.

3.2 Torque Transmission Capability

The performance of Gear I and Gear II is shown in Table 3. Firstly, the torque transmission on both rotors of Gear I and Gear II is simulated by locking the LSR, as shown in Fig. 8(a). Because of the novel PM arrangement, the Gear II exhibits a peak torque of 99.47 Nm, which is 136% higher than the 42.11 Nm of Gear I. Then, the rotational speeds of both rotors are set as designed speeds to investigate the average torque on LSR and HSR. As shown in Fig. 8(b), the average torque of Gear II LSR is 99.35 Nm, which shows an improvement of 138% compared with Gear I LSR. In the meantime, due to the unequal usage of PM materials, the PM volumetric torque density is calculated to perform a fair comparison.

Table 3. Performance of Gear I and Gear II

Configuration	Average torque (Nm)	Torque/PM volume (Nm/L)	Efficiency (%)
Gear I	41.74	100.19	94.47
Gear II	99.35	121.12	96.08

As shown in Table 3, the PM volumetric torque density of Gear II is increased by 20.9% compared with Gear I. Moreover, the torque ripples of the proposed RMG and the conventional RMG are very low, which is also reflected on Fig. 8(b). The torque ripples of the LSRs on both gears are lower than 1.5%. The torque ripples of the Gear II HSR is 1.89%, which is considerably lower than the 5.16% of Gear I HSR.

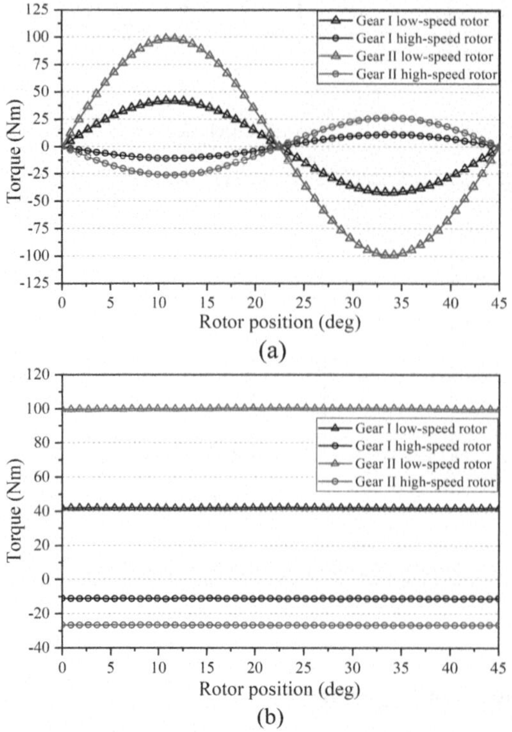

Fig. 8. (a) Torque-load angle characteristic when the LSR is locked. (b) Average torque.

3.3 Loss and Efficiency

The loss of the RMG is composed of two main components, which include the core loss generated in the stator yoke, LSR, and HSR, and the magnet eddy current loss in stator

PMs. The magnet eddy current loss is referred to as solid loss in this study. The loss of Gear I and Gear II at different LSR speeds, ranging from 1000 rpm to 6000 rpm, is shown in Fig. 9(a). The core loss of Gear II at different speeds is significantly larger than Gear I due to the increased magnetic flux density distribution in the iron core materials, which is revealed in Fig. 6. The Gear I and Gear II have a similar solid loss at different rotational speeds as shown in Fig. 9(a).

The efficiency of RMG is expressed as:

$$\eta = \frac{P_{out}}{P_{in}} \times 100\% = \frac{T_L \omega_L}{T_L \omega_L + P_{core} + P_{solid}} \times 100\% \quad (11)$$

where P_{in}, P_{out}, P_{core}, and P_{solid} are the input power, output power, core loss, and solid loss, respectively. T_L is the average torque of LSR.

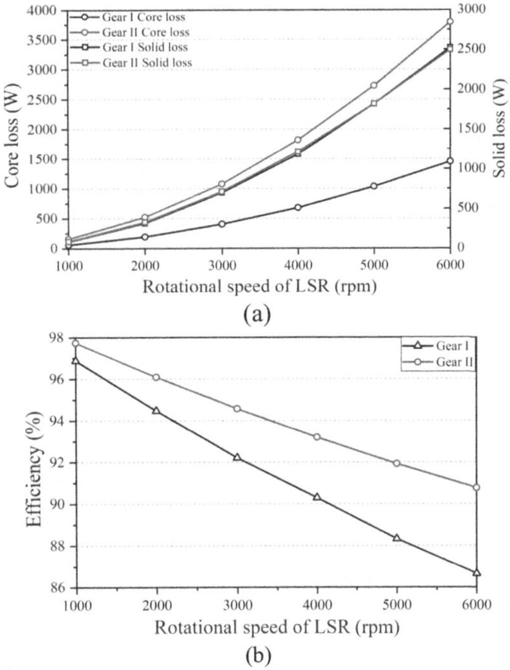

Fig. 9. (a) Core loss and solid loss of Gear I and Gear II. (b) The efficiency of Gear I and Gear II at a different rotational speed of LSR.

The efficiency of Gear I and Gear II at different speeds is shown in Fig. 9(b). The efficiency is reduced with increasing rotational speeds of LSR due to the increase of core loss and solid loss. However, the efficiency of Gear II is higher than Gear I from 1000 rpm to 6000 rpm, which shows that the increase of the average torque of the proposed RMG can overcome the increase in core loss compared with the conventional RMG. As a result, the Gear II can maintain a high efficiency of over 90% when the speed of LSR reaches 6000 rpm.

4 Conclusion

The main contribution of this study is to introduce an optimization method of PM arrangement in electric machines or MGs within a short computing time by using a novel general optimization pattern. The general pattern can generate six different types of PM arrangements. By using the proposed PM pattern in the optimal design of a RMG, the RMG topology with high torque and low torque ripple can be obtained by using the global optimization method. The merits of this optimal design method are revealed by the optimization results of the conventional RMG with surface-mounted PM and the proposed RMG with the general PM pattern. The average torque of the proposed RMG is 138% higher than the conventional RMG, and the PM volumetric torque density is increased by 20.9%. In the meantime, the torque ripple of the proposed RMG HSR is 1.89%, which is greatly lower than the 5.16% of the conventional RMG. The torque ripples of both LSRs are lower than 1.5%, which is significantly low in RMGs. Moreover, due to the enhancement of average torque, the efficiency of the proposed RMG can be maintained over 90% over a large speed range, which is also notably higher than the conventional RMG.

References

1. Yin, X., Pfister, P., Fang, Y.: A novel magnetic gear: toward a higher torque density. IEEE Trans. Magn. **51**(11), 1–4 (2015)
2. Zhu, Z.Q., Howe, D.: Electrical machines and drives for electric, hybrid, and fuel cell vehicles. Proc. IEEE **95**(4), 746–765 (2007)
3. Chau, K.T., Chan, C.C., Liu, C.: Overview of permanent-magnet brushless drives for electric and hybrid electric vehicles. IEEE Trans. Ind. Electron. **55**(6), 2246–2257 (2008)
4. Tlali, P.M., Wang, R.-J., Gerber, S.: Magnetic gear technologies: a review. In: 2014 International Conference on Electrical Machines (ICEM), pp. 544–550. IEEE, Berlin, Germany (2014)
5. Shoaei, A., Wang, Q.: A comprehensive review of concentric magnetic gears. IEEE Trans. Transport. Electrific. 10(3), 5581–5598 (2024). https://doi.org/10.1109/TTE.2023.3317772
6. Shen, J., Li, H., Hao, H., Jin, M., Wang, Y.: Topologies and performance study of a variety of coaxial magnetic gears. IET Electr. Power Appl. **11**(7), 1160–1168 (2017)
7. Jian, L., Chau, K.T., Gong, Y., Jiang, J.Z., Yu, C., Li, W.: Comparison of coaxial magnetic gears with different topologies. IEEE Trans. Magn. **45**(10), 4526–4529 (2009)
8. Wong, H.Y., Baninajar, H., Dechant, B., Southwick, P., Bird, J.Z.: Design and experimental testing of a magnetic gear for an electric aircraft drivetrain. IEEE Trans. Ind. Appl. **59**(4), 3934–3944 (2023)
9. Rasmussen, P.O., Andersen, T.O., Jorgensen, F.T., Nielsen, O.: Development of a high-performance magnetic gear. IEEE Trans. Ind. Appl. **41**(3), 764–770 (2005)
10. Uppalapati, K.K., Bird, J.Z., Jia, D., Garner, J., Zhou, A.: Performance of a magnetic gear using ferrite magnets for low speed ocean power generation. In: 2012 IEEE Energy Conversion Congress and Exposition (ECCE), pp. 3348–3355. IEEE, Raleigh, NC, USA (2012)
11. Uppalapati, K.K., Calvin, M.D., Wright, J.D., Pitchard, J., Williams, W.B., Bird, J.Z.: A magnetic gearbox with an active region torque density of 239 N·m/L. IEEE Trans. Ind. Appl. **54**(2), 1331–1338 (2018)
12. Hasanpour, S., Gardner, M.C., Johnson, M., Toliyat, H.A.: Comparison of reluctance and surface permanent magnet coaxial magnetic gears. In: 2020 IEEE Energy Conversion Congress and Exposition (ECCE), pp. 307–314. IEEE (2020)

13. Li, K., Zhu, Z., Wu, P.: A reluctance magnetic gear for high speed and vibration motor systems. In: 2018 25th International Conference on Mechatronics and Machine Vision in Practice (M2VIP), pp. 1–5. IEEE (2018)
14. Aiso, K., Akatsu, K., Aoyama, Y.: A novel reluctance magnetic gear for high-speed motor. IEEE Trans. Ind. Appl. **55**(3), 2690–2699 (2019)
15. Aiso, K., Akatsu, K., Aoyama, Y.: Reluctance magnetic gear and flux switching magnetic gear for high speed motor system. In: 2017 IEEE Energy Conversion Congress and Exposition (ECCE), pp. 2445–2452. IEEE, Cincinnati, OH (2017)
16. Arani, M.S., Afsari, S.A.: Design of rotor structure for reluctance magnetic gear to improve torque characteristic. In: 2020 11th Power Electronics, Drive Systems, and Technologies Conference (PEDSTC), pp. 1–5. IEEE, Tehran, Iran (2020)
17. Afsari Kashani, S.A.: Design and optimization of coaxial reluctance magnetic gear with different rotor topologies. IEEE Trans. Ind. Electron. **69**(1), 101–109 (2022)
18. Safarpour, R., Pakdelian, S.: Topology optimization of the reluctance coaxial magnetic gear. IEEE Trans. Magn. **58**(8), 1–7 (2022)
19. Fu, W.N., Li, L.: Optimal design of magnetic gears with a general pattern of permanent magnet arrangement. IEEE Trans. Appl. Supercond. **26**(7), 1–5 (2016)
20. Zhao, X., Niu, S.: Design and optimization of a new magnetic-geared pole-changing hybrid excitation machine. IEEE Trans. Ind. Electron. **64**(12), 9943–9952 (2017)

Wind Power Scheduling Deviation Compensation Based on Dual-Frequency Pulse Control

Jianhong Zhu[1], Han Li[1], Juping Gu[2(✉)], Linxin Mao[1], Shaoxuan Chen[1], and Yu He[1]

[1] School of Electrical Engineering and Automation, Nantong University, Nantong 226019, China
[2] Suzhou University of Science and Technology, Suzhou 215009, China
gu.jp@ntu.edu.cn

Abstract. The intermittency of wind power fluctuations and the low reliability of reporting scheduling plans have always affected the friendly access of grid connected wind power. "One machine, one storage" provides an effective solution. An innovative control method for dual-frequency PWM pulses in an improved bidirectional DC/DC Zeta Sepic circuit model is proposed to address the controllability of energy throughput in energy storage systems with dynamic deviation compensation for wind power planning. Firstly, by analyzing the working timing logic mechanism of the energy transfer converter, the feasibility of the energy storage throughput circuit is verified. Then, by analyzing the small signal control model of the bidirectional DC/DC converter during charging and discharging operation, the dual-frequency PWM automatic control of charging and discharging conditions is achieved, and the controllability of power exchange of the energy storage system at the wind turbine grid connection point is achieved. Finally, the model predictive control (MPC) method is adopted to effectively compensate the deviation of wind power plans, which can be accurately tracked. The simulation experiment demonstrates the effectiveness and feasibility of the converter model structure and control algorithm.

Keywords: wind power planning · deviation compensation · dual-frequency PWM · bidirectional energy storage converter · model prediction

1 Introduction

The gap fluctuation of wind power output increases the difficulty of participating in grid dispatch [1]. At the same time, the increasing plan bias with the wind turbine capacity affects the balance of supply and demand, hence, wind power suppliers would suffer economic punishment from electricity prices [2, 3]. Steady output and reliable planning forecast are the goal of scheduling wind power pursuit. In view of the flexible energy throughput characteristics of the storage system, which can dynamically absorb or supplement the unpredictable redundant power or owed power of plan bias. Some scholars have proposed such idea, increase the reliability of the wind power plan, so

K. Li et al. (Eds.): LSMS/ICSEE 2024, CCIS 2218, pp. 160–174, 2025.
https://doi.org/10.1007/978-981-96-0232-2_13

facilitating grid power department to make reasonable scheduling in advance, further ensure the safe and stable operation of the power system. To reduce the influence of uncertainty of new energy generation on grid command tracking, literature [4] proposes a strategy that applies energy storage for two-stage power distribution and three timescales, which effectively controlled energy storage exchange power and improved the tracking ability of new energy for grid command. The literature [5] uses the energy group balance plan of pumping water to solve the uncertainty of wind power output and improve economic benefits. The literature [6] proposes a control strategy of energy storage system based on MPC and dual-layer fuzzy control system tracking wind power planning, avoiding exceeding the limit of charging and discharging action.

In general, most of the researches on wind power plan deviation compensation mentioned in the existing literatures are limited to the upper-layer dispatch model, less on the underlying implementation technology. However, the behavioral control of energy throughput of storage systems not only requires solving the problem of given parameters, but also requires studying the bidirectional converter working mechanism as an "intermediate bridge". Bidirectional converter structural design and control technology have become the study focuses in recent years. There is an urgent need for mature products with high reliability in the field of engineering applications, especially popular in the "one machine, one storage" configuration system.

To converter model structure, literature [7] proposes a bidirectional DC/DC converter with partial switch sharing, which reduces the reactive return loss and achieves high voltage gain. The literature [8] proposes a multi-input soft-switching DC/DC converter for new energy power generation systems, which reduces the switching current stress and can independently transmit new energy sources to the DC grid. The literature [9] proposes a new multi-port DC/DC converter suitable for DC microgrid. The proposed converter has significant flexibility in connection between energy storage and load at different voltage and power levels. The literature [10] proposes a novel non-isolated bidirectional DC/DC converter for hybrid energy storage systems, which has a high voltage conversion ratio and a wide voltage gain range.

To converter pulse control, PWM controls are divided into single-frequency control and dual-frequency control. The literature [11] proposes a unique dual-frequency boost converter that does not require load temporary changes information during current control, increasing efficiency compared to traditional converters. The literature [12] proposes a parallel dual-frequency single-phase and grid connected inverter, which reduces the switching loss and eliminates the current harmonic of the low-frequency unit. In order to improve the transient response in bidirectional DC/DC converter, literature [13] combined pulse width modulation and pulse frequency modulation to design a phase shift control, and verified the effectiveness of the proposed method through experiments. In addition to optimizing the pulse frequency control, literature [14] obtained fast dynamic performance by converting various operating conditions, and realized quickly and accurately control through self-immunity method based on artificial neural network. In order to further expand the disturbance immunity range of the system and improve the robustness of the DC/DC converter to disturbance, literature [15] proposes a second-order linear active disturbance rejection control strategy, with better rapidity, robustness and adaptability.

Based on improved Zeta-Sepic non-isolated bidirectional DC/DC converter analysis, a wind power plan deviation compensation control method of storage system is proposed by dual-frequency PWM. Through the analysis of the switching working state of the bidirectional DC/DC converter, as well the small signal model, working mechanism of model predictive control (MPC) strategy is researched. By simulation, effectiveness of control method is verified. Dual-frequency control makes up for the shortcomings of single-frequency control, and energy storage energy throughput in the plan deviation compensation process can obtain reliable control.

2 Dual-Frequency Pulse DC/DC Converter Control Target

The wind power is rectified and integrated into the DC busbar, and the storage is also integrated into which for scheduling power compensation. The wind/storage system can operate flexibly in both off and online grid modes. To offline operation mode, the energy storage system outputs or absorbs power according to the changes between renewable energy and load, maintains internal power balance. The DC bus voltage is controlled constant to ensure stable and reliable operation of the DC microgrid system. To online mode, the DC microgrid is connected to grid through the DC/AC grid-connected converter. DC microgrid can transmit surplus energy to the grid or be supplemented flexibly.

Assume that the scheduling deviation of wind power to be compensated is ΔP, that is

$$\begin{cases} \Delta P = \Delta P1 + \Delta P2 \\ \Delta P1 = Pg - \mathrm{Pr} \\ \Delta P2 = P\mathrm{load} - Pg \end{cases} \tag{1}$$

In the formula, Pg is the planned power, Pr is the actual wind power, $P\mathrm{load}$ is the load power, $\Delta P1$ and $\Delta P2$ are the power required of energy storage compensation for tracking the wind generation plan and load tracking respectively. When $\Delta P > 0$, the energy storage system is discharged to the DC bus through the bidirectional DC/DC converter, and when $\Delta P < 0$, the energy from the DC bus is charged to the energy storage system through the bidirectional DC/DC converter.

When the wind power planning deviation ΔP is more than or less than zero, the energy storage converter can realize unidirectional energy compensation in a complete switching coordination working cycle, so as to achieve reliable wind power planning.

3 Analysis of Working Mechanism of Converter

3.1 Improve the Topology of Bidirectional DC/DC Circuit Model a Subsection Sample

The non-isolated bidirectional DC/DC converter has the advantages of simple structure, few switching components and convenient control, and is more suitable for wind power generation with low voltage and power levels, as that in distributed power generation systems. The topology is as shown in Fig. 1. $U\mathrm{b}$ is the energy storage side voltage, $U\mathrm{d}$

is the DC bus side voltage, *C*b is the energy storage side capacitance, *C*d is the DC bus side capacitance, *C* is the energy storage capacitance, *L*1 and *L*2 are the inductance, *D* is the anti-parallel diode, S11, S12 and S2 are the fully controlled insulated gate bipolar transistor (IGBT). When the actual wind power output is too much, that is, when the planned deviation $\Delta P < 0$, the bidirectional DC/DC converter works in Buck mode, the energy flows from the DC bus to the energy storage system. Instead, the bidirectional DC/DC converter is in Boost mode, where energy flows from the energy storage system to the DC bus.

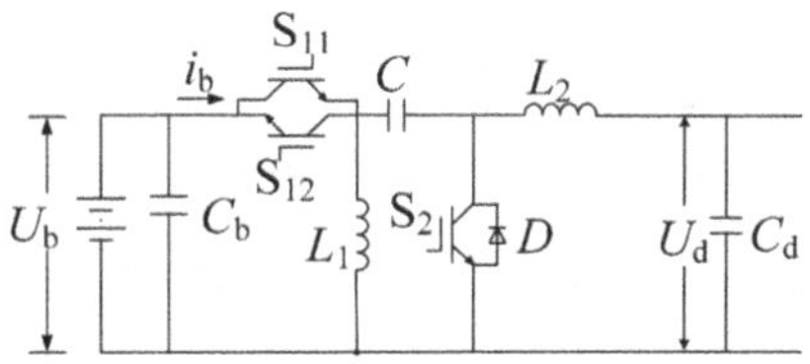

Fig. 1. Non-isolated bidirectional DC/DC converter

In order to avoid the sudden high current and high voltage caused by capacitor voltage mutation and inductive current mutation of energy storage components, 5 switch combination transient switching processes are designed for each complete energy transfer. Wind power planning positive deviation and negative deviation correspond to two different specific 5-mode switch combination switching processes. The specific mode is triggered by positive and negative wind power plan deviation signals to ensure that energy accumulation on the energy storage component L_1 and capacitor *C* will not damage the device.

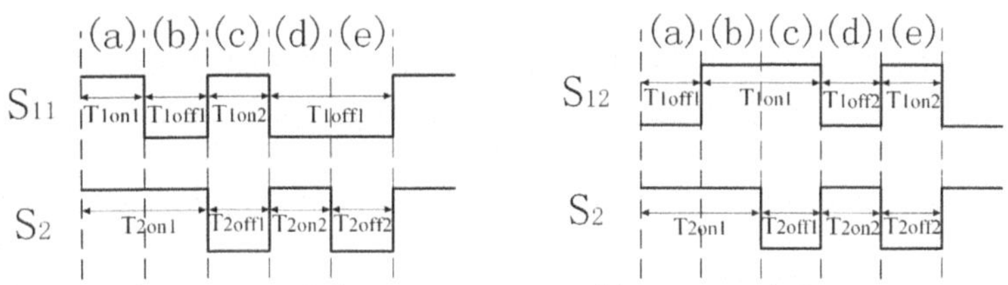

a) □*P* > 0 switch status b) □*P* < 0 switch status

Fig. 2. Schematic diagram of dual-frequency PWM pulse signal control

According to the positive and negative deviation of wind power plan monitored in real time, multi-channel PWM control signal of IGBT in bidirectional DC/DC converter is synchronously started. Figure 2 shows the schematic diagram of dual-frequency PWM pulse signal control. Where, T_{1on1} is the first turn on duration of IGBT S_1 within the complete cycle, T_{1off1} is the first turn off duration of IGBT S_1 within the complete cycle, T_{1on2} is the second turn on duration of IGBT S_1 within the complete cycle, and T_{1off2} is the second turn off duration of IGBT S_1 within the complete cycle. T_{2on1} is the first turn on duration of the IGBT S_2 in the complete cycle, T_{2off1} is the first turn off duration of the IGBT S_2 in the complete cycle, T_{2on2} is the second turn on duration of the IGBT S_2 in the complete cycle, and T_{2off2} is the second turn off duration of the IGBT S_2 in the complete cycle.

When wind power planning deviation $\Delta P > 0$, the bidirectional DC/DC converter is in Boost mode, corresponding to 5 different switching modes of the IGBT, which are equivalent to 5 circuit models, as shown in Fig. 3, where the resistance R_d is the equivalent resistance of the DC bus.

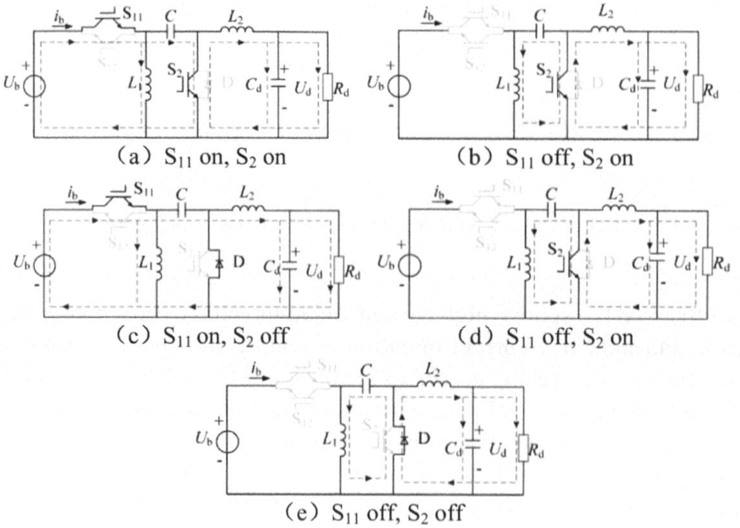

(a) S_{11} on, S_2 on (b) S_{11} off, S_2 on (c) S_{11} on, S_2 off (d) S_{11} off, S_2 on (e) S_{11} off, S_2 off

Fig. 3. Equivalent diagrams of different modes of boost modes

Mode (a) S_{11} is on, S_2 is on, the battery charges the inductor L_1, which stores magnetic field energy, the battery simultaneously charges the capacitor C, which stores electric field energy, and the magnetic field stored on the inductor L_2 flows to the DC bus. Mode (b) S_{11} is off, S2 is on, the magnetic field energy stored on the inductor $L1$ is transferred to the capacitor C, the capacitor C stores electric field energy, the magnetic field on the inductor L_2 continues to flow to the DC bus, mode (c) S_{11} is on, S_2 is off, the battery provides energy to the inductor L_1, the inductor L_1 stores magnetic field energy, the battery charges the inductor L_2 through the capacitor C. And provide energy to the DC bus. Mode (d) S_{11} is off, S_2 is on, the magnetic field on the inductor L_1 can flow to the capacitor C, which stores the electric field energy, and the magnetic field on the inductor L_2 can flow to the DC bus. Mode (e) S_{11} is off, S_2 is off, energy on inductor L_1 continues to flow to capacitor C, capacitor C stores energy, energy on inductor L_2 continues to flow to the DC bus.

When wind power planning deviation $\Delta P < 0$, the bidirectional DC/DC converter adopts Buck mode, corresponding to five different switching modes of the IGBT, which are equivalent to five circuit models, as shown in Fig. 4. Resistor R_b in Fig. 4 is the equivalent resistance of the battery.

Mode (a) S_{12} is on, S_2 is off, and the energy from the DC grid is supplied to the resistor R_b through the inductor L_2 and the capacitor C, and the inductor L_1 is supplied to the resistor R_b. Mode (b) S_{12} is off and S_2 is on, the DC bus energy flows to the inductor L_2, which stores magnetic field energy, the electric field energy in the capacitor C is converted into magnetic field energy and stored in the inductor L_1, and the capacitor C_b maintains the voltage at both ends of the resistor R_b. Mode (c) S_{12} is on, S_2 is on, and the energy in the DC bus continues to flow into the inductor L_2, which stores magnetic field energy, and the capacitor C releases electric field energy to power the resistor R_b. Mode (d) S_{12} is on, S_2 is off, and the energy from the DC bus is supplied to the resistor R_b through the inductor L_2 and the capacitor C, while the inductor L_1 is also supplied to the resistor R_b. Mode (e) S_{12} is off, S_2 is on, and the energy from the DC bus flows to the inductor L_2, which stores the magnetic field energy, the electric field energy in the energy storage capacitor C is converted into the magnetic field energy and is temporarily stored in the inductor L_1, and the capacitor C_b maintains the voltage at both ends of the resistor R_b.

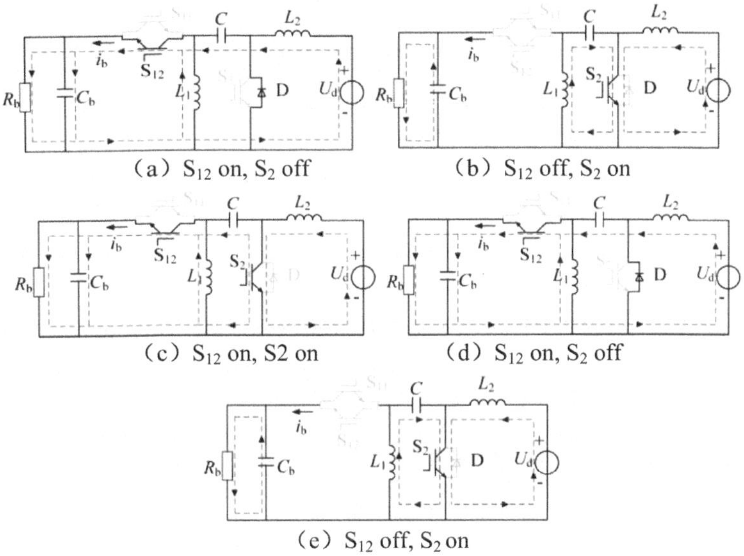

Fig. 4. Equivalent diagrams of different modes of buck modes

3.2 Mathematical Model of Bidirectional DC-DC Converter

The duty cycle of different modes of the bidirectional DC/DC converter during charge and discharge is shown in Table 1.

In steady state, the integral of the voltage U_L at both ends of the inductor L in a period T to the time is zero. Therefore when the wind power plan deviation $\Delta P > 0$,

Table 1. Duty cycle of different modes

Mode	Mode (a)	Mode (b)	Mode (c)	Mode (d)	Mode (e)
Boost mode S_1 duty cycle	$D1T1/T$	$(1-D1)T1/T$	$D2T2/T$	$(1-D2)T2/2T$	$(1-D2)T2/2T$
Buck mode S_1 duty cycle	$D1T1/T$	$(1-D1)T1/T$	$D2T2/2T$	$D2T2/2T$	$(1-D2)T2/T$

the voltage at both ends of L_1 and L_2 is analyzed and the formula (2) is obtained.

$$\begin{cases} \frac{Ud}{Ub} = \frac{D2T2}{(1-D_1)T1+(1-D2)T2} \\ UC = -\frac{D1T1+D2T2}{(1-D_1)T1+(1-D2)T2}Ub \end{cases} \tag{2}$$

Assuming $T1 = n1T2$, $Ud/Ub = \alpha 1$, formula (3) can be obtained.

$$D_1 = \frac{(n1+1)}{n1} - \frac{(\alpha 1+1)}{n1\alpha 1}D2 \tag{3}$$

Similarly, the formula for wind power planning deviation $\Delta P < 0$ can be obtained.

$$\begin{cases} \frac{Ub}{Ud} = \frac{2[(1-D_1)T1+(1-D2)T2]}{2D1T1+D2T2} \\ UC = \frac{2D1T1+D2T2}{(1-D_1)T1+2(1-D2)T2}Ud \end{cases} \tag{4}$$

Assuming $T1 = n2T2$, $Ub/Ud = \alpha 2$, formula (5) can be obtained.

$$D1 = \frac{(1+n2-\alpha 2)}{n2(\alpha 2+1)} - \frac{(2+\alpha 2)}{2n2(\alpha 2+1)}D2 \tag{5}$$

When the wind power plan deviation $\Delta P > 0$, that is, the energy storage discharge mode, the state equations of modes (a)–(e) are shown in Eq. (6)

$$\begin{cases} L1\frac{diL1}{dt} = \frac{(1-D1)T1+(1-D2)T2}{T}uC + \frac{D1T1+D2T2}{T}ub \\ L2\frac{diL2}{dt} = -\frac{D2T2}{T}uC + \frac{D2T2}{T}ub - ud \\ C\frac{duC}{dt} = -iL1 + \frac{D1T1+D2T2}{T}ib \\ Cd\frac{dud}{dt} = iL2 + \frac{1}{Rd}ud \end{cases} \tag{6}$$

In the formula, $T = T1 + T2$.

In the whole working period T, small signal perturbation method and Laplace transform are used to obtain Eq. (7).

$$\begin{cases} sL1\hat{i}L1(s) = M1\hat{u}C(s) + N1\hat{u}b(s) + f1\hat{d}1(s) + f1\hat{d}2(s) \\ sL2\hat{i}L2(s) = -g1\hat{u}C(s) - \hat{u}d(s) + g1\hat{u}b(s) + e1f1\hat{d}2(s) \\ sC\hat{u}C(s) = -\hat{i}L1(s) + N1\hat{i}b(s) + b1Ib\hat{d}1(s) + e1Ib\hat{d}2(s) \\ sCd\hat{u}d(s) = \hat{i}L2(s) + \frac{1}{Rd}\hat{u}d(s) \end{cases} \tag{7}$$

In the formula, $M1 = [(1 - D1)T1 + (1 - D2)T2]/T, N1 = [D1T1 + D2T2]/T$, $b1 = T1/T, e1 = T2/T, g1 = D2T2/T, f1 = Ub - Uc$.

According to formula (7), the transfer function of duty cycle $\hat{d}2$ to inductive current $\hat{i}L2$ in the process of energy storage discharge can be obtained.

$$Gboost(s) = \frac{\hat{i}L2(s)}{\hat{d}2(s)} = \frac{A14s^4 + A13s^3 + A12s^2 + A11s + A10}{B15s^5 + B14s^4 + B13s^3 + B12s^2 + B11s} \quad (8)$$

In the formula, $A14 = C^2CdL1Rde1f1$, $A13 = CL1e1(Cf1 - CdRdg1Ib)$, $A12 = CCdRdM1e1f1(g1Ib + 1) - CL1e1g1Ib$, $A11 = M1e1g1Ib(C - CdRd) + CM1e1f1$, $A10 = -M1e1g1Ib$, $B15 = C^2CdL1L2Rd$, $B14 = C^2L1L2$, $B13 = CRd(CdL2M1 - CL1)$, $B12 = CL2M1$, $B11 = CRdM1$.

When the wind power plan deviation $\Delta P < 0$, that is, the energy storage charging mode, the state equation of modes (a)–(e) is shown in Eq. (9)

$$\begin{cases} L1\frac{diL1}{dt} = \frac{(1-D1)T1+(1-D2)T2}{T}uC - \frac{D1T1+D2T2}{T}ub \\ L2\frac{diL2}{dt} = -\frac{2D1T1+D2T2}{2T}uC - \frac{2D1T1+D2T2}{2T}ub + ud \\ C\frac{duC}{dt} = -\frac{2(1-D1)T1+(2-D2)T2}{2T}iL1 + \frac{2D1T1+D2T2}{2T}iL2 + \frac{D2T2}{2T}ib \\ Cb\frac{dub}{dt} = \frac{D1T1+D2T2}{T}ib + \frac{1}{Rb}ub \end{cases} \quad (9)$$

In the whole working period T, small signal perturbation method and Laplace transform are used to obtain Eq. (10).

$$\begin{cases} sL1\hat{i}L1(s) = M2\hat{u}C(s) - N1\hat{u}b(s) + b2f2\hat{d}1(s) + e2f2\hat{d}2(s) \\ sL2\hat{i}L2(s) = -E2\hat{u}C(s) - E2\hat{u}b(s) + \hat{u}d(s) - b2f2\hat{d}1(s) - \frac{1}{2}e2f2\hat{d}2(s) \\ sC\hat{u}C(s) = -F2\hat{i}L1(s) + E2\hat{i}L2(s) + \frac{1}{2}g2\hat{i}b(s) + b2G2\hat{d}1(s) + \frac{1}{2}e2H2\hat{d}2(s) \\ sCb\hat{u}b(s) = -\frac{1}{Rb}\hat{u}b(s) + e2Id\hat{d}2(s) \end{cases} \quad (10)$$

In the formula, $M2 = [(1 - D1)T1 + (1 - D2)T2]/T, N2 = [D1T1 + D2T2]/T$, $E2 = [2D1T1 + D2T2]/2T, F2 = [2(1 - D1)T1 + (2 - D2)T2]/2T, b2 = T1/T$, $e2 = T2/T, g2 = D2T2/T, f1 = Ub + Uc, G2 = IL1 + IL2, H2 = IL1 + IL2 + I\text{b}$.

According to formula (10), the transfer function of duty cycle $\hat{d}2$ to inductive current $\hat{i}L2$ in the process of energy storage discharge can be obtained.

$$Gbuck(s) = \frac{\hat{i}L2(s)}{\hat{d}2(s)} = \frac{A24s^4 + A23s^3 + A22s^2 + A21s + A20}{B25s^5 + B24s^4 + B23s^3 + B22s^2} \quad (11)$$

In the formula, A24 = −CCbRbe2f2/2, A23 = -CL1e2(CRbE2Ib + CbRbe2E2H2/2), $A_{22} = Ce_2[C_bR_bf_2(1 + F_2M_2/2) + L_1E_2H_2/2],$, $A_{21} = C_bR_be_2H_2M_2(1 - E_2F_2)/2 - CR_be_2E_2F_2I_b(N_2 - M_2) + Ce_2f_2(1 + F_2M_2/2),$, $A20 = e2H2M2(1 + E2F2)/2$, $B25 = C^2CdL1L2Rb$, $B24 = C^2L1L2$, $B23 = CCdRb(L1E2^2 + L2F2M2)$, $B22 = CL1E2^2 + CL2F2M2$.

Formula (12) is obtained by ignoring the power loss and inductance of the IGBT.

$$ubib = udiL2 \quad (12)$$

The transfer function between the inductor L_2 current and the battery current is

$$GiL2\_ib = iL2/ib = ub/ud \quad (13)$$

4 Model Predictive Control

The MPC current inner loop control is established by using Boost mode as an example. The mean value model of the Boost mode of the converter is discretized by the forward Euler method, and Eq. (14) is obtained after sorting.

$$\begin{cases} iL1(k+1) = \frac{Ts}{L1}M1uc(k) + \frac{Ts}{L1}N1ub(k) + iL1(k) \\ iL2(k+1) = -\frac{Ts}{L2}g1uc(k) + \frac{Ts}{L2}g1ub(k) - \frac{Ts}{L2}ud(k) + iL2(k) \\ uc(k+1) = -\frac{Ts}{C}iL1(k) + \frac{Ts}{C}N1ib(k) + uc(k) \\ Cd\frac{ud(k+1)-ud(k)}{Ts} = \frac{Ts}{Cd}iL2(k) + (1 - \frac{Ts}{Cd}\frac{1}{Rd})ud(k) \end{cases} \tag{14}$$

In the formula, T_s is the sampling period.

For the current inner loop control strategy, a stable continuous set model predictive control algorithm is adopted, and the cost function is

$$J = [iL2(k+1) - iL2, \mathrm{ref}]^2 \tag{15}$$

In the formula, $iL2$, ref is the reference value of the current inner loop.

Substitute Eq. (14) into Eq. (15) to get

$$J = \left[-\frac{Ts}{L2}\frac{D2T2}{T}uc(k) + \frac{Ts}{L2}\frac{D2T2}{T}ub(k) - \frac{Ts}{L2}ud(k) + iL2(k) - iL2, \mathrm{ref}\right]^2 \tag{16}$$

Take the partial derivative of D_2 in the above formula and set the partial derivative to zero

$$D2 = \frac{L2}{[ub(k) - uc(k)]Tse1}\left[\frac{Ts}{L2}ud(k) - iL2(k) + iL2, \mathrm{ref}\right] \tag{17}$$

The upper and lower limits of the current of the inductor L_2 are set as $iL2$, max and $iL2$, min respectively, and sorted out.

$$\frac{L2}{Tsg1}\left[iL2, \max + \frac{Ts}{L2}ud(k) - iL2(k)\right] \le D2 \le \frac{L2}{Tsg1}\left[iL2, \min + \frac{Ts}{L2}ud(k) - iL2(k)\right] \tag{18}$$

As can be seen from the above formula, the current of the inductor L_2 can be limited by the duty cycle D_2.

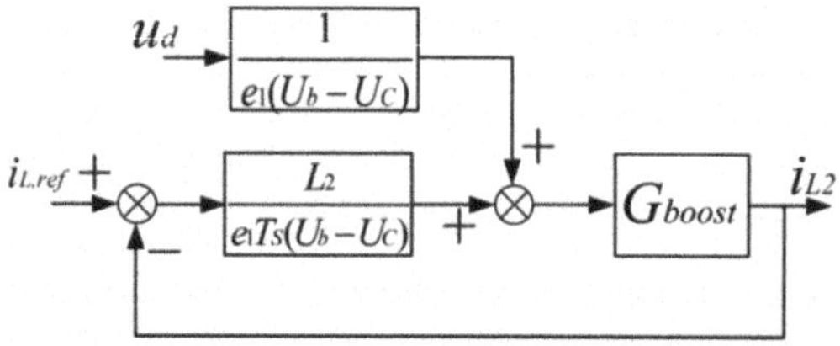

Fig. 5. Block diagram of MPC current inner loop control system in boost mode

As shown in Fig. 5, the block diagram of MPC current inner loop control system in boost mode of converter is shown.

As can be seen from Fig. 5, the closed-loop transfer function from *iL*2, ref to *iL*2 is

$$G_{iL2,iL2\_\mathrm{ref}}(s) = \frac{i_{L2}}{i_{L2,ref}} = \frac{L_2 G_{boost}}{L_2 G_{boost} - e_1 T_s (U_b - U_C)} \tag{19}$$

The closed-loop transfer function from *ud* to *iL*2 is

$$Gud\_iL2(s) = \frac{iL2}{ud} = \frac{TsGboost}{L2Gboost + e1Ts(Ub - UC)} \tag{20}$$

By substituting formula (8) into the above formula, we get

$$G_{iL2,iL2\_\mathrm{ref}}(s) = \frac{L_2\left(A_{14}s^4 + A_{13}s^3 + A_{12}s^2 + A_{11}s + A_{10}\right)}{-e_1 f_1 T_s(B_{15}s^5 + B_{14}s^4 + B_{13}s^3 + B_{12}s^2 + B_{11}s) + L_2(A_{14}s^4 + A_{13}s^3 + A_{12}s^2 + A_{11}s + A_{10})} \tag{21}$$

$$G_{ud\_iL2}(s) = \frac{T_s\left(A_{14}s^4 + A_{13}s^3 + A_{12}s^2 + A_{11}s + A_{10}\right)}{e_1 f_1 T_s(B_{15}s^5 + B_{14}s^4 + B_{13}s^3 + B_{12}s^2 + B_{11}s) + L_2(A_{14}s^4 + A_{13}s^3 + A_{12}s^2 + A_{11}s + A_{10})} \tag{22}$$

According to the control system block diagram shown in Fig. 5, the inductance current *iL*2 can be expressed as

$$iL2, boost(s) = GiL2, iL2\_\,\mathrm{ref}(s)iL2, \mathrm{ref} + Gud\_iL2(s)ud \tag{23}$$

Similarly, the closed-loop transfer function from *iL*2, ref to *iL*2 in buck mode can be obtained as

$$G_{iL2,iL2\_\mathrm{ref}}(s) = \frac{2L_2\left(A_{24}s^4 + A_{23}s^3 + A_{22}s^2 + A_{21}s + A_{20}\right)}{-e_2 f_2 T_s(B_{25}s^5 + B_{24}s^4 + B_{23}s^3 + B_{22}s^2) + 2L_2(A_{24}s^4 + A_{23}s^3 + A_{22}s^2 + A_{21}s + A_{20})} \tag{24}$$

The closed-loop transfer function from *ud* to *iL*2 is

$$Gud\_iL2(s) = \frac{2Ts\left(A24s^4 + A23s^3 + A22s^2 + A21s + A20\right)}{e2f\,2Ts(B25s^5 + B24s^4 + B23s^3 + B22s^2) - 2L2(A24s^4 + A23s^3 + A22s^2 + A21s + A20)} \tag{25}$$

The expression of inductance current *iL*2 is

$$iL2, buck(s) = GiL2, iL2\_\,\mathrm{ref}(s)iL2, \mathrm{ref} + Gud\_iL2(s)ud \tag{26}$$

5 Simulation Verification

In order to verify the effectiveness of the relevant control methods, a bidirectional DC/DC converter simulation model was constructed in Matlab/Simulink environment. The simulation parameters are shown in Table 2.

Figure 6 shows the curve of energy storage changing from 1500 w discharge to 1000 w charge at 0.4 s and the curve of energy storage changing from 1000 w charge to 1000 w discharge at 0.6 s to compensate the wind power deviation. The dashed line

Table 2. Simulation parameters table

Parameter	Symbol	Values
Rated voltage of DC bus/V	U_d	500
Rated voltage of battery/V	U_b	300
Rated charging and discharging power/w	P_N	3000
Initial SOC of battery/%	SOC	50
Battery capacity/Ah	C_1	200
Battery side capacitance C_b/mF	C_b	1.5
Dc bus side capacitance C_d/mF	C_d	4
Energy storage capacitor C/uF	C	2
Inductance L_1/mH	L_1	120
Inductance L_2/mH	L_2	40

represents the simulation result of the double closed-loop PI control method, and the solid line represents the result graph of the MPC control method. It can be seen from Fig. 6(a) that when the converter switches from discharge mode to charge mode, the voltage adjustment time of the double-closed-loop PI control method is 65 ms and the peak voltage is 503.8 V. MPC control method voltage regulation time is 41 ms, the peak voltage is 502.1 V. It can be seen from Fig. 6(b) that when the converter switches from discharge mode to charge mode, the voltage adjustment time of the double-closed-loop PI control method is 75 ms and the peak voltage is 495.5 V. MPC control method voltage regulation time is 62 ms, the peak voltage is 498.8 V. It can be seen from Fig. 6(c) and (d) that compensation of power deviation can be realized quickly when charging and discharging modes are switched. Therefore, under the dual-frequency PWM control, the two control methods can quickly realize the purpose of compensating the wind power deviation under the condition of stabilizing the DC bus voltage, and the MPC control method has better dynamic performance.

Figure 7 shows the curve of energy storage changing from 500 w charging to 2000 w charging at 0.4 s and the curve of energy storage changing from 500 w discharging to 1700 w discharging at 0.8 s to compensate the wind power deviation. The dashed line represents the simulation result of the double closed-loop PI control method, and the solid line represents the result graph of the MPC control method. It can be seen from Fig. 7(a) that when the charging power of the converter changes abruptly, the voltage adjustment time of the double-closed-loop PI control method is 46 ms and the peak voltage is 503 V. The voltage regulation time of MPC control method is 32 ms, and the peak voltage is 500.6 V. It can be seen from Fig. 7(b) that when the discharge power of the converter changes abruptly, the voltage regulation time of the double-closed-loop PI control method is 60 ms and the peak voltage is 499.5 V. The voltage regulation time of MPC control method is 38 ms and the peak voltage is 499.7 V. It can be seen from Fig. 7(c) and (d) that compensation of power deviation can be realized quickly when charging power changes abruptly or discharging power changes abruptly.

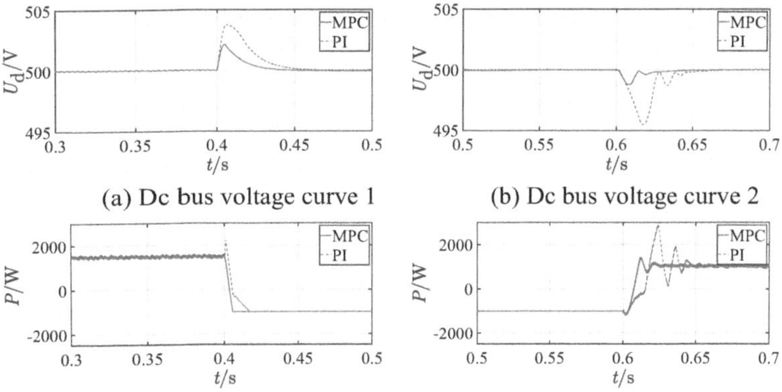

(a) Dc bus voltage curve 1 (b) Dc bus voltage curve 2

(c) Charge and discharge power curve 1 (d) Charge and discharge power curve 2

Fig. 6. Curve of energy storage charging and discharging mode switching

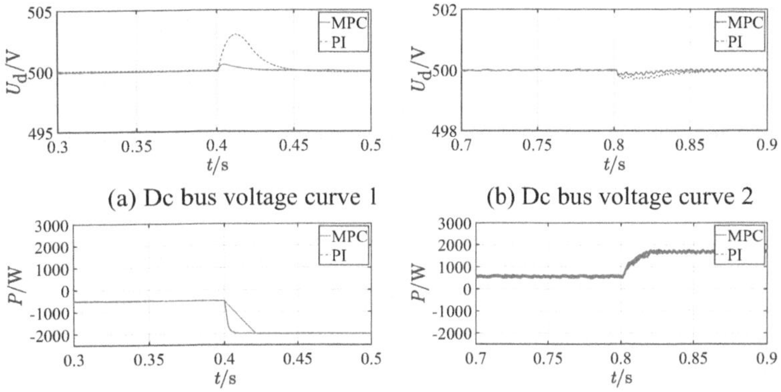

(a) Dc bus voltage curve 1 (b) Dc bus voltage curve 2

(c) Charge and discharge power curve 1 (d) Charge and discharge power curve 2

Fig. 7. Charge and discharge power abrupt curve

Therefore, under the proposed dual frequency PWM control, when the energy storage needs charge and discharge power mutation, the two control methods can realize the purpose of compensating the wind power deviation under the condition of stabilizing the DC bus voltage, and the MPC control method has better dynamic performance, and can maintain the bus voltage at the rated value with faster dynamic response speed.

Table 3 shows the comparison of the performance indexes of the two control methods under different working conditions. It can be seen from the table that dual-frequency PWM has shorter voltage regulation time and smaller peak voltage under MPC control, so wind power planning deviation compensation can be realized quickly and accurately.

Figure 8 is the waveform diagram of the proposed dual frequency PWM in MPC control mode, the converter compensates the power plan deviation, Fig. 8(a) is the waveform diagram of the charge and discharge power, and Fig. 8(b) is the waveform diagram of the energy storage system SOC. By calculating the tracking error, it can be seen that the error of the energy storage is 1.64%when compensating the power

Table 3. Comparison of the performance indicators of the two control methods

Working condition	Parameter			
	PI control method		MPC method	
	Accommodation time t/ ms	Peak voltage/V	Accommodation time t/ ms	Peak voltage/V
ΔP:1500w $\to$ -1000w	65	503.8	41	502.1
ΔP:-1000w $\to$ 1000w	75	495.5	62	498.8
ΔP:-500w $\to$ -2000w	46	503.0	32	500.6
ΔP:500w $\to$ 1700w	60	499.5	38	499.7

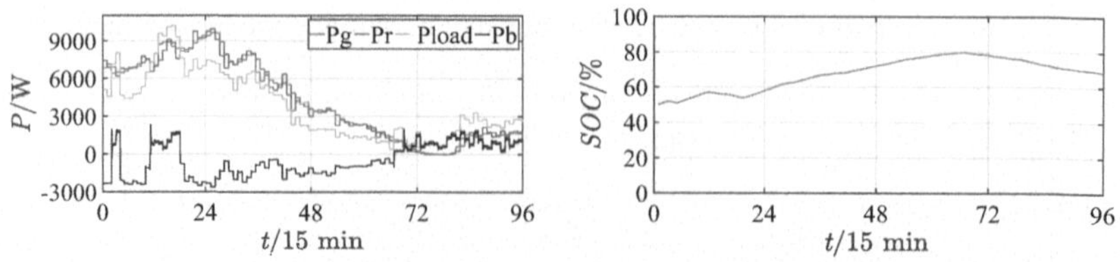

(a) Charge and discharge power curve (b) Energy storage SOC curve

Fig. 8. Converter compensated power plan deviation curve

deviation. It can be seen from the Fig. 8 that when the dual-frequency PWM proposed by MPC control is used to compensate the wind power planning deviation, the power is quickly tracked, and the ability of the energy storage system to compensate the wind power planning deviation is improved.

6 Discussions and Conclusions

The feasibility of the energy storage converter is verified by analyzing the switch combination logic mechanism, and a small signal model of the charge and discharge state of the bidirectional DC/DC converter is established, and then the MPC current inner loop control is established. The dynamic response ability of the converter's dual-frequency PWM control for wind power scheduling deviation compensating is tested by simulation. The method proposed has the characteristics of fast response and stable adjustment process of small peak voltage and is suitable for the application scenario.

However, it must be acknowledged that the research work is not over yet. Firstly, the topic mainly focuses on the control methods of energy transfer, and in terms of physical implementation, it is necessary to fully consider the technical factors of the device itself, and must add correction links in the model. Secondly, due to the complex physical characteristics of the device, the loss of energy transmission exhibits nonlinear characteristics, and the response time of different deviation amplitudes shows nonlinear changes. In actual physical experiments, it is necessary to design virtual moment of inertia with variable damping characteristics for controlling system parameter design.

Finally, consider the transient characteristics of energy storage components, the bidirectional energy conversion and switching process also require the design of reliable buffer circuits.

Acknowledgements. This work is a part of the Innovative Research Group Project of the National Natural Science Foundation of China (Nos. 52377117, 52377104). The authors would like to thank for the supports from both the Ministry of Science and Technology and National Natural Science Foundation of China.

Patents. Chinese invention patents are resulting from the work as follows:

1. A control method for energy storage DC converter for wind power plan deviation compensation. Authorization number: ZL 202010740902.4. Authorization date: 2023/11/14.
2. An energy storage management method for dynamic compensation on wind and solar complementary power plan deviation. Authorization number: ZL 202110863376.5. Authorization date: 2024/01/06.
3. Optimal allocation method of energy storage for day ahead dispatching considering wind power generation and grid load fluctuation. Authorization number: ZL 202210024705.1. Authorization date: 2023/08/25.

References

1. Luo, H., Xu, C., Dai, K., et al.: Balance control of SOC for MMC-BESS with power fluctuation suppression, PCC voltage regulation, and harmonic mitigation in grid-connected wind farm. IEEE Access **10**, 117732–117744 (2022). https://doi.org/10.1109/ACCESS.2022.3218716
2. Rashidizadeh-Kermani, H., Vahedipour-Dahraie, M., Parente, M., et al.: Implementing a flexible penalizing mechanism for wind power producers in the regulating market. IEEE Access **11**, 125204–125216 (2023). https://doi.org/10.1109/ACCESS.2023.3330906
3. Ramirez-Burgueno, L., Sang, Y., Zheng, Y.: Pricing wind power uncertainty in the electricity market. IEEE Access **11**, 58624–58633 (2023). https://doi.org/10.1109/ACCESS.2023.3282575
4. Wei, Y., Cheng, Z., Zhang, H., et al.: Multi-mode tracking strategies for wind-solar-storage hybrid power generation system. In: 2022 2nd International Conference on Electrical Engineering and Control Science (IC2ECS), Nanjing, China (2022). https://doi.org/10.1109/IC2ECS57645.2022.10087947
5. Fujimoto, Y., Inagaki, M., Kaneko, A., et al.: Day-ahead scheduling of wind-hydro balancing group operation to maximize expected revenue considering wind power output uncertainty. IEEE Access **11**, 119200–119218 (2023). https://doi.org/10.1109/ACCESS.2023.3327569
6. Yang, J., Yang, T., Luo, L., et al.: Tracking-dispatch of a combined wind-storage system based on model predictive control and two-layer fuzzy control strategy. Prot. Control Mod. Power Syst. **8**, 1–16 (2023). https://doi.org/10.1186/s41601-023-00334-6
7. Li, Y., Wang, Y., Guan, Y., et al.: Optimized bidirectional DC-DC converter adapted to high voltage gain and wide ZVS range. IEEE Trans. Power Electron. **38**(3), 3486–3499 (2023). https://doi.org/10.1109/TPEL.2022.3221946
8. Sun, Z., Bae, S.: Multiple-input soft-switching DC–DC converter to connect renewable energy sources in a DC microgrid. IEEE Access **10**, 128380–128391 (2022). https://doi.org/10.1109/ACCESS.2022.3227439

9. Saafan, A.A., Khadkikar, V., El Moursi, M.S., et al.: A new multiport DC-DC converter for DC microgrid applications. In: 2021 IEEE Industry Applications Society Annual Meeting (IAS), Vancouver, BC, Canada (2021).https://doi.org/10.1109/IAS48185.2021.9677403
10. Jiang, F., et al.: A zero current ripple bidirectional DC–DC converter with high voltage gain and common ground for hybrid energy storage system EVs. IEEE J. Emerg. Sel. Top. Power Electron. **11**(5), 4882–4894 (2023). https://doi.org/10.1109/JESTPE.2023.3304364
11. Nethra, M., Marimuthu, M., Vijayalakshmi, S.: Design and implementation of double frequency boost converter. In: 2022 3rd International Conference for Emerging Technology (INCET), Belgaum, India, pp. 1–5 (2022). https://doi.org/10.1109/INCET54531.2022.9825420
12. Yang, L., et al.: A novel parallel dual-frequency grid-connected inverter topology and efficiency analysis. In: 2019 10th International Conference on Power Electronics and ECCE Asia (ICPE 2019 – ECCE Asia), Busan, Korea (South), pp. 2350–2355 (2019). https://doi.org/10.23919/ICPE2019-ECCEAsia42246.2019.8797343
13. Yin, J., He, X., Lu, J., et al.: Phase-shift control with unified PWM/PFM for improved transient response in a bidirectional dual-active-bridge DC/DC converter. IEEE Trans. Ind. Electron. **70**(9), 8862–8872 (2023). https://doi.org/10.1109/TIE.2022.3215452
14. Zeng, Y., et al.: Active disturbance rejection control using artificial neural network for dual-active-bridge-based energy storage system. IEEE J. Emerg. Sel. Top. Power Electron. **11**(1), 301–311 (2023). https://doi.org/10.1109/JESTPE.2021.3138341
15. Zhou, S., Guo, X., Zhang, R., et al.: Research on triple interleaved bidirectional DC/DC converter based on second-order linear active disturbance rejection. IEEE J. Emerg. Sel. Top. Ind. Electron. **4**(4), 1074–1083 (2023). https://doi.org/10.1109/JESTIE.2023.3306997

Energy Storage Systems

A Comprehensive Simulation and Validation of Single Particle Aging Mechanism Model for Lithium-Ion Batteries

Kaiyan Shi[1], Tianpeng Lu[1], Xingyu Zhou[2], Tao Sun[1], Yu Wang[2], Xuebing Han[2], and Yuejiu Zheng[1,2(✉)]

[1] University of Shanghai for Science and Technology, Shanghai 200093, People's Republic of China
yuejiu.zheng@usst.edu.cn

[2] State Key Laboratory of Intelligent Green Vehicle and Mobility, Tsinghua University, Beijing 100084, People's Republic of China

Abstract. Aging mechanisms of graphite-$LiFePO_4$ Li-ion batteries and accurately predicting their remaining useful life is crucial for effective battery management in energy storage systems. This study proposes a comprehensive modeling approach based on battery aging side reactions and validates its reliability through durability tests and electron microscopy characterizations. The model, which is based on single particle aging mechanisms, elucidates primary degradation reactions such as solid electrolyte interface layer growth, active material loss, and lithium plating. Durability tests under various cycling and calendar aging conditions confirm the model's accuracy, with errors below 2%. Post-mortem analysis and material characterization tests further validate the model's physical effectiveness using SEM, TEM, and FIB-SEM electron microscopy techniques. The findings provide valuable insights into the aging mechanisms of graphite-$LiFePO_4$ Li-ion batteries and contribute to the development of improved battery management strategies.

Keywords: Li-ion battery · Battery degradation mechanism · Battery mechanism model · Reduced-order model · Aging mechanism verification

1 Introduction

With the increasing penetration of renewable energy sources and China's commitment to peak CO_2 emissions by 2030 and achieve carbon neutrality by 2060, people pay more attention to Battery Energy Storage Systems (BESS). Given the intermittency of renewable energy, BESS plays a pivotal role in balancing supply and demand discrepancies. BESS can store surplus energy during low-demand periods for later use during peak demand, ensuring grid stability. Additionally, BESS provides backup power during grid failures or emergencies, enhancing reliability. In the transportation sector, particularly in Electric Vehicles (EVs) and Hybrid Electric Vehicles (HEVs), battery packs serve as energy storage systems, facilitating emission-free propulsion. Governments worldwide are incentivizing BESS adoption to reduce emissions, enhance energy efficiency, and promote renewable energy integration [1].

K. Li et al. (Eds.): LSMS/ICSEE 2024, CCIS 2218, pp. 177–191, 2025.
https://doi.org/10.1007/978-981-96-0232-2_14

Battery capacity degradation is a crucial challenge in energy storage systems. Predicting and managing battery life can effectively improve battery utilization and safety. Establishing an accurate State of Health (SOH) prediction model that reflects battery aging mechanisms is a key issue that urgently needs to be addressed in the development of energy storage systems.

This study aims to establish a single-particle aging model to predict the lifespan of graphite-$LiFePO_4$ Li-ion batteries by investigating their intrinsic aging side reaction mechanisms. The provided model will be validated against laboratory battery aging test data and further verified through in-situ and dismantling detection of battery aging side reactions to ensure its accuracy and interpretability.

2 Aging Mechanism and Modeling

Lithium-ion batteries inevitably undergo aging processes during operation, which result from complex interactions of various physical and chemical mechanisms [2, 4]. The battery degradation can be influenced by manufacturing characteristics or external conditions. Additionally, battery aging is impacted by specific usage conditions such as ambient temperature, charge/discharge current rates, charge/discharge cutoff voltages, depth of discharge (DOD), assembly pressure, and environmental humidity [6]. A detailed exploration of common battery aging mechanisms and key influencing factors is crucial for selecting appropriate predictive methods and developing more suitable and accurate battery health prediction models [3]. This facilitates the progressive utilization and safe management of batteries. In this section, we will focus on discussing the aging side reactions affecting the lifespan of lithium iron phosphate batteries and model key aging side reactions such as SEI layer growth on the anode surface, active material loss (LAM), and lithium plating.

In battery life prediction, both data-driven models and mechanistic models have their respective advantages and disadvantages [3]. While data-driven models often exhibit high accuracy, they require extensive feature extraction and input, along with multiple model trainings and parameter updates. Furthermore, they cannot effectively incorporate aging mechanisms and adapt to the deteriorative processes stemming from such mechanisms. In contrast, mechanistic models offer clear principles, explaining life degradation phenomena from first principles and reflecting changes in battery capacity through specific aging mechanisms [5]. Many mechanistic models are grounded in physics, considering fundamental governing equations of electrochemical reactions and detailed electrochemical parameters within batteries. They utilize comprehensive models to simulate changes at solid and liquid interfaces within batteries and their implications for battery capacity degradation. Although these comprehensive models provide detailed and precise analyses of battery performance variations, their computational demands pose challenges for practical implementation within energy storage systems.

The Single Particle Model (SPM) is a physics-based reduced-order model commonly employed to simplify battery models [7]. It utilizes two spherical particles to represent the anode and cathodes of the battery, while ignoring the spatial distribution of electrolyte concentration and potential. The SPM comprises ordinary differential equations derived directly from full-order electrochemical models, retaining the most crucial physical characteristics while significantly reducing computational complexity. As a result,

SPM has found widespread application and popularity in online real-time algorithms for battery health analysis [8]. In this chapter, we employ a single particle aging mechanism model to characterize the primary aging side reactions in batteries.

2.1 SEI Layer Growth Aging Mechanism

The Solid Electrolyte Interphase (SEI) layer's growth on a lithium-ion battery's anode is a key factor in battery degradation. This layer, crucial for normal battery function, forms a barrier on the anode's surface, composed of electrolyte byproducts and electrode material decomposition. It enables lithium-ion passage while blocking solvents and solutes, stabilizing the battery interface and extending lifespan. SEI layer formation occurs through anode–electrolyte solute reactions, depositing solid products like lithium salts and polymers. With battery use, solvent components diffuse through the SEI, thickening it and reducing anode porosity, leading to decreased ion de-intercalation rates and increased internal resistance, ultimately impairing battery performance and lifespan [9].

The growth rate of the SEI layer is influenced by input and environmental factors such as current, state of charge (SOC), and ambient temperature. To model the growth rate of SEI, one can construct a growth rate equation based on diffusion and reduction kinetics, as well as the Arrhenius empirical formula (8), i_{SEI} represented as follows:

$$i_{SEI}(t) = nFk_{SEI}C_s(R_n, t)\exp\left(-\frac{\alpha F}{RT}\eta_{SEI}\right) \tag{1}$$

where F is the Faraday's constant, n is the number of electrons reduced, kSEI is the rate constant of the SEI side reaction, ηSEI is the overpotential of the SEI growth side reaction, R is the ideal gas constant, α is the transfer coefficient, Rn represents the radius of the graphite particle. The diffusion of the reactant (S) and reduction product (P) are depicted by Eq. (2).

$$\begin{aligned} \frac{\partial C_s(r,t)}{\partial t} &= D_s\frac{\partial^2 C_s(r,t)}{\partial^2 r} \\ \frac{\partial C_p(r,t)}{\partial t} &= D_p\frac{\partial^2 C_p(r,t)}{\partial^2 r} \end{aligned} \tag{2}$$

The initial conditions indicating the homogeneity of the solution at t = 0, the semi-infinite boundary conditions, and the charge flux density boundary conditions are given by Eq. (3).

$$\begin{aligned} &C_s(r,0) = C_s^* \\ &C_p(r,0) = 0 \\ &\lim_{r\to\infty} C_s(r,t) = C_s^* \\ &\lim_{r\to\infty} C_p(r,t) = 0 \\ &-D_s\left(\frac{\partial C_s(r,t)}{\partial r}\right)_{r=R_s} = \frac{i_{sEI}}{nF} \\ &D_s\left(\frac{\partial C_s(r,t)}{\partial r}\right)_{r\to R_s} = -D_p\left(\frac{\partial C_p(r,t)}{\partial r}\right)_{r\to R_s} \end{aligned} \tag{3}$$

where C* s is the bulk concentration of the solvent reactant at the equilibrium state, Cp is the solvent concentration of the reduction product, Ds is the diffusion coefficient of the reactant product, and Dp is the diffusion coefficient of the reduction product. Rn represents the conditions at the interface of the active material and SEI layer. The SEI growth sub-model can be simplified to the form shown in Eq. (4).

$$Q_{SEI} = \int_0^t i_{SEI}(t)dt = \int_0^t \frac{nFC_s^*\sqrt{D_s}}{\sqrt{\pi t}(1+\lambda\theta)}dt = \int_0^t \frac{k_{SEI}}{2(1+\lambda\theta)\sqrt{t}}\exp(-\frac{E_{SEI}}{RT})dt \quad (4)$$

where $\lambda = C_s^*\sqrt{D_s} \Big/ C_p^*\sqrt{D_i}$, $k_{SEI} = \frac{2nFAC_s^*\sqrt{D_s^0}}{\sqrt{\pi}}$ and $E_{SEI} = \frac{E_s}{2}$ are parameters that need to be adjusted. While λ is determined by the bulk concentration and the diffusion coefficient of both the solvent reactant and reduction product, kSEI is related to the bulk concentration and the diffusion coefficient of the solvent reactant. T is the lumped battery temperature, while θ is computed by:

$$\theta = \frac{C_s(R_n,t)}{C_p(R_n,t)} = \exp\left(\frac{nF}{RT}\eta_{SEI}\right) = \exp\left[\frac{nF}{RT}\left(\eta_m + U_n^{ref} - U_{SEI}^{ref}\right)\right] \quad (5)$$

where η_m is the overpotential of the main battery reaction. U_n^{ref} is the equilibrium potential of the anode, which is the function of the battery SOC, and U_{SEI}^{ref} is the equilibrium potential of the SEI side reactions.

The thickness and resistance of the SEI layer would both enlarge with SEI formation and growth [10]. The growth rate of δ_{SEI} and R_{SEI} are expressed as $\frac{\partial \delta_{SEI}}{\partial t} = \frac{i_{SEI}M}{\rho n F}$, $\frac{\partial R_{SEI}}{\partial t} = \frac{1}{\sigma_{SEI}}\frac{\partial \delta_{SEI}}{\partial t}$, where M, σ_{SEI} and ρ are the molecular weight, ionic conductivity and density of the SEI film, respectively. ω_{SEI} is the volume fraction of SEI in the film. n is the number of exchanged electrons, and F is the Faraday constant.

2.2 LAM Aging Mechanism

The degradation of lithium-ion battery life is largely attributed to the loss of active material in the anode, a consequence of charging and discharging cycles. During discharge, lithium ions intercalate into the anode material, like graphite. Upon charging, they de-intercalate, forming lithiated graphite. This process causes volume changes and stress on the anode, leading to cracks in the solid electrolyte interphase (SEI) film. The continuous expansion and contraction expose new graphite surfaces, generating more SEI films. Stress from lithium-ion de-intercalation can cause particle fracture, reduced stability, and electrode damage. Fractured or detached anode material leads to more SEI film formation, depleting available lithium ions [11].

The stress from lithium-ion de-intercalation leads to a loss of anode active material, reducing lithium-ion attachment and reversible capacity, causing irreversible degradation. Material detachment raises the risk of current collector exposure to electrolyte, potentially corroding it with acidic substances, impacting battery safety. Thus, minimizing active material loss is vital for improving lithium-ion battery performance and safety.

Therefore, when modeling the LAM sub-model, it is crucial to consider the aging principle attributed to the linear accumulation of stress caused by volume expansion of active material particles correlated with the Depth of Discharge (DOD) during each cycle [12]. We can establish an LAM model for stress accumulation fracture based on the "*Basquin*" power law and *Miner's* rule. Through a mechanical model, we can establish the relationship between accumulated damage, D, and the stress amplitude, σ_{ampl}, within each cycle of the particle. The stress amplitude, $\sigma_{ampl,}$ can be computed using the maximum and minimum stresses experienced during each cycle.

$$\sigma_{ampl} = \frac{\sigma_{\max} - \sigma_{\min}}{2} \tag{6}$$

Based on the physical fatigue loss mentioned above, the accumulated material damage D is incorporated into the secondary reaction current of LAM, as shown in the following Eq. (7).

$$i_{LAM}(t) = k_{LAM}(\frac{\sigma_{h,\max}(R_n) - \sigma_{h,\min}(R_n)}{\sigma_{yield}})^{\frac{1}{m}} \tag{7}$$

where k_{LAM} is the factor that quantifies the LAM side reaction current, σ_{yield} is the stress at the maximum yield strength, and m is the slope of the '*Basquin*' power law. k_{LAM}, σ_{yield} and m are parameters to be calibrated from experiments. $\sigma_{h,\max}$ and $\sigma_{h,\min}$ are the maximum and minimum stress within a cycle, and σ_h is calculated by Eq. (8).

$$\sigma_h(R_n) = \frac{\sigma_r(R_n) + 2\sigma_t(R_n)}{3} \tag{8}$$

where σ_r (R_n) is the radial stress and σ_t $(R_n)$is the tangential stress on the surface of the spherical particle. σ_r $(R_n)$and σ_t (R_n) are derived from the core-shell stress-strain model.

2.3 Lithium Plating Aging Mechanism

Lithium plating, a critical issue in battery aging, occurs due to high C-rates, high State of Charge (SOC), low temperatures, and prolonged cycling. During charging, solvated lithium ions excessively deposit metallic lithium on the anode surface instead of intercalating into the material, forming dendrites or mossy structures. Continuous thickening of the Solid Electrolyte Interphase (SEI) film reduces anode porosity, slowing lithium-ion desorption and promoting plating. The cycle of lithium deposition reduces available ions, accelerates plating, and leads to rapid capacity decline and increased internal resistance.

Furthermore, the occurrence of lithium plating can be triggered when anode particles undergo fragmentation. Anode particle fragmentation reduces the capacity for lithium intercalation, causing excess lithium ions to be unable to embed within the graphite layer, thus leading to the deposition of metallic lithium on the surface of the anode [13]. If lithium dendrites deposited in the battery are excessively long and penetrate the separator, encountering the cathode, it can result in internal short circuits, leading to thermal runaway in the battery and severely compromising its safety.

Lithium plating is one of the primary reasons for the transition of battery capacity degradation from linear aging to nonlinear decline. This is because when the anode porosity decreases to a certain threshold, the rate of lithium plating sharply increases.

Lithium plating is charge transfer kinetics-controlled and can be described through the Butler-Volmer equation, as shown in Eq. (9).

$$j_{Li}(t) = j_{0,Li}\left[\exp\left(\frac{\alpha_{a,Li}F}{RT_{bat}}\eta_{Li}\right) - \exp\left(-\frac{\alpha_{c,Li}F}{RT_{bat}}\eta_{Li}\right)\right],\ \eta_{Li} \leq 0 \tag{9}$$

where $j_{0,Li}$ represents the exchange current density, α_a and α_c are the charge transfer coefficients for the anode and cathode, respectively. η_{Li} denotes the overpotential for lithium deposition reactions, as depicted in Eq. (10).

$$\eta_{Li} = \varphi_{s,n} - \varphi_e - U_{Li}^{ref} - R_{sel}i_{Li} \tag{10}$$

where $\varphi_{s,n}$ represents the solid-phase potential, φ_e is the electrolyte potential, and U_{Li}^{ref} is the equilibrium potential for lithium plating reactions. According to Eqs. (9) and (10), lithium plating occurs when $\varphi_{s,n}$ is lower than φ_e. Then, by utilizing Eqs. (9) and (10), the relationship functionΦ between the lithium plating current density j_{Li} and the anode potential difference $\varphi_{s,n} - \varphi_e$ can be derived, where $\varphi_{s,n} - \varphi_e$ is represented as shown in Eq. (11).

$$\varphi_{s,n} - \varphi_e = \eta_m + U_n^{ref} + \frac{R_{SEI}I}{Aa_{s,n}\delta_n} = \eta_m + U_n^{ref} + \frac{R_nR_{SEI}I}{3A\varepsilon_{s,n}\delta_n} \tag{11}$$

where $\varepsilon_{s,n}$ represents the solid-phase volume fraction in graphite particles, which decreases during active material losses, and $a_{s,n}$ is the specific surface area. Therefore, the rate of lithium plating can be calculated based on the charging current and State of Charge (SOC) of the battery, as described in Eq. (12).

$$i_{Li}(t) = k_{Li}A\varepsilon_{s,n}\delta_n\phi\left(\eta_m + U_n^{ref} + \frac{R_nR_{SEI}I}{3A\varepsilon_{s,n}\delta_n} - \varphi_{onset}\right) \tag{12}$$

where k_{Li} is a factor that corrects the lithium plating current causing capacity loss, and k_{Li} and φ_{onset} are calibration parameters.

2.4 Coupling of Aging Side Effects

The degradation of lithium-ion battery lifespan is a result of the coupled interactions among various aging side reactions. These aging side reactions interact with each other, leading to continuous deterioration of battery lifespan [4]. Among them, the growth of the Solid Electrolyte Interphase (SEI) layer reduces the porosity of the battery anode. When the local porosity decreases, the impedance during lithium insertion into the anode increases, leading to a decrease in the potential of the anode's local surface. This increases the likelihood of lithium plating side reactions occurring.

The growth of the SEI layer increases stress between anode particles due to the expanding particle radius. Concurrently, volume changes in the graphite layer during lithium-ion insertion and extraction may rupture the SEI layer lacking elasticity. SEI layer rupture and detachment, alongside localized anode material detachment due to adhesion, can induce Lithium Plating (LAM). Cracks or detachments expose new graphite layers, fostering new SEI layer formation. Active material loss reduces available lithium intercalation capacity, heightening the likelihood of lithium plating side reactions. These interactions collectively degrade lithium-ion battery lifespan.

Therefore, the anode particle radius R_s will increase along with the growth rate of the SEI layer thickness δ_{SEI}. The expansion of R_s further increases the stress σ_h. Similarly, the SEI layer impedance R_{SEI} positively influences the rate of lithium plating side reactions by increasing the overpotential of the SEI layer. Compared to SEI side reactions, the impact of LAM on lithium plating side reactions is much more severe. As the active material begins to crack, the solid-phase volume fraction $\varepsilon_{s,n}$ continuously decreases. The decreasing $\varepsilon_{s,n}$, which is in the denominator of the equation calculating the anode overpotential, sharply accelerates lithium deposition once the anode overpotential reaches the onset potential. Therefore, some parameters of this model, such as δ_{SEI}, R_{SEI} and $\varepsilon_{s,n}$ need to be updated promptly, as shown in Eq. (13).

$$\begin{aligned} \delta_{SEI}(t) &= \delta_{SEI}(t-1) + \frac{\partial \delta_{SEI}}{\partial t}\Delta t \\ R_{SEI}(t) &= R_{SEI}(t-1) + \frac{\partial R_{SEI}}{\partial t}\Delta t \\ \varepsilon_{s,n}(t) &= \frac{\varepsilon_{s,n0}[1 - Q_{LAM}(t)]}{\left(1 - \varepsilon_{s,n0}\right) + \varepsilon_{s,n0}[1 - Q_{LAM}(t)]} \end{aligned} \tag{13}$$

Coupling the three sub-models into a comprehensive lithium-ion battery remaining useful life prediction model, as depicted in Fig. 1.

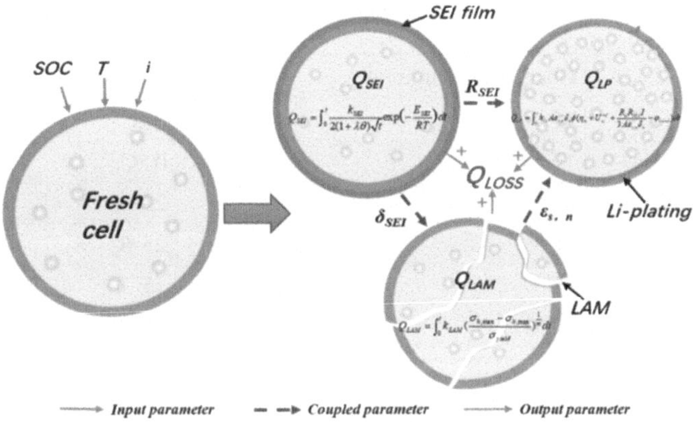

Fig. 1. Aging Mechanisms and Modeling

3 Experiment

3.1 Experimental Design

To investigate the performance of graphite-$LiFePO_4$ Li-ion Batteries under various operating conditions and the need for parameter calibration in the model, we selected prismatic lithium iron phosphate-graphite batteries produced by Gotion High-tech Co., Ltd.

for durability testing. The battery has a cut-off voltage range of 3.65 V to 2.0 V and a nominal capacity of 20 Ah. The durability testing equipment includes a testing computer, test bench, temperature chamber, and battery cell.

Indeed, conducting performance tests under diverse operating conditions provides valuable data for calibrating battery models, thereby improving prediction accuracy. Utilizing experimental data enhances model reliability, enabling better battery life forecasts. Durability experiments offer insights for optimizing battery design and management strategies, ultimately enhancing performance and lifespan to meet the requirements of different application fields more effectively.

We have designed aging experiments under different rates and temperatures, divided into cyclic aging and calendar aging groups, to investigate the influence of different test conditions on battery aging. In the cyclic aging group, each cell cycle consists of 100 full charge-discharge cycles, with reference performance test standard performance test (RPT) conducted at the conclusion of every 100 cycles. The RPT includes standard capacity tests, low-rate capacity tests, Hybrid Pulse Power Characterization (HPPC) tests, and Electrochemical Impedance Spectroscopy (EIS) tests at 50% state of charge (SOC). In the calendar aging group, each cell is left idle for 15 days as one cycle, followed by RPT tests at the end of each cycle.

The durability test arrangement plan is as shown in Tables 1 and 2. Table 1: Cycling Aging Group: Batteries 1–8 undergo cycling aging tests at different rates (1C and 2C) and different temperature gradients (0 °C, 25 °C, 55 °C).

Table 1. Cycling Aging Group

Cell ref	Charge/discharge rate	Temperature
1#	1C	0 °C
2#	1C	0 °C
3#	1C	25 °C
4#	1C	25 °C
5#	1C	55 °C
6#	1C	55 °C
7#	2C	55 °C
8#	2C	55 °C

Calendar Aging Group: Batteries 9–18 undergo calendar aging tests at different temperature gradients (25 °C, 40 °C, 55 °C) and different state-of-charge (SOC) intervals (0%, 50%, 100%). Two batteries within each group are set to the same test conditions to avoid the occurrence of random events.

Table 2. Calendar Aging Group

Cell ref	搁置SOC	环境温度
9#	0%	25 °C
10#	0%	25 °C
11#	50%	25 °C
12#	50%	25 °C
13#	100%	25 °C
14#	100%	25 °C
15#	50%	40 °C
16#	50%	40 °C
17#	50%	55 °C
18#	50%	55 °C

3.2 Analysis of Experimental Results

According to the durability test results, Fig. 2 displays the capacity degradation curves of lithium iron phosphate batteries under 18 different operating conditions. It is observed that within the cycling aging group, the cell 4# maintains optimal capacity under room temperature (25 °C) test conditions, with capacity decreasing to below 80% only after experiencing 7–900 cycles. However, performance of cell 1# is poorest at low temperature (0 °C), with significant capacity degradation occurring after just a few dozen cycles, possibly due to the lithium plating side reaction discussed in our second section. Additionally, increasing the test temperature and increasing the current rate during charging and discharging will both accelerate the battery aging rate.

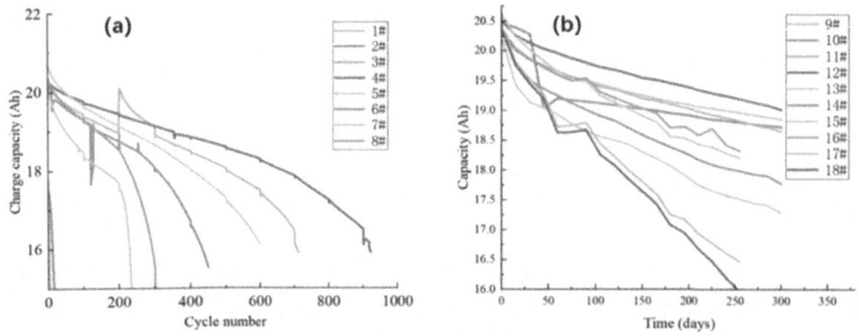

Fig. 2. The capacity retention of batteries under various operating conditions: (a) Capacity retention in the cycling aging group. (b) Capacity retention in the calendar aging group.

In the calendar aging group, it was observed that batteries stored at moderate to low state of charge (SOC) levels (0%, 50%) at room temperature exhibited slower aging rates, with less capacity degradation over 300 days, and minimal differences. Conversely, batteries stored at 100% SOC showed faster capacity degradation rates, accelerating with increasing test environment temperatures. It's worth noting that at 25 °C, the capacity degradation rate of batteries at 100% SOC for cells 13# and 14# was faster than that of cells 15# and 16# stored at 50% SOC at 40 °C. However, at even higher temperatures (55 °C), cells 17# and 18# exhibited the fastest capacity degradation rates.

3.3 Comparison Between Experimental Results and Model Output Results

Calibrating the parameters in the degradation model can significantly improve the model's ability to accurately predict the remaining capacity of the battery [14]. The parameters used internally in the model are obtained through estimation, consistent parameters in the literature, and assumptions used in previous work. The inherent parameters and calibration parameters are listed in reference [8] and Table 3, respectively.

Table 3. Calibrated parameters in the reduced-order model

Parameter	Value	Parameter	Value
λ	86995	k_{LAM}	1.61
σ_{yield}	787.41 MPa	k_{Li}	4.2
k_{SEI}	15.55 $s^{-1/2}$	E_{SEI}	24500 J/mol
m	0.23	φ_{onset}	−5.5 mV

By adjusting parameters, the reduced order mechanism model can accurately predict the remaining useful life of batteries. Figure 3(a) (e) shows a good match between the capacity retention results of the laboratory aging experiment (25 °C, 1C) and the output results of the calibrated reduced order single particle mechanism model. Figure 3(b) (f) shows the error graph between them, and the model output results keep the error within 2%, fully demonstrating the accuracy of the model. Figure 3(c) (g) shows the results of capacity loss caused by each of the three sub models under this input condition, which corresponds to the inherent aging mechanism of the battery.

4 Verification of Reduced Order Model and Aging Mechanism

Through electron microscopy observation, changes in the internal structure of the battery can be directly observed, including the morphology of electrode materials, the formation and changes of SEI films, etc. Thus, providing a more intuitive understanding of the changes that occur during the aging process. By comparing with actual observation results, the accuracy of the theoretical model can be verified, and the model can be further improved and optimized to better predict and understand the aging mechanism of batteries. In this chapter, we mainly verify the aging mechanism of batteries after their end of life by disassembling them, to verify the interpretability of the mechanism

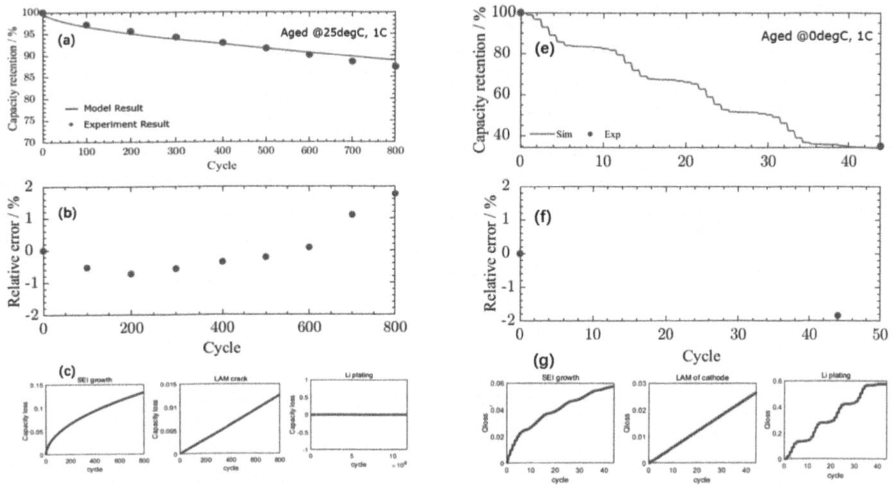

Fig. 3. Aging test results and model output results: (a) (e) Comparison between the capacity retention results of 4 # cell (25 °C, 1C and 0 °C, 1C) and the output results of the mechanism model; (b) (f) Error between experimental and model results; (c) (g) The output results of the three sub models in the reduced order model.

of this reduced order model. Destructive validation involves disassembling the battery and using characterization methods such as material physical and electrochemical analysis to understand the degradation mechanisms of the cathode and anode materials. Validation methods include SEM (Scanning Electron Microscopy), TEM (Transmission Electron Microscopy), FIB (Focused Ion Beam), and EDS (Energy-Dispersive X-ray Spectroscopy), among others.

4.1 Verification of SEI Layer Growth Sub-Model

The growth of the Solid Electrolyte Interphase (SEI) layer cannot be directly observed in-situ and requires disassembly and analysis of the cell after reaching the end of its lifespan. Transmission Electron Microscopy (TEM) allows for the observation of the nanoscale structure of materials inside the battery, enabling analysis of its microstructural changes. In this work, we utilized TEM characterization tests to observe the growth of the anode SEI layer under different operating conditions. During the testing and analysis, we could clearly observe that the anode graphite exhibited a distinct layered structure, while the SEI layer on the graphite surface showed a relatively irregular structure. By evaluating the thickness of the SEI layer structure, we can conduct a semi-quantitative analysis of the SEI layer growth secondary reactions.

Figure 4 shows the TEM results of the anode of fresh LFP batteries and end-of-life batteries under different operating conditions. Comparing Fig. 4(a), it can be observed that the SEI layer of the fresh cell is relatively thin, approximately 2–3 nm. For the 4# cell cycled at 1C and 25 °C, although the SEI layer grows slowly, due to good capacity retention, the cycle life exceeds 900 cycles, with a sufficient usage time resulting in an SEI layer thickness of approximately 8–10 nm at end-of-life (Fig. 4(b)). However, for the

6# cell cycled at 1C and 55 °C, temperature accelerates the growth of the SEI layer, with end-of-life reached after approximately 500 cycles, resulting in an SEI layer thickness of around 6 nm (Fig. 4(c)). In contrast, for the 8# cell cycled at 2C and 55 °C, although the higher rate accelerates SEI layer growth, the cycle life is shorter, reaching end-of-life after only 300 cycles. Consequently, the SEI layer thickness at end-of-life is relatively thinner, approximately 4.6 nm (Fig. 4(d)). Although TEM characterization results may not directly correspond to the output of the mechanism model, the relative trends of SEI layer growth under different conditions are consistent. This characterization test also validates the rationality of the SEI layer growth sub-model within the degradation model.

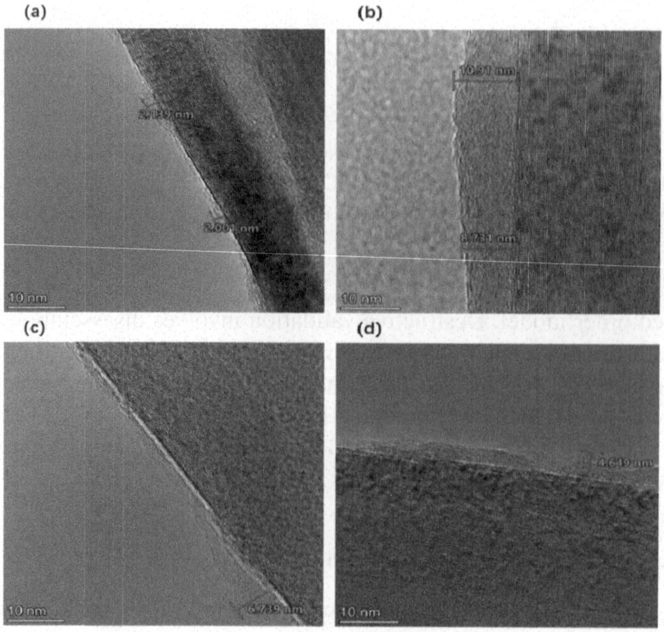

Fig. 4. The TEM results of the anode of fresh LFP batteries and end-of-life batteries under different operating conditions: (a) fresh LFP cell. (b) LFP cell #4. (c) LFP cell #6. (d) LFP cell #8.

4.2 Verification of Lithium Plating Sub-Model

We employed SEM to analyze the disassembled batteries and determine if there were any occurrences of lithium plating on the anode surface. Figure 5 presents the SEM results of LFP batteries, both fresh cell and at the end of life under different conditions. In Fig. 5(a), the fresh cell, serving as the reference, exhibited clear graphite particles without any lithium plating, indicating no cycling had occurred. However, for cell #1 cycled at 1C and 0 °C, SEM results revealed a significant moss-like lithium plating covering the graphite surface, as shown in Fig. 5(b), which aligned perfectly with the predictions of the mechanistic model's lithium plating sub-model. In contrast, for cell #5 cycled at 1C and 55 °C, SEM observation did not reveal any significant lithium plating, as showed in Fig. 5(c), consistent with the conclusions drawn from the mechanistic model's lithium plating sub-model. Regarding LFP cell #8 cycled at 2C and 55 °C, it's noteworthy that this cell experienced capacity fade during durability testing rather than following a linear aging trend. SEM results indicated the presence of numerous dendritic lithium plantings on the graphite surface, as depicted in Fig. 5(d). This suggests that under high-temperature, high-rate cycling conditions, the sharp decline in anode capacity leads to a change in the cathode-to-anode ratio, triggering lithium plating and further exacerbating capacity fade. The significant lithium plating observed in the later stages of cycling under this condition corresponds well with the predictions of the mechanistic model's lithium plating sub-model.

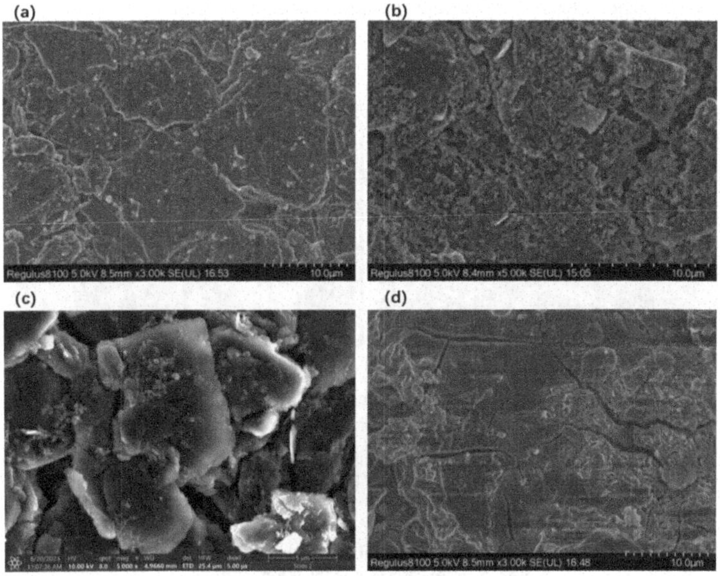

Fig. 5. The SEM results of the anode of fresh LFP batteries and end-of-life batteries under different operating conditions: (a) fresh LFP cell. (b) LFP cell #1. (c) LFP cell #5. (d) LFP cell #8.

4.3 Verification of LAM Sub-Model

This work, we used the methods of disassembled scanning electron microscopy (SEM) and ion beam cutting scanning electron microscopy (FIB-SEM) to qualitatively observe the peeling of anode graphite. Graphite exfoliation typically occurs during battery cycling due to stress accumulation and fatigue damage induced by solvent-mediated lithium-ion intercalation, with the severity closely correlated with cycle count. Through SEM and FIB-SEM observations, the extent of graphite exfoliation was qualitatively assessed, and the operational conditions with the most severe exfoliation were identified. These observational findings will be compared and validated against mechanistic models to enhance understanding of the mechanisms underlying graphite exfoliation and its relationship with cycle count.

Based on the SEM characterization results in Fig. 5, it can be observed that there is no significant anode graphite exfoliation in the newly assembled lithium iron phosphate batteries and most end-of-life batteries under various operating conditions. However, noticeable graphite electrode exfoliation was observed in batteries #3 and #4. These two batteries underwent 1C cycling at 25 °C and exhibited the highest number of cycles and longest duration in the durability test, thus validating each other with the assumption in the mechanistic model regarding the relationship between graphite exfoliation and cycle count. Figure 6 illustrates the graphite electrode exfoliation in these two batteries. Further investigation involved obtaining cross-sections from regions with prominent graphite exfoliation and conducting FIB-SEM characterization, as shown in Fig. 6. It can be observed that the outer layers of the graphite particles indeed exhibit flake-like exfoliation, while the internal structure remains relatively intact.

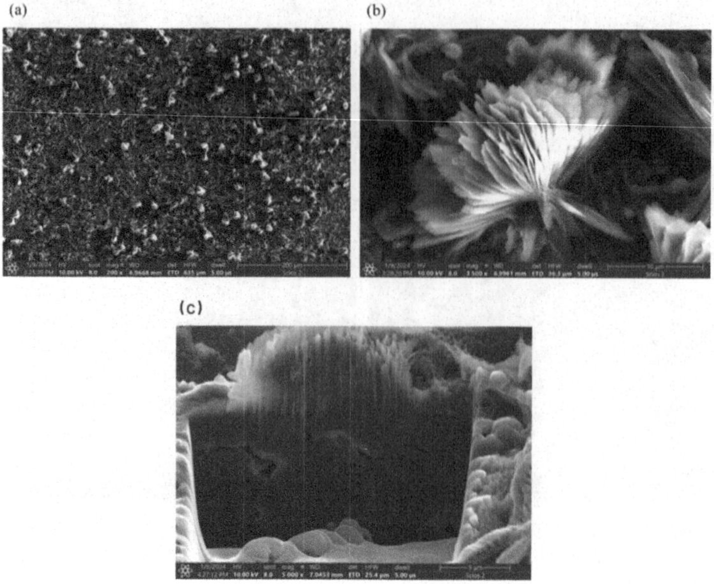

Fig. 6. SEM results of the anode of LFP cell 4 #. (a) Low magnification results. (b) High magnification results. (c) FIB-SEM results.

5 Conclusions

In conclusion, this work comprehensively studied the aging mechanisms of LFP batteries and developed a single-particle aging mechanism model to predict battery life. The model is based on the electrochemical aging mechanisms of LFP batteries, including aging side reactions such as SEI layer growth, loss of active material, and lithium plating, and the proposed model demonstrates high accuracy and interpretability. Validation through durability experiments under different operating conditions and disassembly electron microscopy characterization further confirms the reliability of the model. This research provides a vital theoretical foundation for optimizing battery design and management strategies, thereby enhancing battery performance and lifespan, and advancing energy storage system development. Future research could explore additional factors influencing battery life and optimize the model for improved prediction accuracy. Furthermore, applying the model to practical energy storage systems would further validate its effectiveness and feasibility in real-world applications. Overall, this work contributed to the advancement of battery management technology and established a foundation for more realistic and efficient optimization of battery performance in various application fields.

References

1. Lai, X., Chen, Q., Tang, X., et al.: Critical review of life cycle assessment of lithium-ion batteries for electric vehicles: A lifespan perspective. Etransportation **12**, 100169 (2022)
2. Wang, X., Wei, X., Zhu, J., et al.: A review of modeling, acquisition, and application of lithium-ion battery impedance for onboard battery management. ETransportation **7**, 100093 (2021)
3. Reniers, J.M., Mulder, G., Howey, D.A.: Review and performance comparison of mechanical-chemical degradation models for lithium-ion batteries. J. Electrochem. Soc. **166**(14), A3189–A3200 (2019)
4. Edge, J.S., O'Kane, S., Prosser, R., et al.: Lithium ion battery degradation: what you need to know. Phys. Chem. Chem. Phys. **23**(14), 8200–8221 (2021)
5. Wang, S., Guo, D., Han, X., et al.: Impact of battery degradation models on energy management of a grid-connected DC microgrid. Energy **207**, 118228 (2020)
6. Han, X., Lu, L., Zheng, Y., et al.: A review on the key issues of the lithium ion battery degradation among the whole life cycle. ETransportation **1**, 100005 (2019)
7. Li, J., Landers, R.G., Park, J.: A comprehensive single-particle-degradation model for battery state-of-health prediction. J. Power Sources **456**, 227950 (2020)
8. Wei, Y., Wang, S., Han, X., et al.: Toward more realistic microgrid optimization: experiment and high-efficient model of Li-ion battery degradation under dynamic conditions. ETransportation **14**, 100200 (2022)
9. Li, J., Adewuyi, K., Lotfi, N., et al.: A single particle model with chemical/mechanical degradation physics for lithium ion battery State of Health (SOH) estimation. Appl. Energy **212**, 1178–1190 (2018)
10. Li, D., Danilov, D.L., Zwikirsch, B., et al.: Modeling the degradation mechanisms of C6/LiFePO4 batteries. J. Power Sources **375**, 106–117 (2018)
11. Zhang, Y.Z., Xiong, R., He, H.W., et al.: Aging characteristics-based health diagnosis and remaining useful life prognostics for lithium-ion batteries. ETransportation **1**, 100004 (2019)
12. Fernández, I.J., Calvillo, C.F., Sánchez-Miralles, A., et al.: Capacity fade and aging models for electric batteries and optimal charging strategy for electric vehicles. Energy **60**, 35–43 (2013)
13. Crawford, A.J., Choi, D., Balducci, P.J., et al.: Lithium-ion battery physics and statistics-based state of health model. J. Power Sources **501**, 230032 (2021)
14. Pang, H., Mou, L., Guo, L., et al.: Parameter identification and systematic validation of an enhanced single-particle model with aging degradation physics for Li-ion batteries. Electrochim. Acta **307**, 474–487 (2019)

Semi-supervised State of Health Estimation for Lithium-Ion Batteries via Dynamical Self-encoding

Yifan Shen[1], Xuyang Liu[1], Xingyu Zhou[2], Xuebing Han[2], Yuedong Sun[1], Dongxu Guo[2], and Yuejiu Zheng[1,2](✉)

[1] University of Shanghai for Science and Technology, Shanghai 200093, China
yuejiu.zheng@usst.edu.cn

[2] Tsinghua University, Beijing 100084, PR China

Abstract. Accurate estimation of the State of Health (SOH) of lithium-ion batteries (LiBs) is crucial for performance optimization and safety. However, traditional health features (HFs) extraction methods often fail to capture key HFs in the complex charging data of LiBs, resulting in low model accuracy. The primary contribution of this study is that we successfully derived the HFs of LiBs directly from the external information of the original charging data using a semi supervised architecture and achieved estimation of SOH. To address this issue, we propose a semi-supervised learning approach utilizing Variational Autoencoders (VAE) with Gated Recurrent Unit (GRU). The VAE encoder extracts HFs from unlabeled charging data, enhancing the understanding of data patterns. These HFs are then combined with labeled SOH data to train downstream regression networks for estimation. Experimental results demonstrate that our method performs excellently in practical electric vehicles (EVs), with a maximum Mean Absolute Percentage Error (MAPE) not exceeding 5%. This VAE based semi supervised technique provides a promising solution for SOH estimation of LiBs, demonstrating excellent accuracy and robustness.

Keywords: Lithium-ion Batteries · Electric Vehicles · State of Health · Semi-supervised Learning · Variational Autoencoders (VAE)

1 Introduction

In recent years, the rise of electric ships and electric vehicles (EVs) has played a significant role in social development and environmental protection. Lithium-ion batteries (LiBs) have been widely used and developed as energy storage devices with recyclability, high efficiency, and long lifespan [1–3]. At the same time, the popularity of EVs has brought revolutionary changes to urban transportation methods. However, LiBs use lead to capacity fade [4]. In this context, the estimation of State of Health (SOH) of LiBs is of great significance, as it can help formulate reasonable energy plans and resource allocations to ensure the full environmental and economic benefits of electric boats and new EVs in practical operations.

K. Li et al. (Eds.): LSMS/ICSEE 2024, CCIS 2218, pp. 192–203, 2025.
https://doi.org/10.1007/978-981-96-0232-2_15

The capacity of LiBs is usually considered one of the key indicators of SOH, but it cannot be directly obtained through vehicle sensors. In recent years, with the development of machine learning technology and the large increase in real vehicle data, machine learning algorithms for LiBs' SOH estimation have become one of the mainstream methods [5]. Machine learning-based methods have received considerable attention due to their high accuracy and model independence. Currently, SOH estimation based on maximum likelihood estimation is mainly divided into two steps: first, extracting health features (HFs) reflecting the aging degree from charging data and establishing the relationship between these features and SOH. For example, Song et al. [6] proposed an intelligent SOH estimation framework using actual data collected from a big data platform. Zheng et al. [7] proposed a capacity prediction framework based on a feedforward empirical model and feedback data-driven method. Zhou et al. [8] established a neural network model for estimating the battery capacity of LiFePO4 under multi-stage variable-current fast charging data based on protocol analysis and mechanism research. Hong et al. [9] considered environmental factors and used actual vehicle operation data to predict the battery voltage of electric vehicles based on the Gated Recurrent Unit (GRU) method.

Traditional machine learning-based capacity estimation methods face challenges such as sparse labeled data and difficulty in feature extraction, leading to difficulties in supervised learning and capturing the complex nonlinear relationships of LiBs, thereby limiting the accuracy of capacity estimation. Most machine learning models produce black-box battery state predictions that are difficult to generalize to other battery chemistry fields [10]. However, with the advancement of semi-supervised and unsupervised learning techniques, a new SOH estimation method has gradually become popular. For example, Guo et al. [11] developed a semi-supervised learning technique called partial Bayesian joint training, which enhanced the modeling capability of battery life prediction by extracting hidden information from unlabeled data. Wu et al. [12] designed convolutional autoencoders and recursive autoencoders to automatically extract health features and improved the accuracy of SOH estimation for laboratory data through ensemble learning methods. Qiao et al. [13] proposed a laboratory-to-field capacity estimation framework, which achieved accurate and robust capacity estimation using low-cost fine-tuning techniques and considering various uncertainties. Chen et al. [14] combined denoising autoencoder neural network (DAE-NN) with Gated Recurrent Unit - Recurrent Neural Network (GRU-RNN) and proposed a new battery SOC estimation method DAE-GRU. In summary, hybrid modeling based on self-supervised learning and combining physical models has not only reduced the dependence on labeled data but also improved the accuracy and stability of capacity estimation [15].

This study aims to introduce semi-supervised machine learning methods, especially methods based on the VAE architecture, to address the challenges of sparse labeled data and feature extraction in SOH estimation. We propose a new method that utilizes VAE to extract features from unlabeled charging data and combines it with downstream sub models to achieve accurate estimation of SOH. The significance of this method lies in exploring the potential of semi-supervised learning methods in solving capacity estimation problems. Through this research, we provide a more efficient solution for future battery management and make important contributions to the field of energy storage.

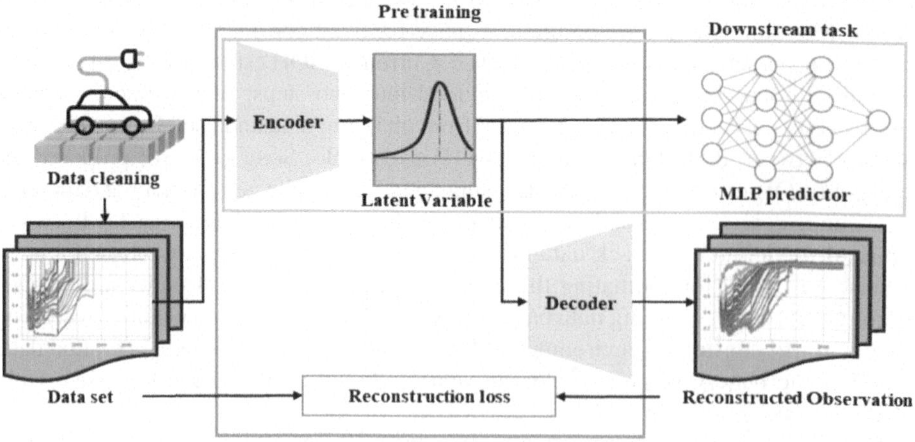

Fig. 1. The overall framework of the proposed method.

2 Dataset Overview

Our dataset consists of one hundred identical model EVs, and all Battery Management System (BMS) recorded data and SOH labels have been validated by the Original Equipment Manufacturer (OEM) to ensure their accuracy. These vehicles have a maximum mileage of up to 260,000 km, exhibiting a noticeable trend of health deterioration over long-term use, as visualized in the heatmap of Fig. 2 (a). The data originates from real-time charging data collected by the BMS. Data from each vehicle is sampled every 10 s. In this approach, we have selected key parameters during the charging process, including voltage, current, State of Charge (SOC), maximum single-cell voltage, minimum single-cell voltage, maximum and minimum temperature of the battery pack, and timestamps. For ease of subsequent processing, we have sliced and trimmed these data into sequences of 128-time steps, equivalent to approximately 21 min of charging, as illustrated in Fig. 2 (b), this also to some extent expanded our dataset. Ultimately, through data cleaning, slicing, and augmentation methods, a total of 349,741 samples were obtained from these 100 vehicles.

For the partitioning of the processed dataset, we have chosen to organize it on a per-vehicle basis. Specifically, we have selected 4 vehicles with the longest mileage from the one hundred as the validation set, while allocating 76 vehicles for the training set and 20 vehicles for the testing set, the ratio of the training set to the test set reached 8:2. This partitioning scheme aims to ensure that the validation set covers a wide range of mileage and maintains sufficient data diversity for training and testing purposes.

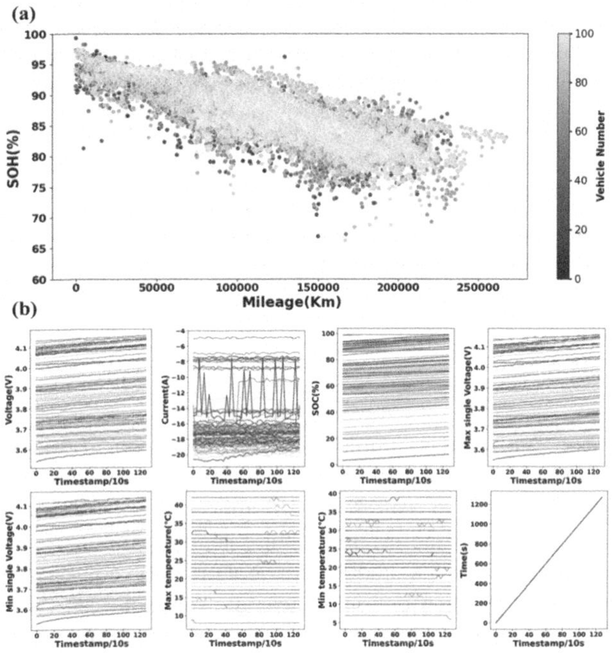

Fig. 2. Introduction to datasets. (a) Heatmap visualization of SOH labels for all vehicles; (b) Samples obtained after performing operations such as cleaning and slicing on the raw data.

3 Health Features Extraction

3.1 Variational Autoencoders

Variational Autoencoder (VAE) is a generative model used for learning the latent representations, also known as latent variables, of data. VAE has wide applications in image generation, data reconstruction, and unsupervised learning, providing effective solutions to various machine learning and deep learning problems. In this section, we will introduce the basic structure and working principle of VAE.

VAE consists of two core components: the encoder and the decoder. The encoder′s role is to map input data to a probability distribution in the latent space, while the decoder maps points sampled from the latent space back to the original data space. The encoder is typically composed of a multi-layer neural network, which maps input data to the mean vector and variance vector in the latent space. Its specific structure is illustrated in Fig. 3.

Assuming the input data is X, the encoder′s output is the mean vector Z_{mean} and the variance vector Z_{logver} in the latent space, which can be seen as abstract representations of the data, and the equations are as follows:

$$Z_{mean}, Z_{mean} = Encoder(X) \tag{1}$$

By sampling from these abstract representations, we can generate new data with features similar to the original data. Therefore, these latent representations not only contain the main features of the original data but also exhibit continuity and interpretability, making them suitable for various machine learning tasks such as classification, clustering, and dimensionality reduction. Subsequently, we sample latent vectors from this probability distribution, the equations are as follows:

$$Z = Z_{mean} + \varepsilon \times e^{0.5 \times Z_{logver}} \tag{2}$$

where ϵ is a random vector sampled from the standard normal distribution $N(0,1)$.

The decoder, also a multi-layer neural network, is responsible for mapping the latent vector Z back to the original data space. The decoder′s output is the reconstructed data X':

$$X' = Decoder(Z) \tag{3}$$

The training process of VAE is completed by maximizing the Evidence Lower Bound (ELBO), which consists of two parts: the reconstruction loss and the regularization term. The reconstruction loss measures the difference between the reconstructed data generated by the decoder and the original data, usually using cross-entropy or mean squared error. This implies that the model learns not only how to generate reconstructed data but also how to make the distribution of the latent space close to the standard normal distribution, making the latent vectors more continuous and interpretable, and thus more suitable as a feature representation of the original data. The regularization term penalizes the difference between points in the latent space and the standard normal distribution, encouraging the latent space to have better continuity and smoothness.

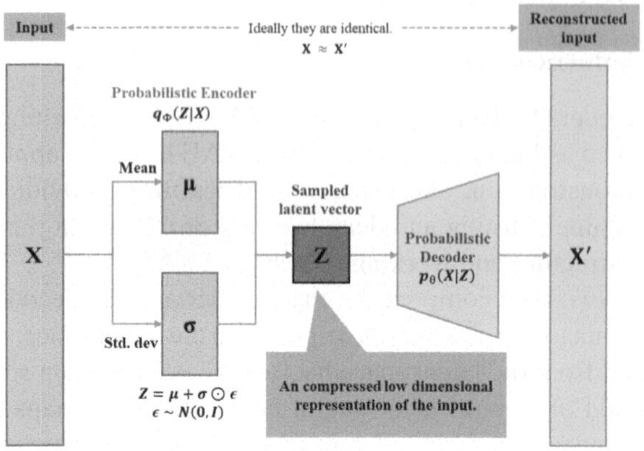

Fig. 3. Overview of VAE Architecture.

3.2 Structure of the GRU Network

Gated Recurrent Unit (GRU) is a variant of recurrent neural networks (RNNs) commonly used for processing sequence data. In this section, we will introduce the structure and characteristics of GRU to better understand its application in our model. Its structure is illustrated in Fig. 4.

Similar to traditional recurrent neural networks, GRU introduces update gates and reset gates to control the flow of information and address the issue of long-term dependencies. Specifically, GRU consists of a hidden state and two gate control units: the update gate and the reset gate. The update gate determines how much information from the previous time step is integrated into the current hidden state. It decides how much of the information from the previous time step is retained and how much new information is accepted. The output of the update gate ranges between 0 and 1, where 0 indicates complete disregard of the previous hidden state and 1 indicates full retention of the previous hidden state. The reset gate determines how much information from the previous time step's hidden state is ignored. It helps the model adapt to different time scales within sequences, thereby enhancing the model's ability to model long sequences. The characteristics of GRU give it an advantage in processing time-series data. Compared to traditional RNNs, GRU can better capture long-term dependencies with fewer parameters and is easier to train. Additionally, GRU's structure is relatively simple, making it computationally efficient and suitable for large-scale data processing.

In our model, we select GRU as the network structure for both the encoder and decoder to better handle input time-series data, extract effective feature representations, and generate accurate reconstruction data.

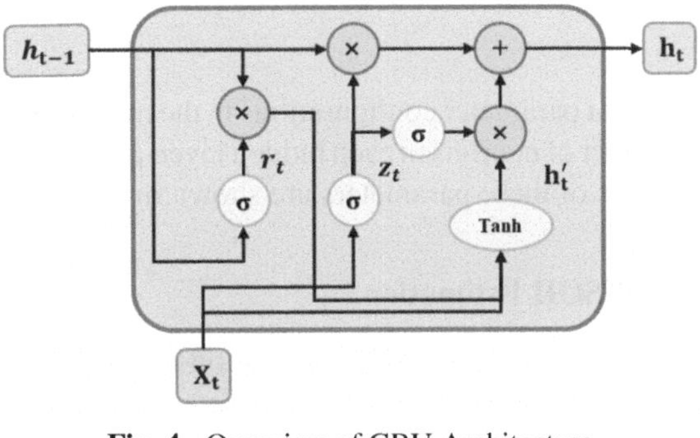

Fig. 4. Overview of GRU Architecture.

4 SOH Estimation Method

4.1 Structure of the MLP Network

Multilayer Perceptron (MLP) is a supervised learning model based on artificial neural networks, commonly used for tasks such as classification and regression. Specifically, MLP learns the relationships between features and patterns in the data to predict and classify labels in the datasets. In our study, MLP is employed as a downstream task, primarily tasked with training using the HFs extracted by VAE and the SOH labels from the datasets.

The specific structure of MLP is illustrated in Fig. 5. It typically consists of one or more hidden layers, with each hidden layer comprising multiple neurons (or nodes). Each neuron is connected to all neurons in the previous layer and has a bias term and an activation function. The neurons in the input layer receive the features extracted by VAE as input, while the neurons in the output layer output the predictions for the labels in the datasets.

Table 1. Hyperparameters of MLP.

Hyperparameter	Value
Input Layer	24
First Hidden Layer	128
Second Hidden Layer	32
Third Hidden Layer	16
Output Layer	1
Number of Hidden Layers	3
Epochs	72
Activation Function	ReLU
Optimizer	Adam

We used a specific set of parameter configurations in the model, including the number of hidden layers, the number of neurons in each hidden layer, and the learning rate, among others. The specific values of these parameters are shown in Table 1.

4.2 The Framework of SOH Estimation

The overall framework of the proposed SOH estimation method is illustrated in Fig. 1, which can be divided into three steps. Firstly, the dataset of EVs is cleaned, sliced, and subjected to data augmentation and normalization, and then divided into training, testing, and validation sets. Next, an unsupervised VAE-GRU method was used to pre train unlabeled samples, and the pre trained encoder was used to automatically extract HFs from the samples. Finally, by retraining the downstream task sub model MLP connected to the encoder, the estimation of SOH can be achieved in just 21 min of charging time.

5 Results and Discussions

5.1 Network Training and Evaluation Index

In this section, we will describe the metrics used for training and evaluating the model performance, focusing on the comparison between predicted results and actual results. We adopt four commonly used evaluation metrics, namely Mean Absolute Percentage

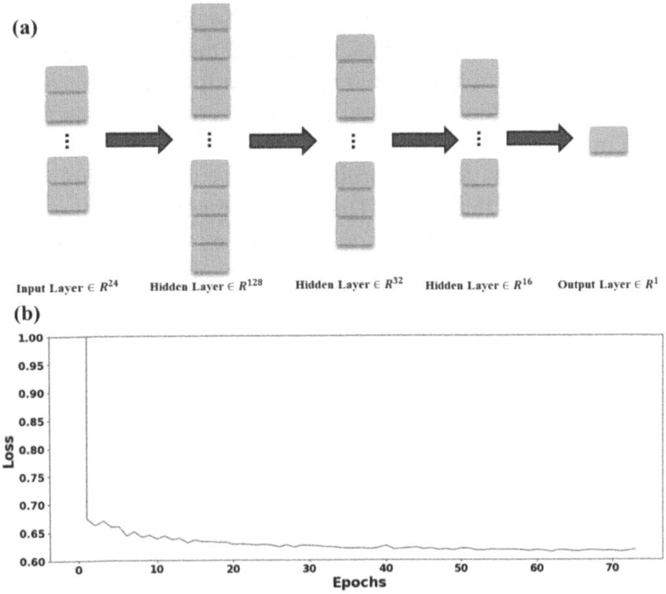

Fig. 5. Faulty module experiment connection and schematic.

Error (MAPE), Mean Absolute Error (MAE), Mean Squared Error (MSE), and Root Mean Squared Error (RMSE), They are defined as follows.

$$MAPE = \sqrt{\frac{1}{n}\sum_{i=1}^{n}(y_i - y_i')^2} \tag{4}$$

$$MAE = \frac{1}{n}\sum_{i=1}^{n}|y_i - y_i'| \tag{5}$$

$$MSE = \frac{1}{n}\sum_{i=1}^{n}\frac{|y_i - y_i'|}{y_i} \tag{6}$$

$$MSE = \frac{1}{n}\sum_{i=1}^{n}(y_i - y_i')^2 \tag{7}$$

These metrics help us evaluate the predictive performance of the model and understand the degree of difference between the predicted values and the actual values. Smaller values of error metrics indicate that the model′s predicted results are closer and more accurate to the actual results.

5.2 Health Features Extraction with the VAE-GRU

In the third section, we introduced the characteristics and capabilities of VAE, and chose GRU as both encoder and decoder. This choice enables us to extract features from preprocessed EV data recorded by BMS. Each original data is transformed into a matrix with 7 channels and 128-time steps, and then dimensionality-reduced to 24 HFs through the encoder. This section will validate whether these features are representative and distinguishable. We randomly selected a set of decoding results from the training dataset, as shown in Fig. 6.

Clearly, the decoded 7 sets of reconstructed curves can accurately track the original curves, with RMSE all less than 0.01. The results indicate that the extracted health features can be accurately restored to charging curves by the corresponding decoder. In other words, these features can represent the charging curves.

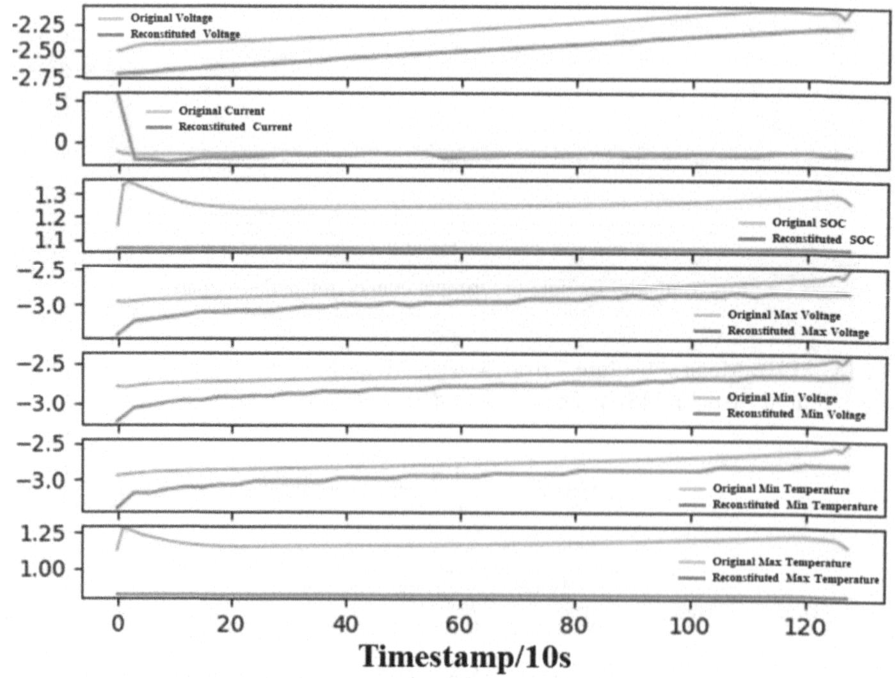

Fig. 6. Pre training results of VAE-GRU.

5.3 SOH Estimation with the MLP Network

After extracting 24 HFs from VAE-GRU, we estimated the SOH of EVs in four validation sets using MLP to validate the performance of the combined method proposed in this section. The SOH estimation results for the four EVs based on the proposed method are shown in Fig. 7. It can be observed that all four EVs have traveled over 100,000 km. Prior to this, the SOH labels certified by the OEM show a certain degree of degradation, as indicated by the dark blue dots. The estimated SOH using our method is represented by light blue dots. Due to preprocessing of the data using the slicing method, individual estimated points exhibit multiple estimation results. It is evident that significant errors tend to occur at locations where the SOH undergoes abnormal jumps. However, the SOH estimated by our method maintains a reasonable degradation trend, with the majority of relative errors concentrated within 3%.

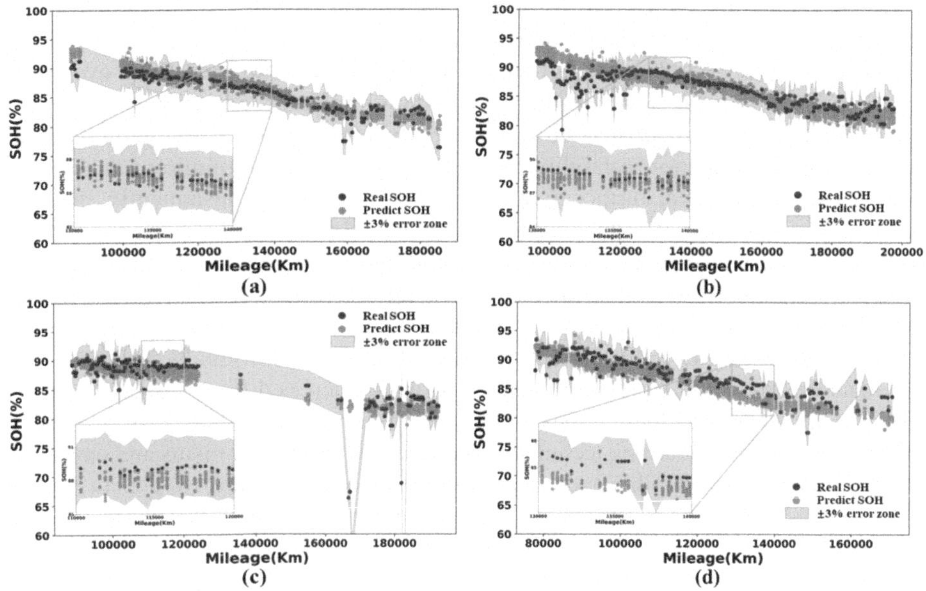

Fig. 7. The SOH estimation results for four EVs (a) EV1, (b) EV2, (c) EV3, (d) EV4.

The estimation errors for the four EVs are presented in Table 2. These results demonstrate that the proposed method performs well in estimating SOH for EVs. Significantly, our data is obtained through slicing, with each segment of original data spanning only 21 min of charging time. This underscores that under our methodology, an accurate estimation of SOH can be achieved with just 21 min of charging data.

Table 2. Comparison of Validation Set Predictions with Labels Using Four Metrics.

	MAPE	MAE	MSE	RMSE
EV1	1.0020	0.0087	0.0001	0.0127
EV2	1.0009	0.0129	0.0003	0.0191
EV3	1.0026	0.0193	0.0019	0.0437
EV4	0.9943	0.0168	0.0004	0.0201

6 Conclusions

This paper proposes a novel semisupervised learning method for estimating SOH in EV's lithium-ion batteries, based on VAE and MLP. Extracting HFs using VAE-GRU encoder and combining it with MLP for SOH estimation, we achieve accurate estimation of battery health status. Experimental results demonstrate the effectiveness of our method across different vehicles, particularly in accurately estimating SOH even with short charging durations. However, there are certain limitations and drawbacks to this method. Firstly, our approach may be sensitive to data fluctuations under certain conditions, resulting in estimation instability. Secondly, this method still requires a substantial amount of labeled data for further training of the sub-models. Consequently, the estimation accuracy of models trained with limited labeled data may be suboptimal. This issue might be addressed in the future through fine-tuning techniques. Lastly, each sample input to the current model requires a minimum record length of 21 min. In cases where the data volume is small, the estimation accuracy may be compromised, thereby limiting the robustness of SOH estimation under this model. Additionally, there are concerns regarding model interpretability and performance across different datasets. In the future, we aim to address these issues and further improve our method. Firstly, we will explore advanced deep learning techniques such as attention mechanisms and ensemble learning to enhance the stability and accuracy of the model. Secondly, we plan to expand the dataset size and introduce more environmental conditions and workloads to enhance the model's generalization ability. Lastly, we will attempt to achieve the same or even better SOH estimation performance by using fine-tuning and linear probe techniques, utilizing only 20% or even less of the currently labeled data. Overall, we believe that the semisupervised learning approach based on VAE and MLP provides an innovative and feasible solution for EV lithium-ion battery state estimation, with broad application prospects.

Acknowledgments. This research is supported by National Natural Science Foundation of China (NSFC) under the Grant number of 52277222, Shanghai Science and Technology Development Fund 22ZR1444500.

Disclosure of Interests. The authors have no competing interests to declare that are relevant to the content of this article.

References

1. Zheng, Y., et al.: A simplification of the time-domain equivalent circuit model for lithium-ion batteries based on low-frequency electrochemical impedance spectra. J. Power. Sources **489**, 229505 (2021)
2. Wu, H., Chen, S., Hong, Y., et al.: Thermal safety boundary of lithium-ion battery at different state of charge. J. Energy Chem. **91**, 59–72 (2024)
3. Aannir, M., Hakkou, R., Levard, C., et al.: Towards a closed loop recycling process of end-of-life lithium-ion batteries: recovery of critical metals and electrochemical performance evaluation of a regenerated LiCoO2. J. Power. Sources **580**, 233341 (2023)

4. Lai, X., Chen, Q., Tang, X., et al.: Critical review of life cycle assessment of lithium-ion batteries for electric vehicles: A lifespan perspective. Etransportation **12**, 100169 (2022)
5. Vidal, C., Malysz, P., Kollmeyer, P., et al.: Machine learning applied to electrified vehicle battery state of charge and state of health estimation: State-of-the-art. IEEE Access **8**, 52796–52814 (2020)
6. Song, L., Zhang, K., Liang, T., et al.: Intelligent state of health estimation for lithium-ion battery pack based on big data analysis. J. Energy Storage **32**, 101836 (2020)
7. Zheng, Y., Cui, Y., Han, X., et al.: A capacity prediction framework for lithium-ion batteries using fusion prediction of empirical model and data-driven method. Energy **237**, 121556 (2021)
8. Zhou, X., Han, X., Wang, Y., et al.: A data-driven LiFePO4 battery capacity estimation method based on cloud charging data from electric vehicles. Batteries **9**(3), 181 (2023)
9. Hong, J., Yang, H., Ma, F.: Multi-forward-step battery voltage prediction for real-world electric vehicles using gated recurrent units. J. Energy Storage **73**, 109056 (2023)
10. Ng, M.F., Zhao, J., Yan, Q., et al.: Predicting the state of charge and health of batteries using data-driven machine learning. Nat. Mach. Intell. **2**(3), 161–170 (2020)
11. Guo, N., Chen, S., Tao, J., et al.: Semi-supervised learning for explainable few-shot battery lifetime prediction. Joule (2024)
12. Wu, J., Chen, J., Feng, X., et al.: State of health estimation of lithium-ion batteries using autoencoders and ensemble learning. J. Energy Storage **55**, 105708 (2022)
13. Wang, Q., Ye, M., Cai, X., et al.: Transferable data-driven capacity estimation for lithium-ion batteries with deep learning: a case study from laboratory to field applications. Appl. Energy **350**, 121747 (2023)
14. Chen, J., Feng, X., Jiang, L., et al.: State of charge estimation of lithium-ion battery using denoising autoencoder and gated recurrent unit recurrent neural network. Energy **227**, 120451 (2021)
15. Hsu, C.W., Xiong, R., Chen, N.Y., et al.: Deep neural network battery life and voltage prediction by using data of one cycle only. Appl. Energy **306**, 118134 (2022)

A Reduce-Order Model to Understand the Multiscale Coupling of Battery Electrochemical-Thermal Performance

Jie Lin[1,2](✉) and Howie N. Chu[2,3]

[1] School of Mechanical and Aerospace Engineering, Queen's University Belfast, Ashby Building, Stranmillis Road, Belfast BT9 5AH, UK
j.lin@qub.ac.uk

[2] Department of Engineering Science, University of Oxford, Parks Road, Oxford OX1 3PJ, UK

[3] ICL Group, Tarrytown New York, 769 Old Saw Mill River Road, New York 10591, USA

Abstract. Inhomogeneous heat generation and temperature distribution in association with non-uniform electrochemical reaction is deemed as a serious problem in large-format lithium-ion batteries. To address this challenge, it requires a thorough understanding of the battery's coupled electrochemical-thermal performance. Therefore, in this paper, we would like to investigate the relevant physical phenomena observed in the battery electrical and thermal responses via a reduce-order 3D electrochemical-thermal model. More importantly, the dominant physical mechanisms of the battery performance are to be discussed. The model has been developed to parameterize large-format Li-ion prismatic pouch cells, and it can be further applied to simulate the irregular temperature distribution on the battery due to the non-uniform electrochemical reaction and heat generation. A scaling analysis of the model is carried out to determine the governing dimensionless numbers of the battery during charging and discharging, based on which the model complexity can be dramatically reduced. The key dimensionless number found in this study are known to be related to the material properties of the battery and the control parameters in the tests.

Keywords: lithium-ion Battery · Model · Thermography · Parameterization · Scaling Analysis

1 Introduction

Owing to their merits of high voltage, high energy density, low self-discharge and absence of memory effects, Li-ion batteries have become the leading energy storage devices for consumer electronics and electric vehicles (EVs) [1]. The specific energy, cycle life, safety and cost of Li-ion batteries have substantially progressed in the past decade [2]. However, there are still challenges for operating batteries in high performance applications. In particular, heat generation during battery operations often leads to high temperature and temperature inhomogeneity, which can cause locally irregular degradation of active electrode materials or even catastrophic thermal runaway [3, 4]. Spatially variable

K. Li et al. (Eds.): LSMS/ICSEE 2024, CCIS 2218, pp. 204–218, 2025.
https://doi.org/10.1007/978-981-96-0232-2_16

internal temperatures are commonly observed in large-format batteries [5], and have been shown to impact performance and cycle life significantly [6]. Therefore, proper thermal management of Li-ion batteries is crucial [7, 8].

Since cell temperature can strongly affect the physical properties and reaction kinetics within a battery, this requires a thorough understanding of the heat generation and transient temperature within a cell during charging/discharging. Hence, an experimentally validated battery model is essential to cultivate the relevant mechanisms in the battery and offers great help to predict and optimize battery design parameters.

Currently, battery thermal simulations are primarily based on equivalent circuit models (ECMs) and electrochemical models [9, 10]. ECMs apply an electrical circuit network, which is comprised of a voltage source, resistors and capacitors, to predict the electrical response of a battery [11, 12]. Analogous to the electrical circuit, a lumped battery thermal model can be developed using a combination of thermal resistance and capacitance [13, 14]. Due to their simplicity, ECMs can be solved quickly, favouring their use in battery management systems (BMSs). For example, Damay [15] employed the Randles circuit to model a large prismatic LFP cell, and the joule heating from the cell was calculated based on the electrical losses in the circuit. The reversible and irreversible heat generation was linked to an equivalent circuit which formulated the general energy balance network. In the equivalent circuit, a few thermal resistances were defined to describe the heat conduction and convection, and the heat capacity of the cell was treated as a capacitor. The proposed method was demonstrated to be well-suited for battery thermal management, after validating with full charging and discharging tests on a battery pack. Nonetheless, an ECM is often oversimplified and fail to capture the key electrochemical and physical processes in the battery. The establishment of an ECM requires a few individual tests to parameterize a specific battery, especially via a wide range of electrochemical impedance spectroscopy (EIS) measurements [16].

In contrast, electrochemical models describe the basic electrochemical and physical processes through detailed local descriptions accounting for the conservation of charge, mass and energy, etc. [9]. The most popular electrochemical Li-ion battery model was developed by Fuller, Doyle, and Newman [17–19], known as the standard 'pseudo-2D' (P2D) model. The model adopts the concentrated solution theory and porous electrode theory to account for the diffusion, migration and convection of Li ions in the electrolyte, as well as intercalation into solid particles [20]. Butler-Volmer kinetics is commonly used to calculate the charge transfer on the electrode-electrolyte interface [21]. Following this, Srinivasan and Wang [22] introduced an energy balance equation into the P2D model to study the thermal behaviour of Li-ion batteries, where ohmic heat, active polarization heat and entropic heat were considered. The well-established P2D model has been extensively applied and adapted to various studies and applications. For example, Zhao et al. [23] coupled the P2D electrochemical model with a 3D thermal model. The results from the P2D model were given as the input to the thermal model, while temperature from the thermal model acted as a feedback to affect the electrochemical reaction. Basu et al. [24] applied a 3D battery electrochemical/thermal model to study the thermal management system of a battery pack. The external flow field and conjugate heat transfer of the battery pack was taken into account with a computational fluid dynamics (CFD) model. Kim et al. [25] proposed a 2D simplified electrochemical/thermal model which only involved the charge balance equations of current collectors and solid electrodes. The model was used to investigate the thermal behaviour of a Li-ion pouch cell, validated with thermal imaging data. Yang et al. [26] supplemented the reaction kinetics for solid electrolyte interphase (SEI) growth and lithium plating into the P2D model. The model was then applied to predict the aging characteristics of Li-ion cells after long-term cycling.

It could be summarized from the literature review that electrochemical models can inclusively describe the major phenomena in Li-ion batteries and have demonstrated a high level of feasibility. However, the P2D model is not able to capture spatially non-uniform in-plane electrochemical reactions, and a number of previous studies do not consider local heat generation in the entire 3D geometry of a cell. On the other hand, existing 3D battery electrochemical/thermal models extended from the P2D model are complex and expensive in computations, especially inefficient to work on practical cells with a multi-layer structure. In addition, to establish an electrochemical model for a specific battery requires a thorough parameterization of the battery, whereas some of its physical parameters are difficult even not possible to obtain experimentally.

Therefore, we would like to propose a simplified streamlined 3D electrochemical-thermal model for large-format Li-ion prismatic pouch cells which achieves computational efficiency by reducing the P2D model to a simpler structure [21]. The model is fast enough for battery parameterization and thermal management while still retaining physical information that can be ascribed to microscopic charge and mass transport phenomena [27]. By constructing a complete single-layer cell, we demonstrate in this study that the model could achieve accurate estimations of ionic and thermal conductivity of the electrodes. Concurrently, the complexity of the model, especially the number of cell layers, can be reduced by asymptotic analysis to make the model more time-efficient. Furthermore, the model can be adopted to investigate and predict the battery electrical and thermal performance, in particular the non-uniformity of both electrical and thermal responses within the cell. A scaling analysis is carried out via non-dimensionalisation of the cell model. Several key dimensionless numbers are shown to govern the dynamic performance of the battery during charging and discharging.

2 Experimental

A lock-in thermography test system has been set up to investigate the thermal performance of large-format Li-ion pouch cells during charging/discharging, as shown in Fig. 1. A type of 20 Ah lithium iron phosphate (LFP) battery from A123 Systems was selected for the experiments. The battery was held vertically with its top edge inserted into a test rig and its bottom edge supported by a cell holder. The positive and negative tabs designed at the top edge of the battery were compressed onto two large copper bars. This helped to minimize the contact resistance and consequent joule heating which were commonly observed in the battery tabs. Large copper cables from the battery tester were connected to the top of the copper bars, where the cell voltage and applied current were measured simultaneously. Furthermore, the battery surfaces, especially the front and rear surfaces, were freely exposed to the ambient to prevent any potential thermal contact. The battery front surface, together with the test rig, was painted black to improve its emissivity.

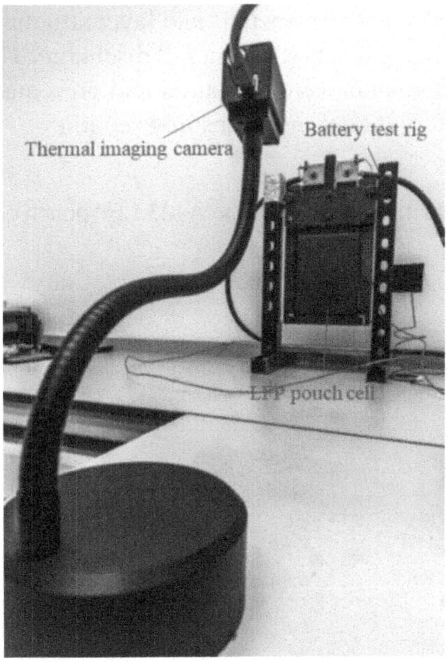

Fig. 1. Test system of the lock-in thermography.

As most of the fundamental battery properties are dependent on state-of-charge (SOC), it is important to maintain constant SOC so that they remain stable. One possible approach is to cycle the battery in a narrow range of SOC with alternating current (AC). Therefore, in this test, the applied current to the battery was set to be a square wave with identical time duration and current for charging and discharging. During cycling, the transient temperature response of the battery front surface was monitored and recorded via a thermal imaging camera (FLIR A35sc). Concurrently, an individual thermocouple was placed near the test rig to measure the ambient temperature and calibrate the thermal imaging camera.

The Li-ion pouch cell was tested at various average SOCs with different C-rates and cycle times of the applied current, as listed in Table 1. Test results obtained from the experiments were adopted to parameterizing the battery via a simplified streamlined electrochemical/thermal model and to predicting the battery performance under other circumstances [27].

Table 1. Operating conditions of the experiment.

Parameter	Nominal value	Test value
SOC	0.30	0.30, 0.50, 0.70
C-rate	4	2, 4
Cycle time (s)	100	50, 100

To obtain the correct physical dimensions and layer structure of the LFP pouch cell, it was further dissembled in a glove box after a full discharge. The geometric parameters of the battery and its components were measured and presented in Table 2, which will be used to establish a battery model in the ensuing section.

Table 2. Geometry of the A123 LFP pouch cell.

Parameter	Symbol	Unit	Value
Battery length	L	mm	200
Battery width	W	mm	150
Battery thickness	δt	mm	7.0
Tab length	$Ltab$	mm	25
Tab width	W_{tab}	mm	48
Tab thickness	δ_{tab}	mm	0.25
Current collector thickness	δ_{cc}	μm	30
Electrode thickness	δ_{el}	μm	52
Separator thickness	δ_{sep}	μm	20
Number of cell layer	N	μm	42

3 Modelling

3.1 Reduce-Order Battery Electrochemical-Thermal Model

Due to the unequal current density appearing in large-format pouch cells, there exists significant non-uniformity of battery state properties (e.g., temperature and SOC). In this case, the commonly adopted P2D battery model is not appropriate and may lead to noticeable errors. Thus the electrochemical reactions and consequent heat generation have to be considered locally. However, extending the conventional 1D cell geometry to 3D without removing the 1D mass balance equation in the electrode particles is rather difficult to program and solve. In order to reduce the computational complexity while still maintaining the dominant characteristics of the battery, a simplified 3D electrochemical-thermal model of a Li-ion battery has been proposed [27]. It has been shown that the model could serve well in extracting the key battery physical properties via optimization of battery electrical and temperature responses. In this study, by reconstructing a complete single-layer cell, including the separator, we would like to demonstrate that the model can be applied to resolve the common challenge in obtaining the correct temperature distributions in the battery. The values of some critical battery properties, especially the ionic and thermal conductivity of the electrodes, can be predicted with more confidence. In addition, an in-depth understanding of the relevant mechanisms in the battery electrical and thermal performance can be facilitated via a quantitative dimensional analysis.

Figure 2 shows the battery model geometry with a single layer of electrodes, separator, current collectors and tabs. The charge balance, species balance and energy balance equations are taken into account for each component in the battery model, as listed in Table 3.

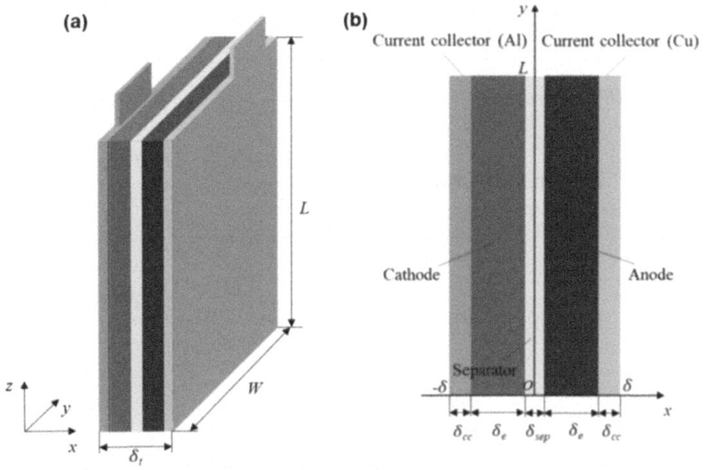

Fig. 2. Model geometry of the simplified battery electrochemical-thermal model.

Table 3. The simplified electrochemical-thermal battery model.

Mechanism	Domain	Governing equations
Charge balance	el	$\nabla \cdot \vec{i_s} = -a_v i_k$, where $\vec{i_s} = -\sigma_s \nabla \phi_s$
		$\nabla \cdot \vec{i_l} = a_v i_k$, where $\vec{i_l} = -\sigma_l \nabla \phi_l$
	sep	$\nabla \cdot \vec{i_l} = 0$, where $\vec{i_l} = -\sigma_l \nabla \phi_l$
	cc	$\nabla \cdot \vec{i_s} = 0$, where $\vec{i_s} = -\sigma_{cc} \nabla \phi_s$
Species balance	el	$\frac{\partial q}{\partial t} = \frac{a_v i_k}{Q_{\max}}$
Energy balance	el	$C_p \frac{\partial T}{\partial t} = \nabla \cdot (k_{el} \nabla T) + \sigma_s \nabla \phi_s \cdot \nabla \phi_s + \sigma_l \nabla \phi_l \cdot \nabla \phi_l + \frac{i_n^2}{\sigma_k} - \frac{a_v T \Delta S}{nF} i_n$
	sep	$C_p \frac{\partial T}{\partial t} = \nabla \cdot (k_{sep} \nabla T) + \sigma_l \nabla \phi_l \cdot \nabla \phi_l$
	cc	$C_p \frac{\partial T}{\partial t} = \nabla \cdot (k_{cc} \nabla T) + \sigma_{cc} \nabla \phi_s \cdot \nabla \phi_s$
Reaction kinetics	el	$i_k = \frac{i_0 nF}{RT} (\phi_s - \phi_l - U_{OCP}) = \frac{i_0 nF}{RT} \eta$ where $\sigma_k = \frac{i_0 nF}{a_v RT}$, $U_{OCP} = U_0 + k_u q + V_{hys} \cdot \mathrm{sgn}(i_k)$

3.2 Dimensionless Model

In order to release the necessity of modelling a full battery geometry and to investigate the dominant factors that control the battery's electrical and thermal responses, a scaling analysis is conducted based on the battery model. It involves nondimensionalizing the independent and dependent variables, as well as all relevant physical properties appearing in the model, expressed as below:

$$\text{Independent variables}: x^* = \frac{x}{\delta},\ y^* = \frac{y}{W},\ z^* = \frac{z}{L},\ t^* = \frac{h\delta t}{C_p L^2} \tag{1}$$

$$\text{Dependent variables}: \phi^* = \frac{nF\phi}{RT_0},\ T^* = \frac{T}{T_0} \tag{2}$$

$$\text{Material properties}: \pi_s = a_v^2\delta^2\frac{\sigma_{k0}}{\sigma_s},\ \pi_l = a_v^2\delta^2\frac{\sigma_{k0}}{\sigma_l},\ \pi_Q = \frac{Q_{\max}R}{nFC_p},\ \pi_i = \frac{a_v i_0 RL^2}{nF\delta h},\ \pi_S = \frac{\Delta S}{R},\ K_u = \frac{nFk_u}{RT_0} \tag{3}$$

$$\text{Control parameters}: \pi_I = \frac{i_{app}}{i_0 a_v \delta},\quad \text{Bi} = \frac{h\delta}{k_{el}} \tag{4}$$

$$\text{where } \sigma_{k0} = \frac{i_0 nF}{a_v RT_0}.$$

Accordingly, a dimensionless form of the equations in Table 3 is presented in Table 4.

3.3 Introduction to Key Dimensionless Numbers

It can be summarized that the governing factors of the battery electrical and thermal performance are the dimensionless numbers discovered in the dimensionless model, summarized as below:

$$Y^* \sim \frac{\delta}{W}, \frac{\delta}{L}, \frac{\delta_{sep}}{\delta}, \frac{\delta_{el}}{\delta}, \pi_s, \pi_l, \pi_Q, \pi_i, \pi_S, K_u, \pi_I, \text{Bi}, \frac{\sigma_{cc}}{\sigma_s}, \frac{k_{sep}}{k_{el}}, \frac{k_{cc}}{k_{el}} \tag{5}$$

where Y^* denotes the objective function, such as ϕ^* and T^*.

On the right side of Eq. (5), the first four dimensionless numbers are geometric aspect ratios, and the last three terms are the ratios of material properties. The physical representations of these dimensionless numbers are clear, so they will not be the focus of the present study. In contrast, the rest of the dimensionless numbers are the key factors that affect the relevant chemical and physical mechanisms taking place in the battery, some of which have been discussed by Newman et al. [21]. Therefore, these dimensionless numbers are of special interests and will be thoroughly investigated in this study. A detailed explanation of them is provided in Table 5, which reveals their physical significance in the battery.

Furthermore, it is known that if the dimensionless numbers in the model are identical for different cells, their objective electrical and thermal responses are expected to be similar. This allows us to simplify the battery model by controlling the dimensionless numbers, where the complexity of the battery geometry can be reduced from multiple layers to single layer. In the single-layer battery model, the thickness of a cell component (e.g., electrode, separator) is equivalent to the sum of all layers in a multi-layer structure.

Table 4. Dimensionless form of the simplified electrochemical-thermal battery model.

Mechanism	Domain	Governing equations
Charge balance	el	$\frac{\partial^2 \phi_{s*}}{\partial x^{*2}} + \frac{\delta^2}{W^2}\frac{\partial^2 \phi_s^*}{\partial y^{*2}} + \frac{\delta^2}{L^2}\frac{\partial^2 \phi_s^*}{\partial z^{*2}} = \pi_s \frac{\eta^*}{T^*}$
		$\frac{\partial^2 \phi_l^*}{\partial x^{*2}} + \frac{\delta^2}{W^2}\frac{\partial^2 \phi_l^*}{\partial y^{*2}} + \frac{\delta^2}{L^2}\frac{\partial^2 \phi_l^*}{\partial z^{*2}} = -\pi_l \frac{\eta^*}{T^*}$
	sep	$\frac{\partial^2 \phi_l^*}{\partial x^{*2}} + \frac{\delta^2}{W^2}\frac{\partial^2 \phi_l^*}{\partial y^{*2}} + \frac{\delta^2}{L^2}\frac{\partial^2 \phi_l^*}{\partial z^{*2}} = 0$
	cc	$\frac{\partial^2 \phi_s^*}{\partial x^{*2}} + \frac{\delta^2}{W^2}\frac{\partial^2 \phi_s^*}{\partial y^{*2}} + \frac{\delta^2}{L^2}\frac{\partial^2 \phi_s^*}{\partial z^{*2}} = 0$
Species balance	el	$\frac{\partial q}{\partial t^*} = \frac{\pi_i}{\pi_Q}\frac{\eta^*}{T^*}$
Energy balance	el	$\frac{\partial T^*}{\partial t^*} = \frac{1}{Bi}\left(\frac{L^2}{\delta^2}\frac{\partial^2 T^*}{\partial x^{*2}} + \frac{L^2}{W^2}\frac{\partial^2 T^*}{\partial y^{*2}} + \frac{\partial^2 T^*}{\partial z^{*2}}\right) +$ $\frac{\pi_i}{\pi_s}\left[\left(\frac{\partial \phi_s^*}{\partial x^*}\right)^2 + \frac{\delta^2}{W^2}\left(\frac{\partial \phi_s^*}{\partial y^*}\right)^2 + \frac{\delta^2}{L^2}\left(\frac{\partial \phi_s^*}{\partial z^*}\right)^2\right]$ $+\frac{\pi_i}{\pi_l}\left[\left(\frac{\partial \phi_l^*}{\partial x^*}\right)^2 + \frac{\delta^2}{W^2}\left(\frac{\partial \phi_l^*}{\partial y^*}\right)^2 + \frac{\delta^2}{L^2}\left(\frac{\partial \phi_l^*}{\partial z^*}\right)^2\right] +$ $\pi_i \frac{\eta^{*2}}{T^*} - \pi_i \pi_S \eta^*$
	sep	$\frac{\partial T^*}{\partial t^*} = \frac{1}{Bi}\frac{k_{sep}}{k_{el}}\left(\frac{L^2}{\delta^2}\frac{\partial^2 T^*}{\partial x^{*2}} + \frac{L^2}{W^2}\frac{\partial^2 T^*}{\partial y^{*2}} + \frac{\partial^2 T^*}{\partial z^{*2}}\right) +$ $\frac{\pi_i}{\pi_l}\left[\left(\frac{\partial \phi_l^*}{\partial x^*}\right)^2 + \frac{\delta^2}{W^2}\left(\frac{\partial \phi_l^*}{\partial y^*}\right)^2 + \frac{\delta^2}{L^2}\left(\frac{\partial \phi_l^*}{\partial z^*}\right)^2\right]$
	cc	$\frac{\partial T^*}{\partial t^*} = \frac{1}{Bi}\frac{k_{cc}}{k_{el}}\left(\frac{L^2}{\delta^2}\frac{\partial^2 T^*}{\partial x^{*2}} + \frac{L^2}{W^2}\frac{\partial^2 T^*}{\partial y^{*2}} + \frac{\partial^2 T^*}{\partial z^{*2}}\right) +$ $\frac{\sigma_{cc}}{\sigma_s}\frac{\pi_i}{\pi_s}\left[\left(\frac{\partial \phi_s^*}{\partial x^*}\right)^2 + \frac{\delta^2}{W^2}\left(\frac{\partial \phi_s^*}{\partial y^*}\right)^2 + \frac{\delta^2}{L^2}\left(\frac{\partial \phi_s^*}{\partial z^*}\right)^2\right]$
Reaction kinetics	el	$i_k = i_0(\phi_s^* - \phi_l^* - K_u q) = i_0 \frac{\eta^*}{T^*}$

When the battery is simplified from N layers to 1 layer, it is found that the electrical and ionic conductivity of the electrodes have to be multiplied by a factor of N^2 yielding

$$\sigma_s^1 = N^2 \sigma_s^N, \quad \sigma_l^1 = N^2 \sigma_l^N, \quad \sigma_{cc}^1 = N^2 \sigma_{cc}^N \tag{6}$$

where the superscript N means the number of cell layers.

Following Eq. (6), the battery model can be simplified to a single-layer geometry which will be employed to investigate a few characteristic phenomena in the battery electrical and thermal responses.

Figure 3 shows a typical example of the cell potential and surface temperatures when subjected to a square wave applied current. It is observed that both voltage and temperature fluctuate with the current, and the temperature non-uniformity is captured by the maximum, average and minimum temperatures. More importantly, several characteristic parameters in the voltage and temperature curves can be defined to reflect the battery intrinsic properties, as highlighted in Table 6.

Table 5. Definition and physical representation of the key dimensionless numbers.

Parameter	Expression	Definition	Physical representation
π_s	$a_v^2\delta^2\frac{\sigma_{k0}}{\sigma_s}$	Ratio of kinetic conductivity to electrical conductivity	Electrochemical reaction vs. electrical resistance
π_l	$a_v^2\delta^2\frac{\sigma_{k0}}{\sigma_l}$	Ratio of kinetic conductivity to ionic conductivity	Electrochemical reaction vs. electrolyte resistance
π_Q	$\frac{Q_{\max}R}{nFC_p}$	Ratio of charge capacity to heat capacity	Relative magnitude of charge capacity to heat capacity
π_i	$\frac{a_v i_0 RL^2}{nF\delta h}$	Ratio of kinetic power to heat transfer coefficient	Electrochemical reaction vs. heat dissipation
π_S	$\frac{\Delta S}{R}$	Normalized entropy change	Effect of entropic heating
K_u	$\frac{nFk_u}{RT_0}$	Normalized OCP gradient with reference to SOC	OCP sensitivity around a certain SOC
π_I	$\frac{i_{app}}{i_0 a_v \delta}$	Normalized applied current	The relative magnitude of applied current
Bi	$\frac{h\delta}{k_{el}}$	Ratio of conductive to convective thermal resistance	Conductive vs. convective thermal resistance

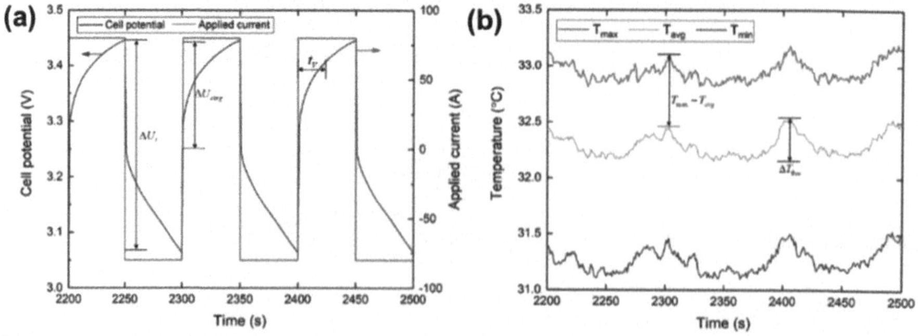

Fig. 3. Illustration of the battery electrical and thermal responses during cycling.

4 Results and Discussion

4.1 Investigation of Cell Layer Simplification

Before the battery properties are parameterized, a simulation is carried out to numerically verify the feasibility of the cell layer simplification based on its dimensionless numbers. Figure 4 shows the temperature results of the battery with different number of cell layers (41, 11, 5 and 1) under identical applied current, i.e., square wave with 100 s cycle time at 4 C-rate. The input of the battery properties in this simulation is a set of initial guesses without validation. The temperature is mapped at t = 2500 s, and a few isothermal lines are plotted on the cell front surface which can be captured by the thermal imaging camera.

Table 6. Definition and physical representation of the objective parameters.

Parameter	Definition	Physical representation
ΔU_t	Total voltage variation in one cycle	The magnitude of ohmic resistance and OCP gradient
ΔU_{chrg}	Voltage variation in charging	The magnitude of OCP gradient
t_V	Time constant in charging	The magnitude of OCP gradient and speed of polarization
T_{avg}	Average surface temperature	The magnitude of heat generation, heat capacity and convective heat transfer
$T_{\max} - T_{\mathrm{avg}}$	Temperature difference between maximum and average	The degree of non-uniform heat generation and magnitude of thermal resistance
ΔT_{fluc}	Temperature fluctuation in one cycle	The magnitude of entropic heating and heat capacity

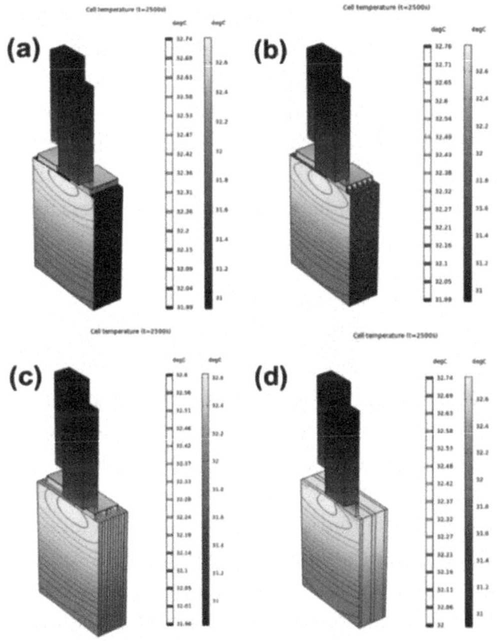

Fig. 4. Battery temperature distribution at $t = 2500$ s with different number of cell layers: (a) 41 layers; (b) 11 layers; (c) 5 layers; and (d) single layer.

When the cell layers are reduced from 41 to 11, 5 and 1, the cell temperature distribution remains similar and there is no significant disagreement in the absolute temperature scales and temperature gradient. A slight discrepancy found in the shape of temperature contours can be attributed to the numerical errors and is thereby negligible. In general, the simplification of the cell geometry is effective and substantially saves the computation time.

4.2 Investigation of Temperature Non-Uniformity

One observation worth noting in the lock-in thermography experiments is the non-uniform temperature distribution on the battery surface. Figure 5 presents the 2D surface temperature field at t = 100, 1000 and 2500 s, respectively. The applied current was 4C with a cycle time of 100 s. At the beginning of the cycling, the cell surface near the top edge was first heated up, due to the intensive current density and chemical reaction rate around the two battery tabs. This phenomenon is commonly seen in most of the existing studies [28, 29]. However, as the test continued, one large hot spot with the highest cell temperature started to appear in the top center of the cell geometry. Heat was then transferred from the hot spot to the surrounding, in particular to the bottom corners where the lowest temperature was observed. The location of the hot spot remained stable as the battery got hotter. This reveals a concentrated current density at this region through which most currents flow between the positive and negative tabs, yielding a large quantity of electrochemical reaction and heat generation. More importantly, the amount of heat generation at the hot spot was more tremendous than the joule heating in the tabs, indicating that it is indeed the critical region of which the heat needs to be appropriately dissipated. This finding differs from the majority of previous work where hot spots were detected in the tabs [30]. It could be explained by the use of highly conductive copper bars in the experiments that improve electrical contact and prevent additional joule heating in the tabs. Consequently, a clear image of the battery temperature development can be focused and captured.

Furthermore, it is noticed that the hot spot was more widely expanded in the vertical direction and the isothermal contours appeared to be convex. Concurrently, a large temperature gradient was found in the horizontal direction. This infers that there may exist anisotropic heat generation and conduction in horizontal and vertical directions, probably due to anisotropic battery physical properties. In order to examine this speculation, the simulation results of the battery model have been verified and are provided in Fig. 5(b), (d) and (f). By applying anisotropic electrical and thermal conductivity to the electrodes, the electric current and heat flows more easily in the vertical direction, leading to greater heat generation and dissipation rates. As a consequence, the simulation is able to predict the transient temperature contour and foresee the locations of the hot and cold spots. Although the shape of the hot spot in the experiment was observed to be slightly irregular, in comparison with the simulated round contour, it is probably due to the uncertainty of the experiment. The difference in absolute temperature and temperature gradient are found to be negligible.

A clearer view of the irregular temperature contours is provided by plotting the horizontal and vertical temperature distributions through the hot spot in Fig. 6. The center of the hot spot was known to be at the middle along the battery width direction and 160 mm battery length from its bottom line. In general, the simulation results agree well with the experimental data in both horizontal and vertical direction and the temperature discrepancy is within $\pm$ 0.3 °C, even though a perfect match of the temperature gradient

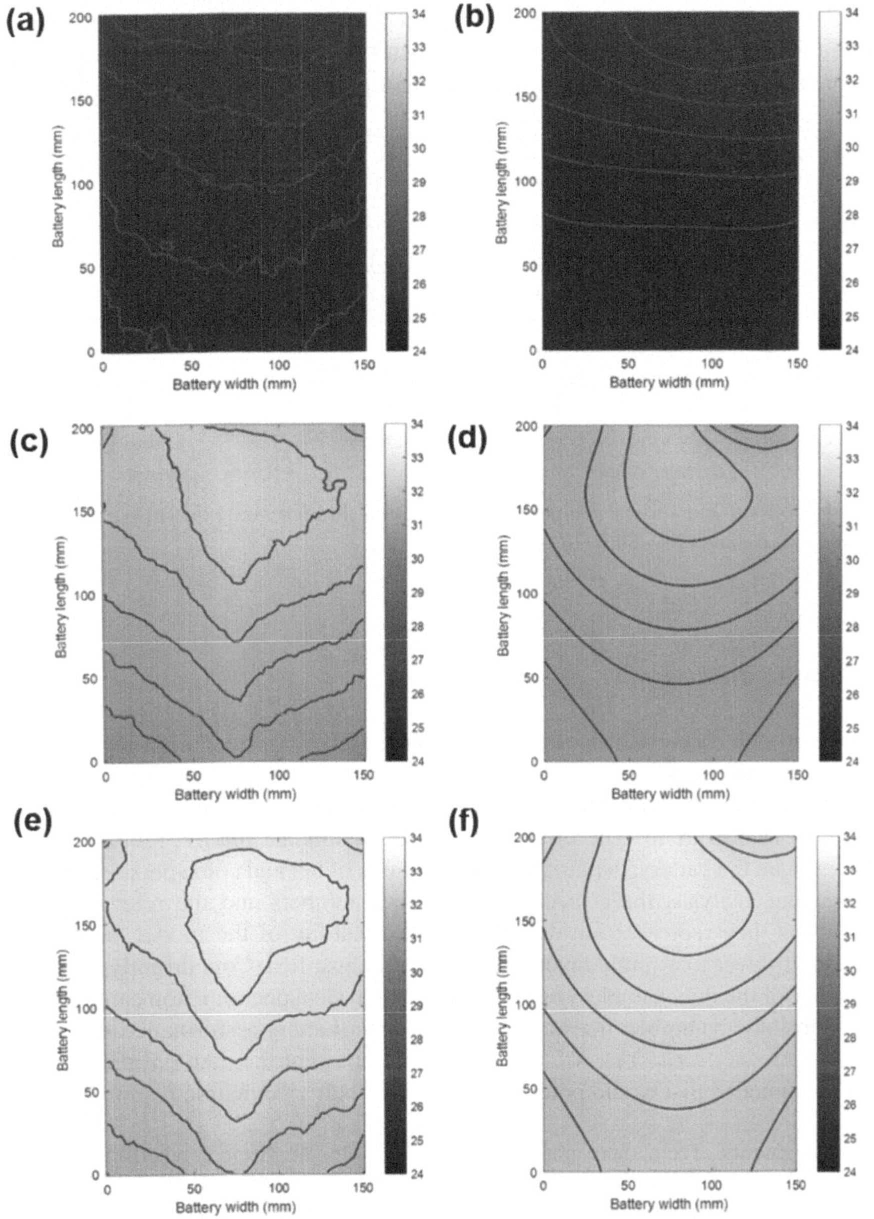

Fig. 5. Experiments and simulations of battery surface temperature response at different time steps: (a) experiment at $t = 100$ s; (b) simulation at $t = 100$ s; (c) experiment at $t = 1000$ s; (d) simulation at $t = 1000$ s; (e) experiment at $t = 2500$ s; and (f) simulation at $t = 2500$ s.

has been difficult. This indicates that the anisotropic in-plane electrical and thermal conductivity of the cell could explain the spatially non-uniform current and temperature distributions. Therefore, it is demonstrated that the battery model can be employed to capture the heat generation and temperature distribution characteristics of large-format pouch cells, which will provide insights into their proper thermal management.

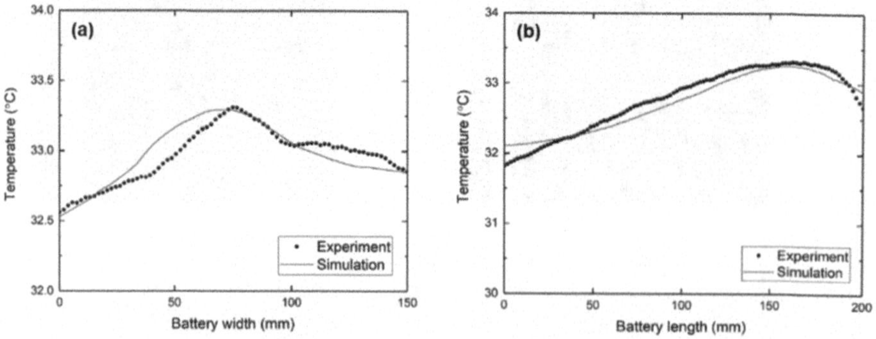

Fig. 6. Horizontal and vertical temperature distribution through hot spot: (a) battery length = 160 mm; and (b) battery width = 75 mm.

5 Conclusion

The coupled electrochemical-thermal performance of the large-format Li-ion pouch cells is examined in this study. A 3D electrochemical-thermal model, which is simplified from the P2D model while still maintaining the important charge and mass balances has been established in our group. The model is applied to simulate the irregular temperature distribution on the battery, in particular the locations of hot and cold spots. Furthermore, via a scaling analysis, the critical dimensionless numbers and the relevant physical mechanisms they represent are determined. The validity of the model simplification from a multi-layer to a single-layer structure is then justified. Concurrently, the relative importance of the dimensionless numbers in each physical phenomenon can be decided. The dimensionless numbers that have an impact on the battery performance are identified to be π_s, π_l, π_Q, π_i, π_S, K_u, π_I, Bi. They represent the material properties and control parameters that would potentially affect the battery dynamic response.

Acknowledgments. The authors gratefully acknowledge funding from (1) the EPSRC Translational Energy Storage Diagnostics (TRENDs) project (ref EP/R020973/1); (2) the STFC Futures Early Career Award under the STFC Batteries Network (ST/R006873/1); (3) the QUB Agility Fund; and (4) RS-NSFC International Exchange Program (IEC\NSFC\233361).

Disclosure of Interests. The authors have no competing interests to declare that are relevant to the content of this article.

References

1. Guo, X., He, H., Zhao, S., Dong, H., Shearing, P.R., Jervis, R., et al.: Unveiling aqueous lithium-ion batteries via advanced modelling and characterisation: a review. Energy Storage Mater. 103505 (2024)
2. Lin, J., Zhao, S., Jervis, R., Shearing, P.: Probing the Electrochemical processes of niobium pentoxides (Nb2O5) for high-rate lithium-ion batteries: a review. ChemElectroChem **11**(6), e202300581 (2024)
3. Feng, X., Ouyang, M., Liu, X., Lu, L., Xia, Y., He, X.: Thermal runaway mechanism of lithium ion battery for electric vehicles: a review. Energy Storage Mater. **10**, 246–267 (2018)
4. Liu, H., Wei, Z., He, W., Zhao, J.: Thermal issues about Li-ion batteries and recent progress in battery thermal management systems: a review. Energy Convers. Manage. **150**, 304–330 (2017)
5. Lin, J., Chu, H.N., Howey, D.A., Monroe, C.W.: Multiscale coupling of surface temperature with solid diffusion in large lithium-ion pouch cells. Commun. Eng. **1**(1), 1 (2022)
6. Paul, P.P., Thampy, V., Cao, C., Steinrück, H.G., Tanim, T.R., Dunlop, A.R., et al.: Quantification of heterogeneous, irreversible lithium plating in extreme fast charging of lithium-ion batteries. Energy Environ. Sci. **14**, 4979–4988 (2021)
7. Xia, G., Cao, L., Bi, G.: A review on battery thermal management in electric vehicle application. J. Power. Sources **367**, 90–105 (2017)
8. Lin, J., Chu, H.N., Thu, K., Wojtala, M., Gao, F., Chua, K.J.: Novel battery thermal management via scalable dew-point evaporative cooling. Energy Convers. Manage. **283**, 116948 (2023)
9. Plett, G.L.: Battery Management Systems: Battery Modeling: Artech House (2015)
10. Jokar, A., Rajabloo, B., Désilets, M., Lacroix, M.: Review of simplified pseudo-two-dimensional models of lithium-ion batteries. J. Power. Sources **327**, 44–55 (2016)
11. Hu, X., Li, S., Peng, H.: A comparative study of equivalent circuit models for Li-ion batteries. J. Power. Sources **198**, 359–367 (2012)
12. He, H., Xiong, R., Guo, H.: Online estimation of model parameters and state-of-charge of LiFePO4 batteries in electric vehicles. Appl. Energy **89**(1), 413–420 (2012)
13. Fleckenstein, M., Bohlen, O., Roscher, M.A., Bäker, B.: Current density and state of charge inhomogeneities in Li-ion battery cells with LiFePO4 as cathode material due to temperature gradients. J. Power. Sources **196**(10), 4769–4778 (2011)
14. Lin, J., Chu, H.N., Monroe, C.W., Howey, D.A.: Anisotropic thermal characterisation of largge-format lithium-ion pouch cells. Batteries Supercaps **5**(5), e202100401 (2022)
15. Damay, N., Forgez, C., Bichat, M.P., Friedrich, G.: Thermal modeling of large prismatic LiFePO 4 /graphite battery. coupled thermal and heat generation models for characterization and simulation. J Power Sources **283**, 37–45 (2015)
16. Merla, Y., Wu, B., Yufit, V., Martinez-Botas, R.F., Offer, G.J.: An easy-to-parameterise physics-informed battery model and its application towards lithium-ion battery cell design, diagnosis, and degradation. J. Power. Sources **384**, 66–79 (2018)
17. Fuller, T.F., Doyle, M., Newman, J.: Simulation and optimization of the dual lithium ion insertion cell. J. Electrochem. Soc. **141**(1), 1–10 (1994)
18. Fuller, T.F., Doyle, M., Newman, J.: Relaxation phenomena in lithium-ion-insertion cells. J. Electrochem. Soc. **141**, 982–990 (1994)
19. Doyle, M., Fuller, T.F., Newman, J.: Modeling of galvanostatic charge and discharge of the lithium/polymer/insertion cell. J. Electrochem. Soc. **140**(6), 1526–1533 (1993)
20. Newman, J., Tiedemann, W.: Porous-electrode theory with battery applications. AlChE J. **21**(1), 25–41 (1975)

21. Newman, J.S., Tobias, C.W.: Theoretical analysis of current distribution in porous electrodes. J. Electrochem. Soc. **109**(12), 1183–1191 (1962)
22. Srinivasan, V., Wang, C.Y.: Analysis of electrochemical and thermal behavior of Li-Ion cells. J. Electrochem. Soc. **150**(1), A98 (2003)
23. Zhao, R., Liu, J., Gu, J.: The effects of electrode thickness on the electrochemical and ther-mal characteristics of lithium ion battery. Appl. Energy **139**, 220–229 (2015)
24. Basu, S., Hariharan, K.S., Kolake, S.M., Song, T., Sohn, D.K., Yeo, T.: Coupled electrochemical thermal modelling of a novel Li-ion battery pack thermal management system. Appl. Energy **181**, 1–13 (2016)
25. Kim, U.S., Yi, J., Shin, C.B., Han, T., Park, S.: Modelling the thermal behaviour of a lithium-ion battery during charge. J. Power. Sources **196**, 5115–5121 (2011)
26. Yang, X.G., Leng, Y., Zhang, G., Ge, S., Wang, C.Y.: Modeling of lithium plating induced aging of lithium-ion batteries: transition from linear to nonlinear aging. J. Power. Sources **360**, 28–40 (2017)
27. Chu, H.N., Kim, S.U., Rahimian, S.K., Siegel, J.B., Monroe, C.W.: Parameterization of prismatic lithium–iron–phosphate cells through a streamlined thermal/electrochemical model. J. Power. Sources **453**, 227787 (2020)
28. Kim, U.S., Shin, C.B., Kim, C.S.: Modeling for the scale-up of a lithium-ion polymer battery. J. Power. Sources **189**, 841–846 (2009)
29. Robinson, J.B., Engebretsen, E., Finegan, D.P., Darr, J., Hinds, G., Shearing, P.R., et al.: Detection of internal defects in lithium-ion batteries using lock-in thermography. ECS Electrochem. Lett. **4**, A106–A109 (2015)
30. Robinson, J., Shearing, P., Brett, D.: Thermal imaging of electrochemical power systems: a review. J. Imaging **2**(1), 2 (2016)

RUL Prediction for Lithium-Ion Batteries Based on Sequence Decomposition and Transformer Networks

Min Zheng(✉), Mengfan Ruan, Junhao Yuan, and Quan Liu(✉)

School of Mechatronic Engineering and Automation, Shanghai University, Shanghai 200072, China

zhengmin203@shu.edu.cn, shuliushijin@163.com

Abstract. Accurate remaining useful life (RUL) prediction for lithium-ion batteries is crucial for battery health monitoring and early fault warning systems. This paper proposes a comprehensive predictive framework that integrates ICEEMDAN, XGBoost, and the deep learning model Transformer. This method first utilizes the ICEEMDAN technique for efficient multi-scale decomposition of battery capacity degradation data, extracting intrinsic mode functions (IMFs) that reflect local capacity fluctuations and a residual sequence (RES) that indicates the overall degradation trend. Subsequently, the XGBoost algorithm assesses the relationship between each IMF and RES component and the battery capacity degradation, assigning appropriate weights for subsequent predictions. Finally, the Transformer model independently predicts each component and combines these predictions, weighted by the XGBoost evaluation, to achieve a more accurate RUL prediction. Using aging experiment data from different batteries, the performance of the ICEEMDAN-XGBoost-Transformer model is compared with that of the ICEEMDAN-Transformer and Transformer models. Results indicate that this method offers high prediction accuracy, with average capacity prediction errors of RMSE, MAPE, and MAE being 0.01834, 0.02164, and 0.01319, respectively. When the prediction starting point is set at cycle 40, the absolute error in RUL prediction is less than 7 cycles, outperforming other similar models.

Keywords: Batteries · RUL · ICEEMDAN · XGBoost · Transformer

1 Introduction

In recent years, with the rapid development of renewable energy and electric vehicles, lithium-ion batteries, as the core energy storage device, have received widespread attention. The reliability and efficiency of batteries directly affect the performance and safety of the entire energy system [1]. Therefore, accurately monitoring the capacity degradation and predicting the Remaining Useful Life (RUL) of batteries has become a hot research topic.

After a comprehensive review of domestic and international literature and related algorithm research, battery RUL prediction methods can mainly be categorized into

K. Li et al. (Eds.): LSMS/ICSEE 2024, CCIS 2218, pp. 219–232, 2025.
https://doi.org/10.1007/978-981-96-0232-2_17

model-based methods and machine learning-based methods [2]. Model-based methods require an understanding of the internal physicochemical reaction mechanisms of batteries to establish aging models, combining filters and observers to capture battery degradation trends for RUL prediction [3]. Common models include electrochemical models [4], equivalent circuit models [5], and empirical models [6], among others. Model-based prediction methods are often characterized by simple parameters, small computational loads, and high accuracy. Liu et al. [7]. Proposed a new particle filtering (PF) framework based on an electrochemical model for predicting the RUL of lithium batteries, which requires a simplified electrochemical model (SEM) and its parameters as state variables in the PF algorithm. However, specially designed current excitation is needed to identify these parameters. The literature [8] compares various battery models used in energy management for electric vehicles, emphasizing that the DP model with two RC networks performs best among the seven models studied, balancing accuracy and computational efficiency, making it suitable for real-time battery management system applications. However, complex models that require various data for parameter setting can lead to inconsistencies in parameters, so model construction needs to balance complexity and accuracy.

Machine learning has the significant capability to compute multivariate datasets, discover patterns, and unlock applications on a large scale [9]. By using various machine learning and deep learning methods to train a large amount of battery aging data, the battery degradation information is extracted, establishing the mapping relationship between input data and output prediction, i.e., the battery degradation model, thus more accurately predicting the RUL of lithium batteries. Several machine learning methods have been successfully applied in the field of battery prediction, such as support vector machines (SVM) [10], neural networks [11], Gaussian process regression (GPR) [12], Bayesian models [13], and more. Zhou [14] et al. used the SVR algorithm to implement an online RUL prediction method, considering the dynamic changes in battery parameters. Zhang [15] et al. captured health indicators highly correlated with RUL from the voltage-time curves during the battery discharge stage and established a mapping based on GPR between RUL and health indicators to achieve RUL prediction. Song [16] et al., based on historical actual capacity samples of batteries, proposed the GRU-RNN algorithm to model the capacity degradation trend and predict the RUL of batteries, with high accuracy and robustness. However, as these methods fail to effectively consider the phenomenon of capacity regeneration, they struggle to fully describe the trend of capacity degradation. Some studies often use signal decomposition algorithms to denoise battery capacity data. Zhou [17] used EMD and ARIMA models to predict the future capacity and RUL of lithium batteries. Yi [18] et al. proposed a prediction method based on CEEMDAN and Long Short-Term Memory neural networks (LSTM), which outperformed the standalone LSTM model. However, EMD and CEEMDAN have issues with mode mixing, residual white noise, and pseudo-components. Methods based on Attention Mechanism (AM) can effectively capture the most critical features in a sequence, hence receiving significant attention in the field of lithium battery RUL prediction. Zhu [19] et al. constructed a hybrid network model for lithium-ion battery capacity estimation using Convolutional Neural Networks (CNNs), Bidirectional Long Short-Term Memory (Bi-LSTM), and AM, and further calculated the SOH and RUL estimation results.

Although attention mechanisms were incorporated, networks based on RNN require a high training time cost and are difficult to compute in parallel.

Considering the limitations of the above algorithms, this paper proposes a fusion algorithm based on ICEEMDAN, XGBoost, and Transformer. The ICEEMDAN algorithm decomposes capacity into multiple components, reducing the impact of local data fluctuations on subsequent work, then uses XGBoost to evaluate the importance scores of each component, and finally, the Transformer model trains and predicts the overall capacity and RUL for each component separately. The main contributions of this paper are as follows:

(1) To consider the phenomenon of capacity regeneration and improve the effect of original data decomposition, ICEEMDAN is used to perform multi-scale decomposition of capacity data, obtaining IMF components representing capacity regeneration and residual signals reflecting the overall trend of capacity degradation, improving prediction accuracy.
(2) IMF components and residual sequences as features of capacity degradation play different roles in the overall capacity change. Therefore, this paper uses the XGBoost algorithm to assess the degree of correlation between each IMF component and residual with capacity degradation, serving as the weight for subsequent component prediction.
(3) In practical applications, it is crucial to accurately predict the RUL of batteries at an early stage. Using a fully AM-based Transformer model, even with limited training samples, it can effectively capture key features and deeply learn the changing rules of each battery component, thus achieving higher early prediction performance. Moreover, the structure of this model allows for parallel processing of input sequences, which can speed up training and save time.
(4) Experiments were conducted using four NASA battery datasets, setting early prediction points, and the results were compared with other models to verify the effectiveness and robustness of our model.

2 Proposed Method

2.1 Iceemdan

Due to issues like mode mixing, white noise residue, and false modes associated with EMD, EEMD, CEEMD, and CEEMDAN, Colominas [20] et al. introduced the Improved Complete Ensemble Empirical Mode Decomposition with Adaptive Noise (ICEEMDAN) algorithm. Unlike other EMD variants, ICEEMDAN focuses on the accurate estimation of the local mean of the signal. This objective is achieved by creating new extrema to obtain the true residue, and each real IMF in the ICEEMDAN algorithm is derived from the difference between the previous residue and the current one. To prevent false modes caused by the overlapping of white noise, ICEEMDAN uses noise added by EMD decomposition at each stage, rather than directly using white noise. The computational process of ICEEMDAN is as follows:

(1) The prediction model $P_k(u(n))$ takes the input feature $u(n)$, the predicted value $y(j)$ plus an error ε_0 plus $P_k(u(n))$, where $k = 1,2,3,\ldots,M$, $u(n)$ represents the decomposed component of the battery capacity, ε_0 represents the error term, and M represents the number of decomposition components.

(2) The error $y(k)$ from the prediction model is evaluated, and a mean value is calculated to obtain the first error r_1:

$$r_1 = \frac{1}{N}\sum_{j=1}^{N} L(y(j)) \tag{1}$$

(3) Next, the error term r_1 is subtracted from the original prediction to give a new error term, r_1, and the same process is applied to obtain the first IMF. The next error term d_1 is:

$$d_1 = y - r_1 \tag{2}$$

(4) To extract the next IMF, perform the following iterative process:

$$d_k = r_{k-1} - r_k \tag{3}$$

where during each iteration, r_k is the average error of the battery capacity (at time k- 1)predicted by the model to the actual battery capacity. The next r_k is given by:

$$r_k = \frac{1}{T}\sum_{i=k}^{N} (L((r_{k-1} + e_k + P_k(u(n))))) \tag{4}$$

2.2 XGBoost

Xtreme Gradient Boosting (XGBoost) is an advanced ensemble learning algorithm, specifically designed for classification and regression tasks. Decision trees are the core of the XGBoost algorithm, which split the data into smaller categories based on different thresholds of input features, forming a tree-like structure. However, decision trees are prone to overfitting. Chen [21] et al. developed XGBoost, an optimized implementation of the Gradient Boosting algorithm that combines multiple decision tree models (referred to as "weak learners") into a powerful predictive model (or "strong learner") and incorporates several additional features such as regularization and tree pruning to mitigate overfitting, thereby enhancing the model′s predictive performance and accuracy.

Hyperparameter tuning in XGBoost is a crucial process that involves carefully selecting and combining parameters to improve model performance and predictive accuracy. The tuning process involves multiple parameters, each of which directly impacts the model′s learning capability and generalization performance. These parameters include the learning rate, number of trees, maximum depth of trees, minimum child weight, Gamma value, and more. Common methods for automatic hyperparameter tuning in XGBoost include grid search and random search. A typical drawback of grid search is that the search can be very time-consuming, especially for grid models. Random search, on the other hand, tests combinations arbitrarily chosen from a list of parameters, providing a relatively higher chance of finding the optimal parameters and requiring less time. Therefore, this study employs the random search method.

2.3 Transformer

The Transformer model is a novel deep learning method for sequence modeling that is entirely based on the attention mechanism [22]. It has achieved success in key tasks such as speech recognition, language modeling, and machine translation, outperforming recurrent neural networks and convolutional neural networks.

Figure 1 illustrates the architecture of the Transformer model, which includes an encoder and a decoder. The encoder and decoder primarily consist of embeddings, positional encoding, multi-head attention mechanisms, layer normalization, and feed-forward neural networks. The encoder uses an input layer to transform input data and incorporates positional encoding with sine and cosine functions to embed sequence information. The multi-head self-attention layer assigns weights to parts of the input, allowing the model to learn the relationships between input segments. The encoder stack processes the input sequence, while the decoder generates the output for translation tasks. The feed-forward networks and normalization layers enhance the performance of the architecture and stabilize training. Finally, the output layer predicts the next parts of the sequence or generates output. The innovation of this architecture lies in its ability to capture complex dependencies within sequences, making it highly effective for various language tasks.

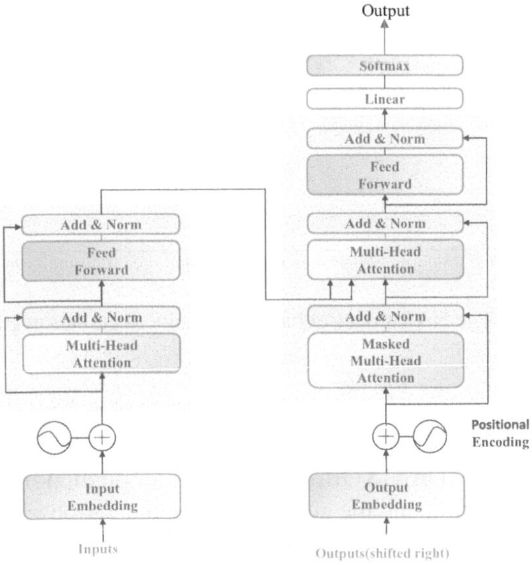

Fig. 1. Transformer structure.

3 Prediction Model Construction

3.1 Datasets

This experiment uses the NASA lithium-ion battery aging dataset, selecting four batteries tested at a temperature environment of 24 °C, specifically B0005, B0006, B0007, and B0018, as research subjects. The rated capacity is 2Ah. In the battery charging experiments, charging first occurs under a constant current (CC) mode at 1.5A until the battery voltage reaches 4.2V, followed by constant voltage (CV) mode charging until the charging current decreases to 20mA. For the discharging experiments, discharging takes place under a constant current (CC) mode at 2A until the voltage of batteries B0005, B0006, B0007, and B0018 drops to 2.7V, 2.5V, 2.2V, and 2.5V, respectively. These cycling charge and discharge tests aim to simulate the battery's accelerated aging process, with battery end of life (EOL) when the capacity reduces to 70% of the rated capacity (from 2Ah to 1.4Ah) (Fig. 2).

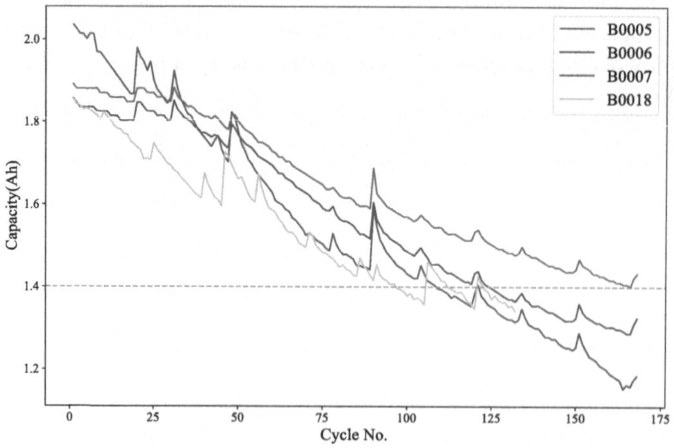

Fig. 2. The NASA dataset battery capacity degradation curve.

3.2 Prediction Process

The process of the ICEEMDAN-XGBoost-Transformer prediction method proposed in this paper is shown in Fig. 3, and the algorithm implementation steps are as follows:

(1) The ICEEMDAN technique is utilized to perform multi-scale decomposition on the battery capacity data, ultimately separating out several Intrinsic Mode Functions (IMFs) and a residual sequence (RES). Each IMF captures local fluctuation characteristics within the data, while the RES indicates the overall degradation trend of the battery capacity.

(2) The IMFs and RES are used as feature variables, with the pre-decomposition battery data serving as the response variable. The XGBoost algorithm quantifies the correlation of each feature with battery capacity degradation, evaluates the importance of each feature, and assigns weights for subsequent predictions based on this.

(3) The Transformer model independently trains and predicts each IMF and RES, combining the prediction results of each component weighted according to the evaluation results of the XGBoost model to obtain the overall future capacity of the battery. Prediction of RUL begins when the predicted capacity falls below the End of Life (EOL) threshold.
(4) The model's capacity prediction performance is evaluated using Root Mean Square Error (RMSE), Mean Absolute Error (MAE), and Mean Absolute Percentage Error (MAPE) metrics, and the model's RUL prediction performance is assessed using Relative Error (RE), comparing it with other models to verify the effectiveness of this algorithm.

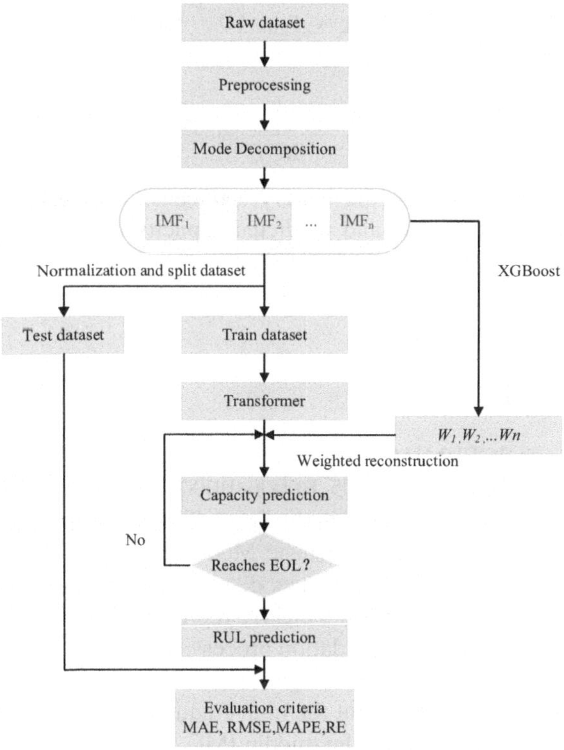

Fig. 3. The prediction process based on the ICEEMDAN-XGBoost-Transformer model.

3.3 Evaluation Criteria

The evaluation metri cs used in this paper are Mean Absolute Error (MAE), Root Mean Square Error (RMSE), Mean Absolute Percentage Error (MAPE), and Relative Error (RE). The formulas are as follows:

$$MAE = \frac{1}{n}\sum_{i=1}^{n}\left|y_{\text{real}}(i) - y_{\text{pred}}(i)\right| \tag{5}$$

$$RMSE = \sqrt{\frac{1}{n}\sum_{i=1}^{n}\left|y_{\text{real}}(i) - y_{\text{pred}}(i)\right|^2} \tag{6}$$

$$MAPE = \frac{1}{n}\sum_{i=1}^{n}\left|\frac{y_{\text{pred}}(i) - y_{\text{real}}(i)}{y_{\text{real}}(i)}\right| \tag{7}$$

$$RE = \frac{\left|RUL^{\text{pred}} - RUL^{\text{real}}\right|}{RUL^{\text{real}}} \tag{8}$$

where n represents the length of the battery sequence. y_{real} and y_{pred} respectively represent the actual capacity data and predicted capacity data, RUL^{real} and RUL^{pred} respectively represent the measured value and predicted value.

4 Results and Discussion

4.1 Decomposition of the Datasets

This paper employs the ICEEMDAN and CEEMDAN methods to decompose the capacity data of NASA dataset batteries B0005, B0006, B0007, and B0018. The decomposition results for battery B0005 are illustrated in the Fig. 4. The algorithm successfully decomposed the battery data into multiple Intrinsic Mode Functions (IMFs). Among these, the last IMF, acting as the residual sequence, displays the primary trend of the battery capacity decline, while the earlier IMFs clearly represent different frequency fluctuations during the battery capacity degradation process. The IMFs from the ICEEMDAN algorithm decomposition exhibit increasingly longer fluctuation periods as the decomposition level increases, indicating that the high-frequency components are effectively captured and separated. Particularly, the middle IMFs, which exhibit prominent peaks, may represent critical moments of battery performance decline. Next, we will use these decomposed components for detailed regression analysis and further construct predictive models to estimate the RUL of the battery.

4.2 XGBoost

After decomposing the battery capacity data through ICEEMDAN, multiple Intrinsic Mode Functions (IMFs) and a Residual Sequence (RES) with varying frequencies were obtained. Subsequently, the XGBoost regression analysis method was used to evaluate the correlation of each component with future battery capacity. To better fit the data,

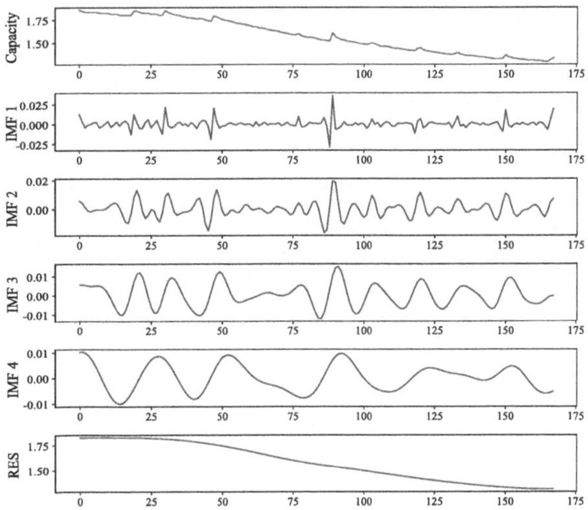

Fig. 4. B0005 Capacity decomposition results.

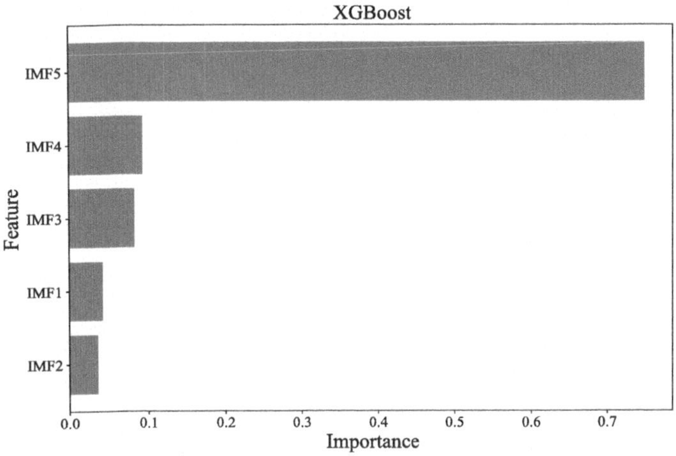

Fig. 5. IMFs regression analysis based on XGBoost.

this paper adopted a random search algorithm to tune the hyperparameters of XGBoost. Ultimately, through multiple iterations of search, a set of optimized hyperparameters was determined, which achieved coefficient of determination scores above 0.99.

After parameter optimization, the XGBoost model provided correlation values for each IMF. The Fig. 5 more intuitively shows the contribution of each IMF component to the overall capacity of battery B0005. It is evident that the lowest frequency fluctuation component (RES) accounts for a significantly greater proportion compared to all preceding components, as the RES reflects the overall degradation trend of the battery capacity. However, this does not imply that higher frequency IMFs are less correlated with the original capacity; for instance, the highest frequency IMF1 of battery B0005 scored higher than IMF2, indicating the necessity of using XGBoost to assign appropriate weights to each component.

4.3 Capacity Prediction

To evaluate the predictive performance and robustness of our proposed ICEEMDAN-XGBoost-Transformer model, we compared it with the ICEEMDAN-Transformer model and the Transformer model at the same prediction starting point (set at 90). To ensure the fairness of the comparison experiment, the same NASA dataset was used to train and test all three models, while keeping the structure and hyperparameters of the Transformer model unchanged. Figure 6 and Table 1 show the prediction results and error metrics of the different models.

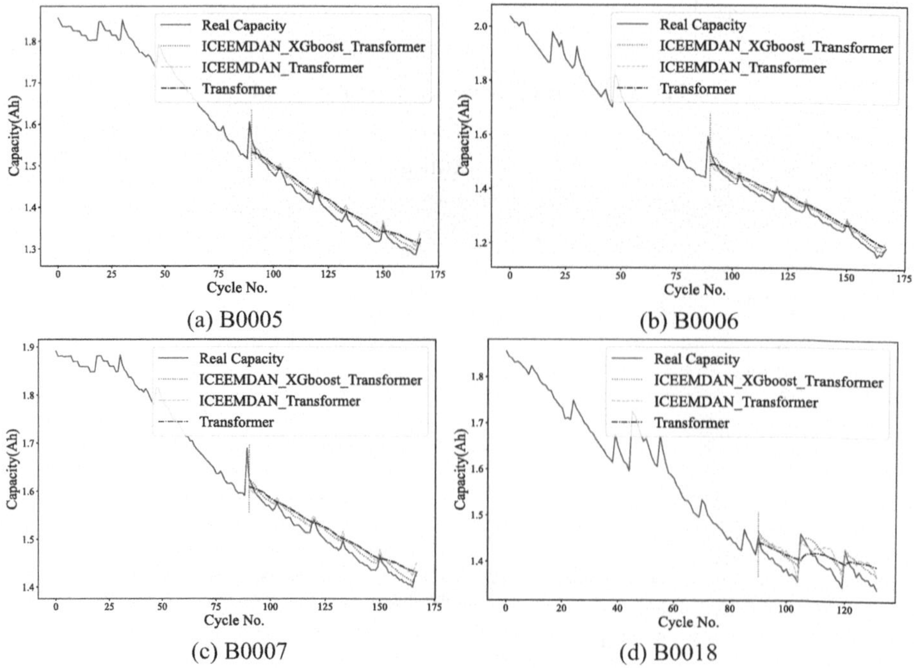

(a) B0005 (b) B0006

(c) B0007 (d) B0018

Fig. 6. Capacity prediction results for different model.

Figure 6 visually demonstrates the capacity prediction performance of each model. Compared to the other methods, the Transformer model produces a smoother prediction curve, indicating that although it successfully captures long-term dependencies in time-series data, it merely fits to a smooth curve, which introduces errors in future capacity predictions, especially when dealing with capacity regeneration phenomena. The ICEEMDAN-Transformer model, due to its multi-scale decomposition of the original capacity data, can account for capacity regeneration issues but may also be overly sensitive to each capacity fluctuation, which might be unnecessary. Compared to these two models, the method proposed in this paper stands out in predicting future capacity. This method not only focuses on capacity regeneration phenomena but also uses the XGBoost method to assess the importance of each fluctuation component in capacity degradation. Features that reflect the overall degradation trend of capacity are given priority attention, while noise or less relevant fluctuations are filtered out. Therefore, the model's predictions are not overly sensitive to every minor change in battery status, thereby enhancing the stability and robustness of the model's predictions.

Table 1. Error indicators for different methods.

Battery	Method	MAE	RMSE	MAPE
B0005	Transformer	0.02197	0.02742	0.01566
	ICEEMDAN-Transformer	0.02085	0.02277	0.01498
	ICEEMDAN-XGBoost-Transformer	**0.01487**	**0.01706**	**0.0107**
B0006	Transformer	0.02789	0.03634	0.02115
	ICEEMDAN-Transformer	0.02241	0.02715	0.01685
	ICEEMDAN-XGBoost-Transformer	**0.01886**	**0.02379**	**0.01416**
B0007	Transformer	0.02142	0.02498	0.01431
	ICEEMDAN-Transformer	0.02038	0.02189	0.01363
	ICEEMDAN-XGBoost-Transformer	**0.01399**	**0.01577**	**0.00935**
B0018	Transformer	0.02853	0.03196	0.02046
	ICEEMDAN-Transformer	0.02751	0.03373	0.01985
	ICEEMDAN-XGBoost-Transformer	**0.02565**	**0.02992**	**0.01854**

A comparison of the error metrics between the ICEEMDAN-Transformer model and the Transformer model clearly shows a reduction in prediction errors after incorporating the ICEEMDAN decomposition step. For example, for battery B0005, the MAE of the ICEEMDAN-Transformer decreased from 0.02197 to 0.02085, and the RMSE reduced from 0.02742 to 0.02277. The reason is that the IMFs produced by the ICEEMDAN decomposition capture dynamic changes across different time scales, which helps reveal long-term trends and cyclical changes hidden beneath noise and short-term fluctuations, effectively enhancing the Transformer model's ability to capture battery capacity changes. Of course, when compared to the ICEEMDAN-XGBoost-Transformer model, the latter shows the lowest error values across MAE, RMSE, and MAPE metrics. This indicates that under the same testing conditions, this model provides more accurate future capacity predictions. For instance, the MAE of the ICEEMDAN-XGBoost-Transformer model for battery B0005 decreased by 32.3% and 28.6% compared to the Transformer model and ICEEMDAN-Transformer model, respectively. In the prediction for battery B0006, the RMSE of the ICEEMDAN-XGBoost-Transformer model decreased by 34.6% and 12.4% compared to the Transformer model and ICEEMDAN-Transformer model, respectively.

4.4 RUL Prediction

Accurate RUL prediction and robust uncertainty management are crucial for lithium-ion batteries, and a battery is considered to have reached its End of Life (EOL) when its capacity falls to 80%. Early prediction of future RUL during initial stages of battery usage is essential for practical applications, hence this paper chooses the 40th cycle as the prediction start point. To study the predictive performance and robustness of the model in this paper, all NASA batteries were tested and compared with the ICEEMDAN-Transformer and Transformer models. Since battery B0007 did not reach EOL during the entire cycle experiment, Fig. 7 only shows the RUL prediction error results for the other three batteries.

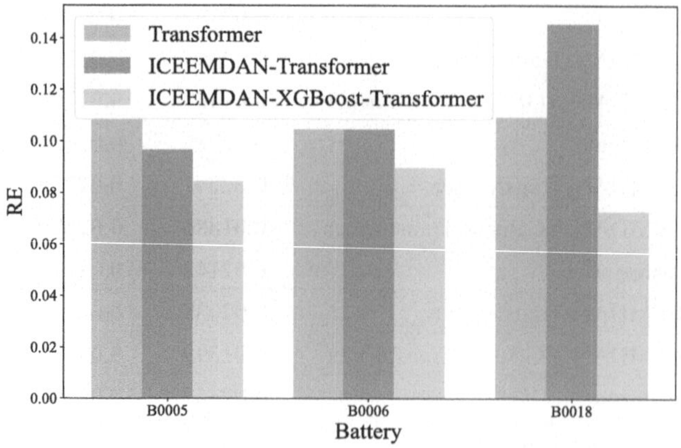

Fig. 7. RUL error indicators for different methods.

Starting from the prediction point, the actual RULs of batteries B0005, B0006, and B0018 were 83, 67, and 55 cycles, respectively. The ICEEMDAN-XGBoost-Transformer model's prediction errors (RE) were all below 0.1, outperforming other models. For example, battery B0005 dropped below EOL at cycle 123, and starting from cycle 40, the battery's RUL was 83 cycles. The ICEEMDAN-XGBoost-Transformer model predicted an RUL of 76 cycles, only 7 cycles short of the actual value, with RE values 22% and 12.5% lower than those of the Transformer model and ICEEMDAN-Transformer model, respectively. For battery B0018, the advantages of the ICEEMDAN-XGBoost-Transformer model were even more significant, with RE value decreasing from 0.10909 in the Transformer model to 0.07273, a drop of about 33%. This indicates significant improvements in the model's ability to capture the end-of-life behavior of the battery by combining ICEEMDAN decomposition and XGBoost feature selection.

In summary, compared to other deep learning algorithms, the method proposed in this paper demonstrated optimal performance, indicating that the model extracted very useful hidden information from the input aging feature sequences. Particularly in the prediction metric RE, successfully reducing it below 10%, it shows lower errors compared to other methods, demonstrating that the ICEEMDAN-XGBoost-Transformer network consistently focuses on key features for better prediction of battery RUL.

5 Conclusions

This paper introduces a hybrid algorithm based on ICEEMDAN, XGBoost and Transformer for predicting the capacity and RUL of lithium batteries. In this study, compared to other popular intelligent algorithms, the hybrid method shows advantages in handling non-stationary, nonlinear time series.

Experiments utilized a degradation dataset of four NASA battery discharge capacities, and the prediction results of other methods were compared with those of this method to verify the effectiveness and robustness of the proposed method. Specifically, in terms of capacity prediction, the proposed ICEEMDAN-XGBoost-Transformer model had average MAE, RMSE, and MAPE values of 0.01834, 0.02164, and 0.01319, respectively, which are lower than the average errors of other models. For RUL prediction, the average RE of this model was 0.08221, while the average RE of other models was greater than 0.1. Cross-comparison results indicate that the proposed hybrid method can effectively predict the future capacity and RUL of lithium batteries.

The Transformer model proposed in this paper has many hyperparameters, resulting in huge model and huge storage resource consumption. This not only increases computational costs, but may also limit the application of the model in resource-constrained environments. Therefore, future research needs to explore more efficient model simplification techniques to improve the efficiency of training and deployment while ensuring model performance. In addition, the dataset used in this paper is obtained under the conditions of constant current and constant voltage charge and discharge, and the working condition is relatively simple. Future studies need to combine battery data under more complex operating conditions, such as charging and discharging data for electric vehicles, which will help verify and optimize the effectiveness of the model in practical applications.

References

1. Liu, C., Zhuo, J., Zhao, D., et al.: A review on the utilization of energy storage system for the flexible and safe operation of renewable energy microgrids. Proc. CSEE **40**(1), 1–18 (2020)
2. Song, K., Hu, D., Tong, Y., et al.: Remaining life prediction of lithium-ion batteries based on health management: a review. J. Energy Storage **57**, 106193 (2023)
3. Sadabadi, K.K., Jin, X., Rizzoni, G.: Prediction of remaining useful life for a compo-site electrode lithium-ion battery cell using an electrochemical model to estimate the state of health. J. Power. Sources **481**, 228861 (2021)
4. Shao, S., Zhu, Z., Peng, W., Dai, J., Wu, H.: Variation and mechanism of lithium-ion concentration in the liquid phase during charging and discharging cycles. Energy Storage Sci. Technol. **10**(3), 1187–1195 (2021)
5. Dai, N.C., Bae, S.J.: Equivalent circuit simulated deep network architecture and transfer learning for remaining useful life prediction of lithium-ion batteries. J. Energy Storage **71**, 108042 (2023)
6. Bloom, I., Cole, B.W., Sohn, J.J., et al.: An accelerated calendar and cycle life study of Li-ion cells. J. Power. Sources **101**(2), 238–247 (2001)
7. Liu, Q., Zhang, J., Li, K., et al.: The remaining useful life prediction by using electro-chemical model in the particle filter framework for lithium-ion batteries. IEEE Access **8**, 126661–126670 (2020)

8. He, H., Xiong, R., Guo, H., et al.: Comparison study on the battery models used for the energy management of batteries in electric vehicles. Energy Convers. Manage. **64**, 113–121 (2012)
9. Lombardo, T., Duquesnoy, M., El-Bouysidy, H., et al.: Artificial intelligence applied to battery research: hype or reality? Chem. Rev. **122**(12), 10899–10969 (2021)
10. Nuhic, A., Terzimehic, T., Soczka-Guth, T., et al.: Health diagnosis and remaining useful life prognostics of lithium-ion batteries using data-driven methods. J. Power. Sources **239**, 680–688 (2013)
11. Zhou, J.B.: Research on Lithium-ion Battery remaining useful life estimation with relevance vector machine. Harbin Inst. Technol. (2013)
12. Yuanchang, D., Xiaoqiong, P., Jianfang, J.I.A., et al.: Remaining useful life prediction of lithium-ion batteries based on SVD-SAE-GPR. Energy Storage Sci. Technol. **12**(4), 1257 (2023)
13. Wang, L., Cao, H., Ye, Z., et al.: Bayesian large-kernel attention network for bearing remaining useful life prediction and uncertainty quantification. Reliab. Eng. Syst. Saf. **238**, 109421 (2023)
14. Zhou, J., Liu, D., Peng, Y., et al.: Dynamic battery remaining useful life estimation: An on-line data-driven approach. In: 2012 IEEE International Instrumentation and Measurement Technology Conference Proceedings, pp. 2196–2199. IEEE (2012)
15. Zhang, S., Guo, X., Zhang, X.: Multi-objective decision analysis for data-driven based estimation of battery states: a case study of remaining useful life estimation. Int. J. Hydrogen Energy **45**(27), 14156–14173 (2020)
16. Song, Y., Li, L., Peng, Y., et al.: Lithium-ion battery remaining useful life pre-diction based on GRU-RNN. In: 2018 12th International Conference on Reliability, Maintainability, and Safety (ICRMS), pp. 317−322. IEEE (2018)
17. Zhou, Y., Huang, M.: Lithium-ion batteries remaining useful life prediction based on a mixture of empirical mode decomposition and ARIMA model. Microelectron. Reliab. **65**, 265–273 (2016)
18. Yi, L., Liu, B., Wang, Y., et al.: Online prediction method of RUL of lithium battery based on CEEMDAN-ILSTM. In: 2022 IEEE 6th Conference on Energy Internet and Energy System Integration (EI2), pp.1824–1829. IEEE (2022)
19. Zhu, Z., Yang, Q., Liu, X., et al.: Attention-based CNN-BiLSTM for SOH and RUL estimation of lithium-ion batteries. J. Algorithms Comput. Technol. **16** (2022)
20. Colominas, M.A., Schlotthauer, G., Torres, M.E.: Improved complete ensemble EMD: a suitable tool for biomedical signal processing. Biomed. Sig. Proc. Control **14**, 19–29 (2014)
21. Chen, T., Guestrin, C., Xgboost: A scalable tree boosting system. In: Proceedings of the 22nd Acm Sigkdd International Conference on knowledge Discovery and Data Mining, pp. 785–794. Association for Computing Machinery (2016)
22. Vaswani, A., Shazeer, N., Parmar, N., et al.: Attention is all you need. Adv. Neural Inf. Proc. Syst. **30** (2017)

An Estimation Method of State of Health of Sodium-Ion Batteries Based on TCN

Xikang Wang[1,5], Hui Dai[2], Jiaolong Ye[2], Dianan Liu[3], Weijie Lin[4], Guanhao Du[4], and Yuanjun Guo[5(✉)]

[1] Southern University of Science and Technology, Shenzhen 518055, China
[2] CRRC Zhuzhou Locomotive Co., Ltd., Zhuzhou 412000, China
[3] Shendianneng Technology Group Co., Ltd., Shanghai 518131, China
[4] Shenzhen Hithium Energy Storage Technology Co., Ltd., Shenzhen 518000, China
[5] Shenzhen Institutes of Advanced Technology, Chinese Academy of Sciences, Shenzhen 518055, China
yj.guo@siat.ac.cn

Abstract. With the widespread use of new energy vehicles and the large-scale application of energy storage, lithium-ion batteries are facing resource shortages. Sodium-ion batteries, due to their affordability and abundant resources, are considered a promising energy storage technology. Accurately assessing their health status is crucial to ensuring the efficient performance and long lifespan of batteries. This study combines the techniques of time-series convolutional neural networks and unscented Kalman filtering to propose a novel health status estimation framework. The time-series convolutional neural network leverages its high parallelism and ability to capture long-term dependencies to provide strong support for estimation, while the unscented Kalman filter ensures the robustness of state estimation. Through experimental validation on three sodium-ion batteries, this method demonstrates excellent estimation results, with an average absolute error of 4.52%, average root mean square error of 4.09%, and model fitting up to 93.63%. This research not only provides an efficient estimation method for the health status of sodium-ion batteries but also offers valuable insights for battery management and maintenance in practical applications, aiming to further enhance battery performance, safety, and lifespan.

Keywords: Sodium-ion Battery · SOH · TCN · UKF · BMS

1 Introduction

Currently, global energy demand is growing exponentially to meet the needs of modern production and daily life [1]. Lithium-ion batteries, as a clean energy storage technology, have emerged as one of the most promising energy storage solutions for electric vehicles and energy storage systems [2]. However, with the large-scale commercialization of electric vehicles, the scarcity of lithium resources is becoming increasingly prominent. Facing the potential risk of lithium depletion, the development of new types of batteries utilizing low-cost metal materials abundant in resources has become a significant focus of current battery research [3, 4].

K. Li et al. (Eds.): LSMS/ICSEE 2024, CCIS 2218, pp. 233–243, 2025.
https://doi.org/10.1007/978-981-96-0232-2_18

Due to the high abundance and low cost of sodium, sodium-ion batteries are widely regarded as a promising new battery technology that could potentially replace lithium-ion batteries [5]. At the same time, sodium is more readily available, and its existence in various compounds allows for its direct use in battery production [6]. In addition, sodium-ion batteries also demonstrate high power density output capabilities. With the optimization of electrode materials design, they are expected to achieve similar volume and weight power density as commercial lithium-ion batteries [7]. In summary, sodium-ion batteries have the potential to become the next generation of high-performance batteries widely used in fields such as electric vehicles [8].

Currently, research on state estimation of sodium-ion batteries is still limited. FENG et al. [9] achieved the estimation of State of Health (SOH) for sodium-ion batteries for the first time and proposed a dual-exponential particle filter combined with Gaussian process regression based on wavelet analysis. Meanwhile, WEI et al. [10] achieved the estimation of Remaining Useful Life (RUL) for sodium-ion batteries for the first time. To address the research gap in the field of health estimation for sodium-ion batteries, this study proposes a method that combines Temporal Convolutional Network (TCN) and Unscented Kalman Filtering (UKF) to achieve accurate estimation of State of Health (SOH) for sodium-ion batteries.

2 SOH Estimation Model

The definition of SOH can typically be classified from several different perspectives, mainly including capacity-based SOH, impedance-based SOH, and open-circuit voltage-based SOH:

$$\mathrm{SOH} = \frac{\mathrm{Q_i}}{\mathrm{Q_0}} \times 100\% \tag{1}$$

$$\mathrm{SOH} = \frac{\mathrm{R_i}}{\mathrm{R_0}} \times 100\% \tag{2}$$

$$\mathrm{SOH} = \frac{\mathrm{OCV_i}}{\mathrm{OCV_0}} \times 100\% \tag{3}$$

In the equations, Qi represents the current battery capacity, Q0 represents the rated battery capacity; Ri represents the current battery internal resistance, R0 represents the rated battery internal resistance; OCVi represents the current battery open-circuit voltage, OCV0 represents the rated battery open-circuit voltage.

The SOH definition adopted in this study is based on battery capacity calculation. This section primarily introduces the basic structure of TCN-UKF and the construction of the SOH estimation model.

2.1 Temporal Convolutional Network (TCN)

So far, the topic of sequence modeling in the context of deep learning has been mainly related to recursive neural network architectures (e.g., LSTMs and GRUs).S. Bai et al.

[11] argued that this way of thinking has become obsolete and that Convolutional Networks should be considered as one of the main candidates when modeling sequence data. They were able to show that convolutional networks can achieve better performance than RNNs in many tasks, while avoiding the common pitfalls of recursive models, such as the gradient explosion/vanishing problem or the lack of memory retention. In addition, using convolutional networks instead of recurrent networks improves performance because it allows parallel computation of the output. Their proposed architecture is called Temporal Convolutional Network (TCN) and will be explained in the following section.

TCN refers to Temporal Convolutional Networks, a novel algorithm that can be used to solve time series prediction.

The algorithm was first proposed in 2016 by Lea et al. [12], when they were doing research on video action segmentation, and in general this routine consists of a two-step process: firstly, low-level features are computed using a CNN that (usually) encodes spatio-temporal information, and secondly, these low-level features are fed into a classifier that uses (usually) captures high-level temporal domain information. The main drawback of this approach is the need for two separate models.

TCNs provide a unified approach to capturing all two levels of information hierarchically. a TCN can take a series of arbitrary lengths and output them as the same length. In the case of a one-dimensional fully convolutional network architecture, causal convolution is used. A key feature is that the output at time t is only convolved with the elements that occurred before t.

The structure of causal convolution and dilated convolution is shown in Fig. 1 and Fig. 2. Causal convolution is a special variant of one-dimensional convolution. It differs from conventional convolutional networks in that this network is not a fully connected network, does not utilize upsampling and downsampling, and does not include pooling layers. Dilated convolution refers to the dilation size of the convolution, which is the sampling interval of the convolution. By increasing the dilation size, the network can remember more past information with fewer layers. Generally, higher layers use larger dilation sizes.

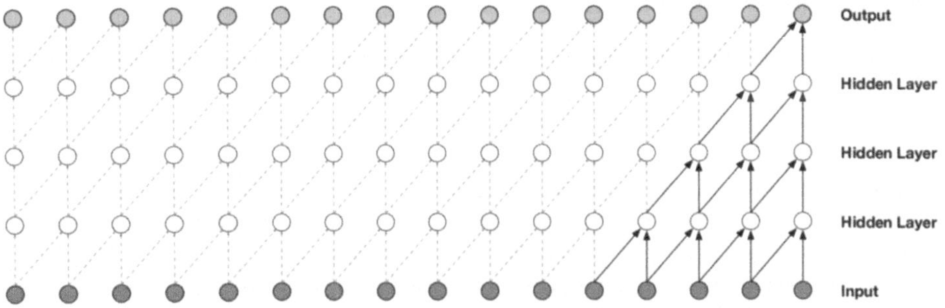

Fig. 1. Visualization of a stack of causal convolutional layers

In dilated convolution, the elements in the convolution kernel are no longer directly adjacent, but are expanded by inserting holes (or "dilation") within the kernel to enlarge the receptive field. The size of the receptive field depends on the network's architecture

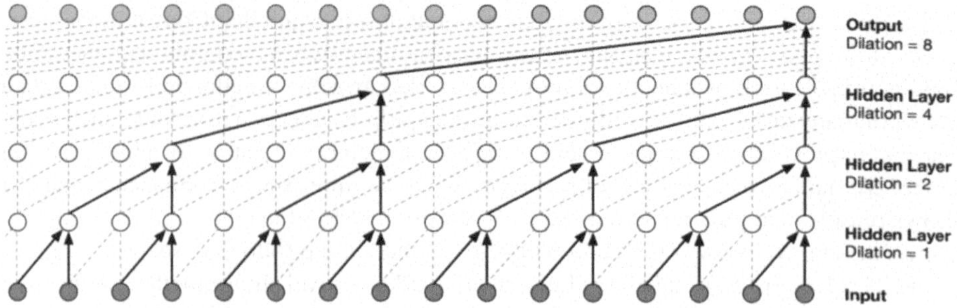

Fig. 2. Visualization of a stack of dilated causal convolutional layers

and parameter settings, which can typically be adjusted through the size of the convolution kernel, stride, and the number of layers in the network. The formula for receptive field width is:

$$w = 1 + \sum_{i=0}^{n-1} (k - 1) \cdot b^{i} = 1 + (k - 1) \cdot \frac{b^{n} - 1}{b - 1} \tag{4}$$

In the equation, k represents the kernel size, b represents the dilation base, n represents the number of layers, and i represents the layer below the new layer.

TCN has four main advantages over traditional recurrent neural networks (RNN): Firstly, it can process input sentences in parallel rather than sequentially, thereby improving processing efficiency. Secondly, TCN's receptive field size can be flexibly adjusted according to task requirements, enhancing the model's versatility. Thirdly, due to the absence of parameter sharing across different time steps, TCN exhibits more stable gradients, reducing the issues of vanishing and exploding gradients. Finally, as TCN shares convolutional kernels within layers, it consumes less memory compared to RNN, making it more memory-efficient

2.2 Unscented Kalman Filtering (UKF)

Unscented Kalman filtering is a sampling strategy to approximate the nonlinear distribution of the filtering method, the core of the UKF is the use of the Unscented Transform (UT), using the framework of the Kalman filtering algorithm. UKF uses the UT transform to determine the sampling in the vicinity of the estimation point, converting a state estimation point into multiple estimation points, and according to the difference in the weights of the state estimation points, the value of the state estimation point is passed to the later observation value. According to the different weights, the values of the state estimation points are passed to the later observation values, and according to the errors of the observation values and the real measurement values, the corresponding weights are constantly corrected through feedback, so as to approximate the probability density distribution of the function, which can make the mean and covariance after the nonlinear transformation more accurate, and ultimately get the relatively more ideal state values. The computational complexity of UKF is comparable to the Extended Kalman Filter (EKF), but it outperforms EKF in terms of performance[13].

The system's state equation and measurement equation can be expressed as:

$$x_k = f(x_{k-1}, u_k) + w_k \tag{5}$$

$$z_k = h(x_k) + v_k \tag{6}$$

In the equations, x_k represents the state vector of the system, u_k represents the control input of the system, z_k represents the measurement vector, f and h are nonlinear state transition and measurement functions, w_k and v_k represents process noise and measurement noise.

The overall process of UKF filtering for predicting TCN values involves:

(1) Initialization

state variables: $E[x(0)] = \hat{x}(0)$, error covariance matrix: $E[(\hat{x}_0 - x_0)(\hat{x}_0 - x_0)^T] = P_0$, Q is process noise covariance matrix, R is measurement noise covariance matrix.

(2) Sampling 2n + 1 Sigma points

$$\begin{cases} \chi_i(k-1) = \hat{x}(k-1), i = 0 \\ \chi_i(k-1) = \hat{x}(k-1) + (\sqrt{(n+\lambda)P(K-1)})_i, i = 1 - n \\ \chi_i = \hat{x}(k-1) - (\sqrt{(n+\lambda)P(K-1)})_{i-n}, i = n+1 - 2n \end{cases} \tag{7}$$

(3) Time update

Sigma point nonlinear transformation:

$$\chi_i(k|k-1) = f[\chi_i(k-1)] \tag{8}$$

Mean after Sigma point nonlinear transformation:

$$\hat{x}(k|k-1) = \sum_0^{2n} \omega_m^i \chi_i(k|k-1) \tag{9}$$

State error covariance matrix:

$$P(k|k-1) = \sum_0^{2n} \omega_c^i [\chi_i(k|k-1) - \hat{x}(k|k-1)][\chi_i(k|k-1) - \hat{x}(k|k-1)]^T + Q \tag{10}$$

In the equation, the weights are:

$$\omega_m^0 = \frac{\lambda}{n+\lambda}, \omega_c^0 = \frac{\lambda}{n+\lambda} + (1 - \alpha^2 + \beta)$$
$$\omega_m^i = \omega_c^i = \frac{\lambda}{2(n+\lambda)}, i = 1 - 2n$$

(4) Prediction observation:

$$y_i(k|k-1) = h[\chi_i(k|k-1)] \tag{11}$$

(5) Calculate the mean and covariance of observation predictions:

$$\hat{y}(k|k-1) = \sum_{i=0}^{2n} \omega_{\mathrm{m}}^{\mathrm{i}} y_i(k|k-1) \tag{12}$$

$$\begin{aligned} P_{yy} &= \sum_{i=0}^{2n} \omega_{\mathrm{c}}^{\mathrm{i}} [y_i(k|k-1) - \hat{y}(k|k-1)][y_i(k|k-1) - \hat{y}(k|k-1)]^T + R \\ P_{xy} &= \sum_{i=0}^{2n} \omega_{\mathrm{c}}^{\mathrm{i}} [\chi_i(k|k-1) - \hat{x}(k|k-1)][y_i(k|k-1) - \hat{y}(k|k-1)]^T \end{aligned} \tag{13}$$

(6) Calculate Kalman gain:

$$K(k) = P_{xy} P_{yy}^{-1} \tag{14}$$

(7) Update state covariance:

$$\begin{aligned} \hat{x}(k) &= \hat{x}(k|k-1) + K(k)[y(k) - \hat{y}(k|k-1)] \\ P(k) &= P(k|k-1) - K_k P_{yy} K_k^T \end{aligned} \tag{15}$$

2.3 TCN-UKF Estimation Framework

In this study, a fused SOH estimation framework based on Temporal Convolutional Network (TCN) and Unscented Kalman Filter (UKF) is developed. The framework skillfully combines the advantages of deep learning and classical filtering techniques, aiming to significantly improve the accuracy and efficiency of battery state of health estimation.

Firstly, TCN has significant advantages in handling time series data. Its ability to effectively capture long-term dependencies enables it to perform well in various timing prediction tasks. TCN is able to handle long time-span data through its special convolutional architecture, while avoiding the common gradient explosion and gradient vanishing problems of recurrent neural networks. This makes TCN particularly suitable for battery SOH prediction, as the health state of a battery is a complex process that gradually changes over time and contains a lot of time-dependent information.

Secondly, the framework uses Untraced Kalman Filtering (UKF) to deal with the nonlinearities present in the battery SOH estimation process. UKF is a state-of-the-art nonlinear filtering method that provides higher accuracy when dealing with nonlinear system state estimation. It can estimate the nonlinear state of the system more accurately and correct the prediction bias in real time by means of the traceless transformation technique. The introduction of this filtering method not only improves the stability and accuracy of the prediction results, but also enhances the robustness of the model to noise and uncertainty.

By combining TCN and UKF, the fusion framework proposed in this study fully utilizes the advantages of both techniques. TCN is responsible for capturing the long-term dependencies and complex patterns in the battery SOH variations, whereas UKF deals with the nonlinear part of the data and performs real-time correction during the prediction process. In this way, the model not only predicts battery SOH more accurately, but also maintains high stability in the face of uncertainties and variations in real applications.

The proposed estimation framework is shown in Fig. 3, which demonstrates how the combination of TCN and UKF can work together in the battery SOH estimation process. With this innovative fusion approach, this study provides a new and efficient solution for battery state of health estimation, which has important practical application value and theoretical significance.

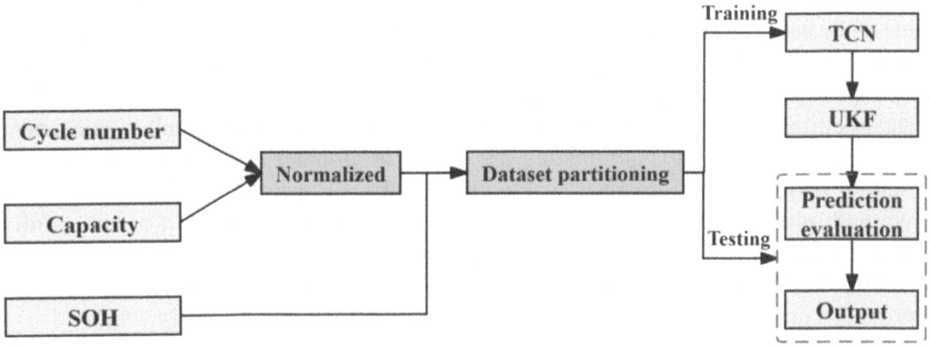

Fig. 3. The architecture of TCN-UKF

3 Experiment

3.1 Experimental Setup

This study utilized an in-house sodium-ion battery dataset for experimentation to validate the proposed method for estimating the State of Health (SOH) of sodium-ion batteries. The experimental dataset comprises three CR2032 button cells, differing in cathode material and charge-discharge current, all operated under constant current conditions. The cathode material of Cell 1 is cerium, with a charge-discharge current of 5C and a voltage range of 1.5–4.1 V. The cathode material of Cell 2 is iron, with a charge-discharge current of 1C and a voltage range of 1.5–4.1 V. The cathode material of Cell 3 is also cerium, with a charge-discharge current of 1C and a voltage range of 1.5–4.1 V. All cells underwent a total of 500 charge-discharge cycles, with external observables such as voltage, current, and capacity being recorded throughout the process.

The prediction process was divided into three times. Each prediction uses the entire life cycle data of two batteries as a training set to predict the SOH of the other battery. The input features are the number of cycles and the capacity (mAh), which is a direct reflection of the physical and chemical state of the battery. The reasons for choosing these two features are as follows:

First, the number of cycles is an important indicator of a battery's lifetime. The more cycles a battery has, the more its performance and state of health (SOH) usually decreases gradually. Therefore, by analyzing the cycle count, we can capture the changing pattern of the battery over time and usage, which is crucial for predicting the future health of the battery.

Second, the capacity (mAh) directly reflects the battery's power storage capacity, which is one of the core indicators of the battery's state of health. With the use of the battery, its capacity will gradually decay and can no longer maintain the original level of power storage. By monitoring changes in capacity, we can accurately assess the current state of the battery and possible future performance degradation. This is an important reference value for battery life prediction and performance evaluation.

The comprehensive analysis of these characteristics can effectively predict the SOH value of the battery, thus evaluating the service life and performance of the battery. These two variables are not used as characteristics since the battery is tested under constant temperature and constant current conditions. This is because constant temperature and constant current conditions ensure that the environment and current remain consistent during battery testing, thus eliminating the effects of ambient temperature and current variations on battery performance, making the number of cycles and capacity the most direct and reliable characteristics to reflect the health of the battery. In order to fully measure the precision and accuracy of prediction, three indicators are used: root mean square error (RMSE), mean absolute error (MAE) and R-squared. The mathematical expressions are:

$$\begin{cases} RMSE = \sqrt{\frac{1}{n}\sum_{i=1}^{n}(X_{pred}^{i}, X_{true}^{i})} \\ MAE = \frac{1}{n}\sum_{i=1}^{n}\left|X_{pred}^{i} - X_{true}^{i}\right| \\ R^2 = 1 - \frac{\sum_{i=1}^{n}\left|X_{pred}^{i} - X_{true}^{i}\right|^2}{\sum_{i=1}^{n}(X_{pred}^{i}, \overline{X}_i)^2} \end{cases} \tag{16}$$

3.2 Result Analysis

To verify the effectiveness and superiority of the proposed method for estimating State of Health (SOH) of sodium-ion batteries, four algorithms including TCN, RNN, LSTM, and SVR were compared. The evaluation metrics and prediction curves for the comparative experiments of Cell1 are shown in Table 1 and Fig. 4; those for Cell2 are presented in Table 2 and Fig. 5; and those for Cell3 can be found in Table 3 and Fig. 6.

Table 1. Evaluation of Cell 1 prediction results

Evaluation	TCN	RNN	LSTM	SVR	TCN-UKF
R^2	0.9491	0.8341	0.2698	0.9181	0.9818
MAE	0.0252	0.0435	0.0827	0.0321	0.0129
RMSE	0.0262	0.0473	0.0994	0.0332	0.0157

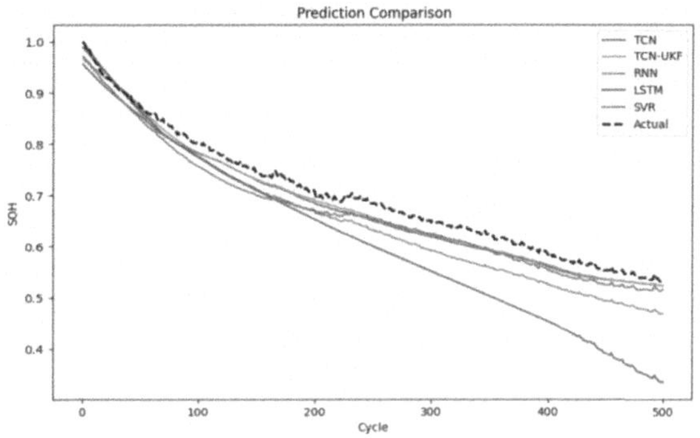

Fig. 4. The estimation curve of Cell 1

Table 2. Evaluation of Cell 2 prediction results

Evaluation	TCN	RNN	LSTM	SVR	TCN-UKF
R^2	0.8169	0.3381	0.2473	0	0.8872
MAE	0.0899	0.1541	0.1831	0.3186	0.0755
RMSE	0.1042	0.1981	0.2113	0.3575	0.0817

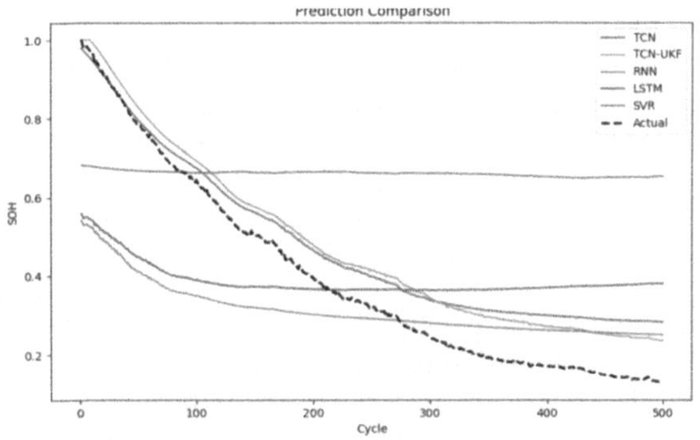

Fig. 5. The estimation curve of Cell 2

Table 3. Evaluation of Cell 3 prediction results

Evaluation	TCN	RNN	LSTM	SVR	TCN-UKF
R^2	0.9001	0.8957	0.9157	0.9446	0.9401
MAE	0.0445	0.0484	0.0411	0.0313	0.0345
RMSE	0.0494	0.0505	0.0454	0.0368	0.0383

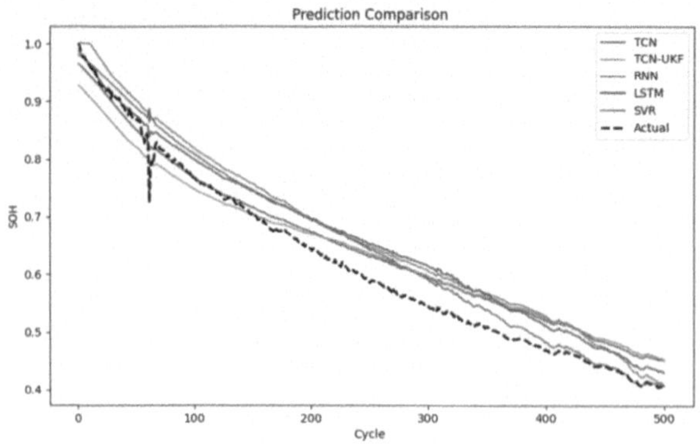

Fig. 6. The estimation curve of Cell 3

From the above three sets of graphs, it can be seen that the TCN-UKF prediction model proposed in this paper has the smallest error and the highest level of fit between the estimated SOH values and the ground truth values in all the experiments. Specifically, the TCN-UKF model performs very well in different experiments, and the deviation between its estimated and actual values is significantly lower than that of other models. The average RMSE is about 0.0452, with the maximum value not exceeding 0.0817, while the average MAE is about 0.0409, with the maximum value not exceeding 0.0755.In addition, the TCN-UKF model also performs excellently in terms of the fitting level. As can be seen from the graphs, the model is able to fit the trend of battery SOH very well, and the prediction results have a very high level of agreement with the ground truth values. This high fitting level is not only reflected in the low RMSE and MAE values, but also in the high consistency between the predicted and real curves.

4 Conclusion

This paper proposes a sodium-ion battery State of Health (SOH) estimation framework based on TCN and UKF. The model fully leverages the unique advantages of both in handling time-series data and state estimation, thereby significantly improving the accuracy of SOH estimation and ensuring its robustness. Through experiments on batteries under different operating conditions, the model demonstrates excellent adaptability and flexibility. Experimental validation on sodium-ion batteries under three different conditions shows accurate estimation results. The experimental results not only confirm the superiority and effectiveness of this method but also demonstrate outstanding performance in SOH estimation accuracy, model fitting, and other evaluation metrics compared to existing methods. This estimation method provides a crucial guarantee for the reliability and long-term durability of new energy storage technologies by ensuring accurate and robust estimation. Accurately estimating battery health status enables effective monitoring and management, thereby further extending their lifespan. Overall, this study provides a highly effective and practical technical strategy for the health management system of sodium-ion batteries, which holds profound significance for advancing the application and development of sustainable energy technologies.

Acknowledgement. This study was funded by the Science and Technology Project of Shenzhen (grant numbers GJHZ20210705141401005, CJGJZD20220517141405012 and KCXST20221021111402006), the "Nanling Team Project" of Shaoguan City, the Shenzhen Excellent Innovative Talents (grant number RCYX20221008093036022) and the Shenzhen-Hong Kong Joint Funding Project (A) (grant number SGDX20230116092053005).

References

1. Qi, Q., Tao, F., Hu, T., et al.: Enabling technologies and tools for digital model. J. Manuf. Syst. **58**, 3–21 (2021)
2. Yang, S., Zhang, C., Jiang, J., et al.: Review on state-of-health of lithium-ion batteries: characterizations, estimations and applications. J. Clean. Prod. **314**, 128015 (2021)
3. Li, H., Zhang, X., Zhao, Z., et al.: Flexible sodium-ion based energy storage devices: recent progress and challenges. Energy Storage Mater. **26**, 83–104 (2020)
4. Xu, Z., Wang, J.: Toward emerging sodium-based energy storage technologies: from performance to sustainability. Adv. Energy Mater. **12**(29), 2201692 (2022)
5. Vaalma, C., Buchholz, D., Weil, M., et al.: A cost and resource analysis of sodium-ion batteries. Nat. Rev. Mater. **3**(4), 1–11 (2018)
6. Abraham, K.M.: How comparable are sodium-ion batteries to lithium-ion counterparts? ACS Energy Lett. **5**(11), 3544–3547 (2020)
7. Fang, Y., Xiao, L., Chen, Z., et al.: Recent advances in sodium-ion battery materials. Electrochemical Energy Rev. **1**, 294–323 (2018)
8. Wang, Q., Zhu, X., Wang, T., et al.: The development status of sodium-ion batteries. Mod. Car **2023**(12), 129–131 (2023)
9. Yifeng, F., Jiani, S., Haiying, C., et al.: State of health prediction for sodium-ion batteries. Energy Storage Sci. Technol. **10**(4), 1407 (2021)
10. Wei, M., Balaya, P., Ye, M., et al.: Remaining useful life prediction for 18650 sodium-ion batteries based on incremental capacity analysis. Energy **261**, 125151 (2022)
11. Bai, S., Kolter, J.Z., Koltun, V.: An empirical evaluation of generic convolutional and recurrent networks for sequence modeling. arXiv preprint arXiv:1803.01271 (2018)
12. Lea, C., Flynn, M.D., Vidal, R., et al.: Temporal convolutional networks for action segmentation and detection. In: Proceedings of the IEEE Conference on Computer Vision and Pattern Recognition, pp. 156–165 (2017)
13. Pan, Q., Yang, F., Ye, L., et al.: A review of a class of nonlinear filters: unscented kalman filtering (UKF). Control and Decision (05), 481–489+494 (2005)
14. Van Den Oord, A., Dieleman, S., Zen, H., et al.: Wavenet: A generative model for raw audio. arXiv preprint arXiv:1609.03499 (2016)

Internal Short-Circuit Diagnosis Method in Lithium-Ion Battery Based on Temporal Convolutional Network Voltage Prediction

Shiwen Zhao[1], Qiao Peng[2], Jingyang Fang[1], Kang Li[3], Liwang Ye[4], and Kailong Liu[1](✉)

[1] School of Control Science and Engineering, Shandong University, Jinan, China
Kailong.Liu@email.sdu.edu.cn
[2] Information Technology, Analytics and Operations Group, Queen's University Belfast, Belfast BT9 5EE, UK
[3] School of Electronic and Electric Engineering, University of Leeds, LS2 9JT Leeds, UK
[4] PetroChina Shenzhen New Energy Research Institute Co., Ltd, Shenzhen, China

Abstract. Internal short circuit failure may cause thermal runaway, which poses a huge threat to the safe operation of lithium-ion batteries. Therefore, it is crucial to conduct research on internal short circuit diagnosis. This paper introduces an internal short-circuit diagnosis method comprising a voltage prediction approach based on temporal convolutional network and a residual evaluation method based on sliding window cumulative summation. Fault diagnosis can be accomplished by assessing the residual discrepancy between the predicted and measured voltages. This method only requires normal battery data to train the model without high-quality fault data, which is practical and feasible. Experimental verification confirms that the method can swiftly and accurately detect faults.

Keywords: Lithium-ion battery · internal short circuit · fault diagnosis · voltage prediction · temporal convolutional network

1 Introduction

Lithium-ion batteries are extensively utilized in a variety of electronic devices and transportation vehicles, including mobile phones, laptops, electric cars, and energy storage stations [1–3]. Their key advantages, such as high energy density and long cycle life, contribute significantly to their status as one of the most commonly used battery types in modern electronic devices and transportation vehicles [4–6]. However, given their involvement in complex and potentially hazardous electrochemical reactions, lithium-ion batteries inherently carry higher safety risks as energy storage devices [7–9]. Among the various failure modes of lithium batteries, internal short circuit (ISC) failure is acknowledged as the most detrimental [10, 11]. Minor ISC faults contribute to increased capacity loss, disruption of the battery pack's consistency, and accelerated battery aging. And as the fault progress, they may escalate to thermal runaway, culminating in serious

K. Li et al. (Eds.): LSMS/ICSEE 2024, CCIS 2218, pp. 244–253, 2025.
https://doi.org/10.1007/978-981-96-0232-2_19

safety incidents [12, 13]. This has seriously affected the popularity and application of lithium-ion batteries in fields such as electric vehicles [14, 15]. Therefore, diagnosing ISC faults within lithium-ion batteries is an urgently needed area of research.

Diagnosing ISC faults primarily involves two approaches: model-based methods and data-based methods [16]. The primary process of the model-based method involves constructing a model based on the battery mechanism. Subsequently, the residual signal is obtained by comparing the measurable signal with the signal generated by the model. Finally, the residual is evaluated to determine the diagnostic result [17]. For example, Meng et al. [18] proposed a weighted function self-adjusting state and fault estimator based on 2RC-network equivalent circuit model. However, model-based methods have drawbacks, including extensive computational requirements and susceptibility to interference, rendering them challenging to apply in practical scenarios.

In the data-based diagnosis method, diagnosis is conducted directly using measured voltage, current, and temperature data without the need for modeling [19]. Then, ISC fault diagnosis is achieved by analyzing the data to derive rules and detect changes in those rules using statistical or machine learning techniques. Among methods based on statistical techniques, Wu et al. [20] proposed an ISC diagnosis method based on voltage cosine similarity. This technique diagnoses ISC by magnifying voltage inconsistencies into cosine similarity between vectors. However, those methods exhibit low accuracy and is vulnerable to interference from inconsistencies among module batteries. In contrast, machine learning-based methods achieve greater diagnostic accuracy despite slightly higher computational demands. For example, Zhu et al. [21] introduced an ISC diagnosis method based on multi-machine learning fusion. They conducted transfer learning on a significant volume of simulation data, resulting in a high level of diagnostic accuracy. However, most contemporary machine learning methods demand a significant quantity of high-quality fault data that is challenging to obtain, in order to effectively train models and attain high-precision diagnosis [22].

Therefore, to address the aforementioned challenges, this paper introduces a voltage prediction method based on temporal convolutional network (TCN) for diagnosing lithium-ion battery ISC faults. Leveraging TCN's efficient time series processing and parallel computing capabilities, the TCN model can utilize voltage and current data to swiftly and precisely predict the normal operating voltage for the upcoming 60 s. Then, the sliding window cumulative summation method is employed to assess the residual between the predicted voltage and the measured voltage for fault diagnosis, demonstrating high accuracy and sensitivity.

2 Methodology

2.1 Voltage Prediction Model a Subsection Sample

The structure of the predicted voltage model consists of three layers of TCN followed by a final fully connected layer, as depicted in Fig. 1. TCN is a deep learning framework that employs convolutional networks for sequence modeling and prediction [23]. It excels at capturing local dependencies within sequence data. In contrast to recurrent neural networks commonly employed for sequence prediction, TCNs offer the advantage of parallel computation, leading to enhanced performance and resolution of issues like exploding or vanishing gradients [24].

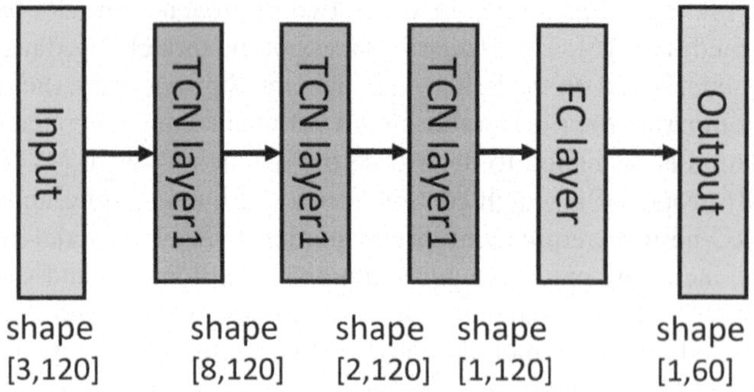

Fig. 1. Model structure.

The structure of TCN is shown in Fig. 2. The TCN comprises multiple residual blocks akin to ResNet. Each residual block contains four structures: dropout layer, ReLU activation function, weight normalization layer and one-dimensional dilated causal convolution. It is worth noting that the one-dimensional (1-D) dilated causal convolution used by TCN serves as the cornerstone of sequence modeling and prediction. The causal convolution is to output y_t at time t, and its input can only be t and the time before t. It can be described by the following formula:

$$y_t = f(x_1, x_2, \ldots x_t), t = 1, 2, \ldots N \tag{1}$$

where N is the length of sequence data, and *f(.)* represents *1-D* convolution calculation.

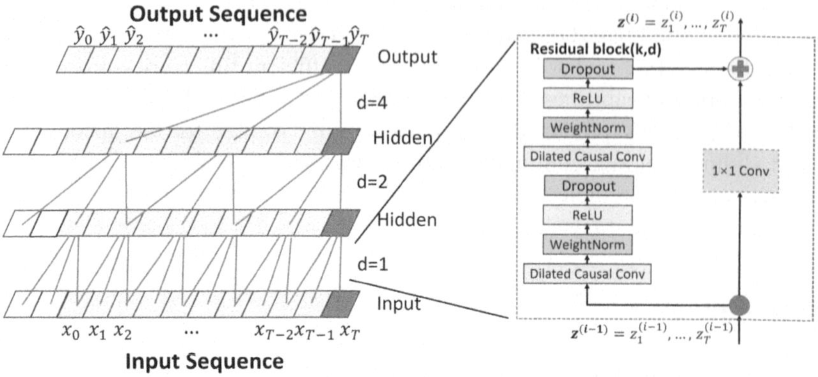

Fig. 2. TCN structure.

What's more, to enhance the network's receptive field without increasing the depth, TCN employs dilated convolution. This technique enables the distance between convolution kernels to exponentially increase with the depth of the layer, facilitated by a dilation factor (d).

2.2 Data Preprocessing and Model Training

When using TCN for prediction, three characteristic data need to be used, namely current voltage, current and set future current. The data needs to be processed before it can be fed into the model. First, the voltage sequence data is normalized:

$$X' = \frac{X - \min(X)}{\max(X) - \min(X)} \tag{2}$$

where X is the original data, and X' is the normalized data. Then the sliding window is used to extract the sequence data to obtain the samples input to the model. The samples obtained are related to the length of the sliding window and the moving step size.

When training the model, The 'Adam' optimizer is selected to optimize the parameters of the model. The optimization objective is to minimize the mean square error (MSE) loss function:

$$MSE = \frac{1}{n}\sum\nolimits_{i=1}^{n}\left(Y_i - \hat{Y}_i\right)^2 \tag{3}$$

where n is the number of input samples, Y_i, $\widehat{Y}_i$ are the true target value and model prediction value respectively.

2.3 Residual Evaluation

The residual formula, expressing the difference between the predicted voltage and the measured voltage, can be formulated as follows:

$$e = y_{predict} - y_{measure} \tag{4}$$

where e is the residual, $y_{predict}$ is the predicted voltage. And $y_{measure}$ is the measured voltage. When a fault occurs, the battery voltage experiences a sudden change, resulting in an increase in the residual between it and the predicted normal operating voltage. Therefore, short circuit faults can be diagnosed based on residual changes. However, due to errors in the model's prediction of voltage, the residuals may exceed a constant threshold even in normal cells. During normal battery operation, the residual fluctuates around zero, with both positive and negative values occurring evenly. Therefore, a sliding window is used to sum the residuals over a period of time to eliminate the interference of normal errors.

3 Experiments

The experimental platform comprises an Arbin battery tester, a temperature control box, a main control computer, and a battery module, as illustrated in Fig. 3. The 18650 specification ternary lithium battery used in the experiment has a capacity of 2Ah and an operating voltage range of 2.5–4.2 V. During the experiment, six batteries were connected in series to create a module. Subsequently, a resistor was connected in parallel with the No. 4 battery to simulate a short-circuited battery. There are three types of short-circuit resistors(R_{SC}) selected: 1 Ω, 5 Ω, and 10 Ω. The Urban Dynamometer Driving Schedule is used as the discharging mode for the battery pack. Then, at the 600-s mark of the experiment, the short-circuit resistor was introduced into the module.

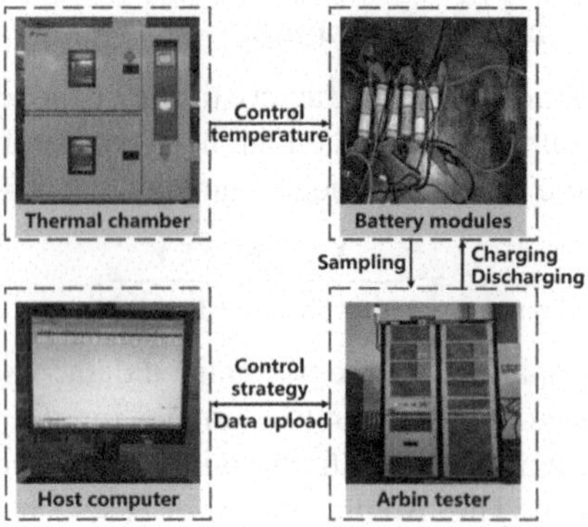

Fig. 3. Experiment platform

Under normal conditions, the voltage of each battery in the module is displayed in Fig. 4. The figure illustrates that the voltages of each battery are relatively close, with minor differences present. Figure 5 displays the voltages of each battery under ISC fault conditions, with vary resistors connected to battery 4. When battery 4 is connected in parallel with a resistor, its voltage drops compared to the other batteries. It's worth noting that there are no obvious changes in voltage when the 10 Ω resistor is connected in parallel to battery 4.

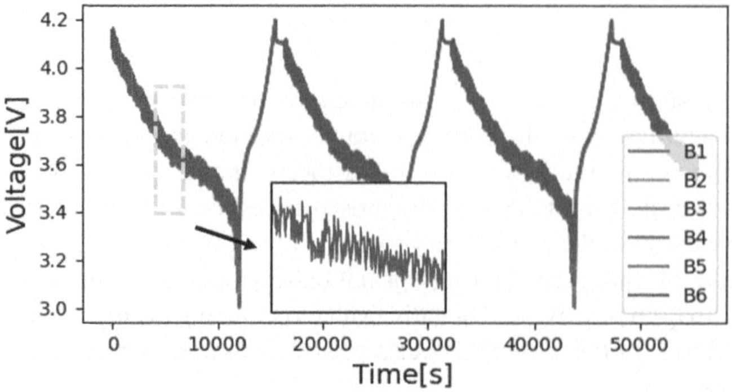

Fig. 4. Normal operating voltage of each battery in the module

4 Results and Analysis

4.1 Training Model

The data from five out of the six batteries in the module are utilized as a training set to train the model, while the data from the remaining battery is used as a test set to evaluate the training effectiveness. Among them, the five-cell battery data used as the training set are concatenated along the time dimension, which can improve the generalization

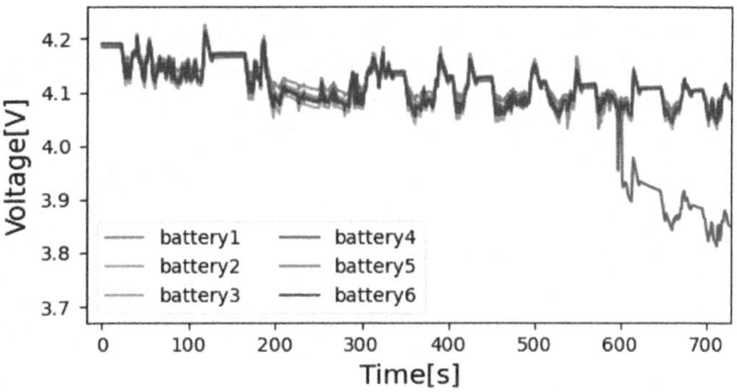

(a)

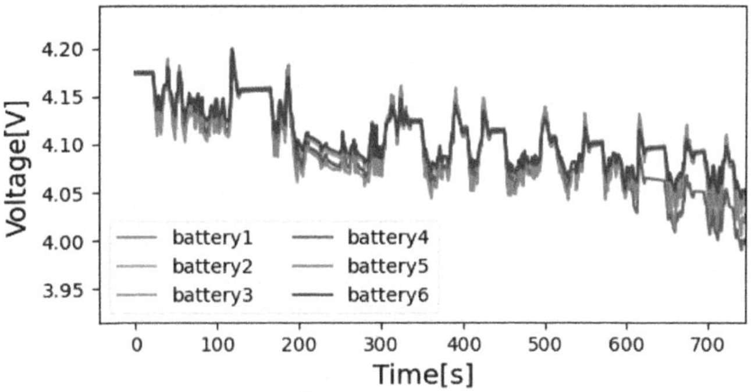

(b)

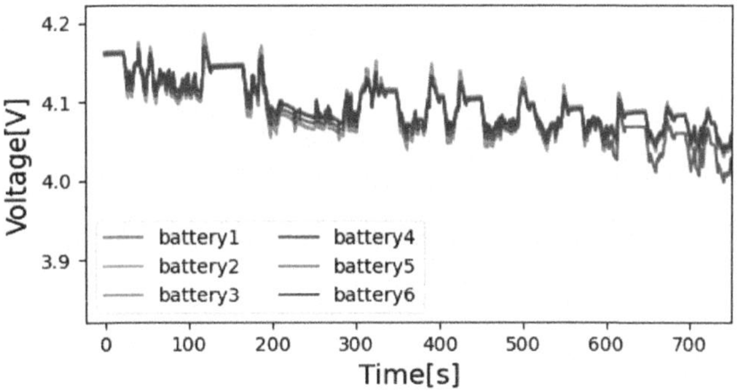

(c)

Fig. 5. Each battery voltage of the module with internal short circuit battery. (a) $R_{SC} = 1\ \Omega$. (b) $R_{SC} = 5\ \Omega$. (c) $R_{SC} = 10\ \Omega$

ability of the model. Next, the data underwent preprocessing using the method outlined in Sect. 2.2. When using a sliding window to extract data, the window length is selected as 120 and the step size is 1. The extracted data is used as input to the model to predict the voltage after 60 s. Figure 6 shows the prediction results of the trained model. It is evident that the predicted voltage aligns more closely with the actual voltage. Figure 6(b) illustrates the discrepancy between the predicted voltage and the measured voltage. From the figure, it's evident that the residual error is minimal and evenly fluctuates around zero. Furthermore, sharp errors may directly exceed the set threshold. Therefore, establishing the threshold value for residuals directly for fault diagnosis could result in misdiagnosis.

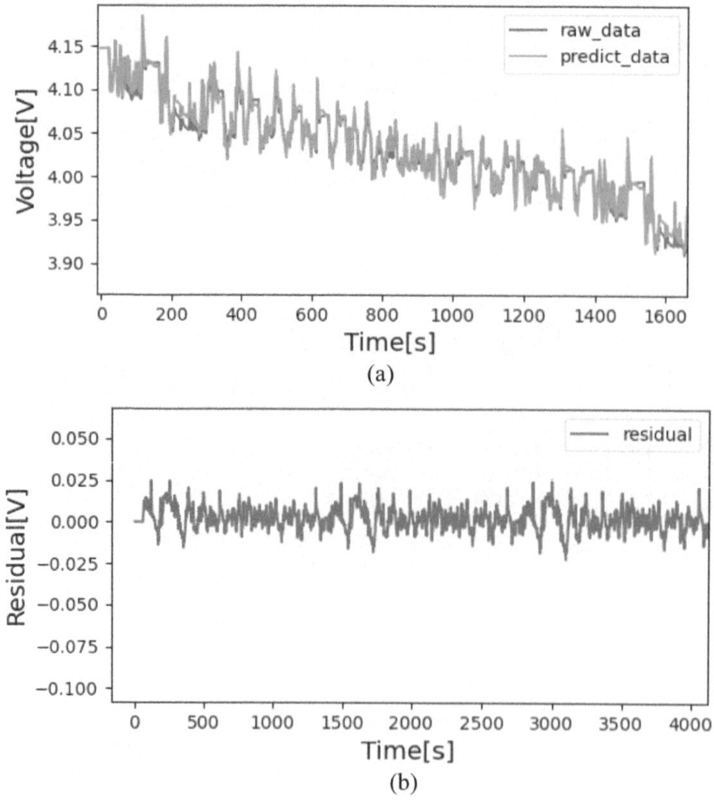

Fig. 6. Voltage prediction results and residual

4.2 Fault Diagnosis Results and Analysis

According to the residual between the predicted voltage and the measured voltage, the cumulative sum is calculated by the method in Sect. 2. And the length of the sliding window used to calculate the cumulative sum is 100, and the step size is 1. The diagnostic results obtained for ISC faults caused by various short-circuit resistances are depicted in Fig. 7. It's evident that when there's no fault occurrence, the cumulative sum is close to zero. When a fault occurs, it can be seen that the cumulative sum of battery 4 exceeds the threshold. Specifically, when a 1 Ω resistor is connected in parallel to battery 4 at 600 s, the fault is detected at 603 s. When the 5 Ω and 10 Ω resistors are connected to the No.4 battery, fault detection is achieved at 615 s and 654 s respectively. Hence, ISC faults of varying severity can be diagnosed quickly and accurately using the proposed method.

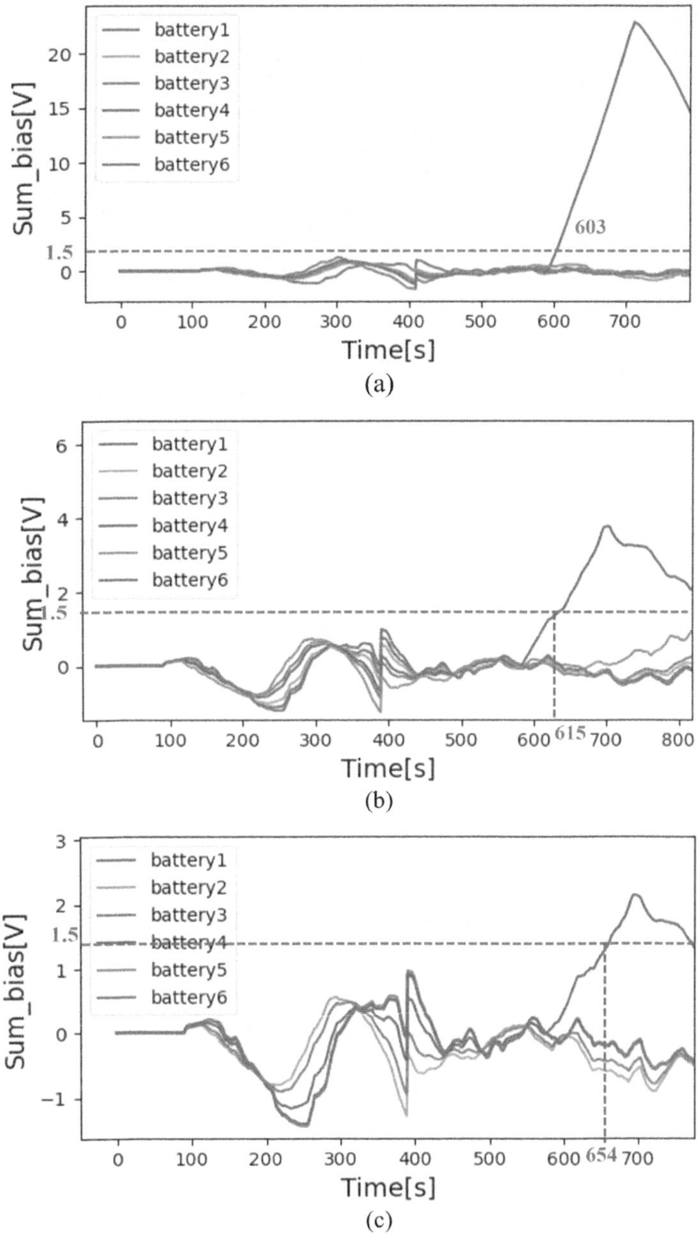

Fig. 7. Diagnostic result. (a) $R_{SC} = 1\ \Omega$. (b) $R_{SC} = 5\ \Omega$. (c) $R_{SC} = 10\ \Omega$

5 Conclusion

This paper proposes an ISC fault diagnosis method based on TCN voltage prediction. Firstly, employ TCN to forecast the forthcoming normal voltage by utilizing voltage and current data, then compare it with the measured voltage to derive the residual. Subsequently, a sliding window technique is applied to aggregate the residuals, thereby mitigating the impact of normal errors. Finally, the effectiveness of the proposed method was verified through experiments. The experimental results substantiate that faults can be diagnosed promptly and accurately, even in instances where there are no discernible changes in voltages.

Acknowledgments. This work was supported by National Natural Science Foundation of China under Grant 62373224, and Natural Science Foundation of Shandong Province under Grant ZR2024JQ021, and Shenzhen Science and Technology Program under Grant GJHZ20240218113404009.

Disclosure of Interests. The authors declare that they have no known competing financial interests or personal relationships that could have appeared to influence the work reported in this paper.

References

1. Zhu, T., Cruden, A., Peng, Q., Liu, K.: Enabling extreme fast charging. Joule **7**(12), 2660–2662 (2023)
2. Peng, Q., Li, W., Fowler, M., Chen, T., Jiang, W., Liu, K.: Battery calendar degradation trajectory prediction: data-driven implementation and knowledge inspiration. Energy, 130849 (2024)
3. Liu, K., et al.: Transfer learning for battery smarter state estimation and ageing prognostics: Recent progress, challenges, and prospects. Adv. Appl. Energy **9**, 100117 (2023)
4. Li, X., Wang, Z.: A novel fault diagnosis method for lithium-Ion battery packs of electric vehicles. Measurement **116**, 402–411 (2018)
5. Hu, X., Zhang, K., Liu, K., Lin, X., Dey, S., Onori, S.: Advanced fault diagnosis for lithium-ion battery systems: a review of fault mechanisms, fault features, and diagnosis procedures. IEEE Ind. Electron. Mag. **14**(3), 65–91 (2020)
6. Liu, K., Peng, Q., Teodorescu, R., Foley, A.M.: Knowledge-guided data-driven model with transfer concept for battery calendar ageing trajectory prediction. IEEE/CAA J. Automatica Sin. **10**(1), 272–274 (2023)
7. Zhang, K., et al.: An early soft internal short-circuit fault diagnosis method for lithium-ion battery packs in electric vehicles. IEEE-ASME Trans. Mechatron. **28**(2), 644–655 (2023)
8. Fan, C., Liu, K., Ren, Y., Peng, Q.: Characterization and identification towards dynamic-based electrical modeling of lithium-ion batteries. J. Energy Chem. (2024)
9. Zhang, C., et al.: State of Health Estimation for Lithium-ion Batteries Based on Mechanism Fundamental Learning under Variable Charging Strategies (2023)
10. Wang, G., Zhao, G., Xie, J., Liu, K.: Ensemble learning based correlation coefficient method for robust diagnosis of voltage sensor and short-circuit faults in series battery packs. IEEE Trans. Power Electron. (2023)

11. Lai, X., et al.: Mechanism, modeling, detection, and prevention of the internal short circuit in lithium-ion batteries: recent advances and perspectives. Energy Storage Mater. **35**, 470–499 (2021)
12. Qiao, D., Wang, X., Lai, X., Zheng, Y., Wei, X., Dai, H.: Online quantitative diagnosis of internal short circuit for lithium-ion batteries using incremental capacity method. Energy **243**, 123082 (2022)
13. Huang, L., et al.: A review of the internal short circuit mechanism in lithium-ion batteries: Inducement, detection and prevention. Int. J. Energy Res. **45**(11), 15797–15831 (2021)
14. Schmid, M., Kleiner, J., Endisch, C.: Early detection of internal short circuits in series-connected battery packs based on nonlinear process monitoring. J. Energy Storage **48**, 103732 (2022)
15. Hu, J., He, H., Wei, Z., Li, Y.: Disturbance-immune and aging-robust internal short circuit diagnostic for lithium-ion battery. IEEE Trans. Industr. Electron. **69**(2), 1988–1999 (2021)
16. Yuan, H., Cui, N., Li, C., Cui, Z., Chang, L.: Early stage internal short circuit fault diagnosis for lithium-ion batteries based on local-outlier detection. J. Energy Storage **57**, 106196 (2023)
17. Qiao, D.: Quantitative diagnosis of internal short circuit for lithium-ion batteries using relaxation voltage. IEEE Trans. Ind. Electron. (2024)
18. Meng, J., Boukhnifer, M., Delpha, C., Diallo, D.: Incipient short-circuit fault diagnosis of lithium-ion batteries. J. Energy Storage **31**, 101658 (2020)
19. Yuan, H., Wang, G., Cui, N.: Internal short circuit fault diagnosis for lithium-ion battery using a novel voltage prediction method. In: 2021 China Automation Congress (CAC), pp. 3252–3257. IEEE (2021)
20. Wu, X., Wei, Z., Wen, T., Du, J., Sun, J., Shtang, A.A.: Research on short-circuit fault-diagnosis strategy of lithium-ion battery in an energy-storage system based on voltage cosine similarity. J. Energy Storage **71**, 108012 (2023)
21. Zhu, G., Sun, T., Xu, Y., Zheng, Y., Zhou, L.: Identification of internal short circuit faults in lithium-ion batteries based on a multi-machine learning fusion. Batteries **9**(3), 154 (2023)
22. Wang, Y., Zhang, C., Hu, J., Zhang, P., Zhang, L., Xu, Z.: Research on internal short circuit detection method for lithium-ion batteries based on battery expansion characteristics. J. Power. Sources **587**, 233673 (2023)
23. Fan, J., Zhang, K., Huang, Y., Zhu, Y., Chen, B.: Parallel spatio-temporal attention based TCN for multivariate time series prediction. Neural Comput. Appl. **35**(18), 13109–13118 (2023)
24. Gong, L., Yu, M., Jiang, S., Cutsuridis, V., Pearson, S.: Deep learning based prediction on greenhouse crop yield combined TCN and RNN. Sensors **21**(13), 4537 (2021)

State of Charge Estimation of Lithium-Ion Battery Based on a Nonlinear Equivalent Circuit Model

Chuanxin Fan[1(✉)], Chunfei Gu[2], Qingyuan Li[1], Xinyu Lu[1], Wenwen Qin[1], and Xinxiang Tian[2]

[1] School of Automation, Nanjing Institute of Technology, Nanjing, China
Chuanxin.fan@njit.edu.cn

[2] School of Electric Power Engineering, Nanjing Institute of Technology, Nanjing, China

Abstract. Accurate battery mathematical models are essential to estimate the lithium-ion battery state of charge (SOC). The conventional equivalent circuit model, however, does not describe the actual electrochemical nonlinear dynamic response of a lithium-ion battery. In this work, a novel nonlinear equivalent circuit model (NLECM) is established, which is based on pulse-multisine signal for parameter estimation. Based on the established NLECM model, a window variational adaptive extended Kalman filter (WVAEKF) is first applied for SOC estimation. The designed WVAEKF can identify variations in the error innovation sequence distribution and modify the window's length. The experimental results demonstrate that the WVAEKF algorithm's estimation of SOC has an error limit of 1%. Compared with the traditional AEKF algorithm with a fixed window length of 6, the RMSE of SOC estimation based on WVAEKF algorithm is reduced by 23%.

Keywords: Li-ion Battery · Nonlinear Equivalent Circuit Model · State of Charge Estimation · Window Variational Adaptive Extended Kalman Filter

1 Introduction

Lithium-ion (Li-ion) batteries are extensively utilized in electric vehicles (EVs) because of their exceptional energy density, extended lifespan, and superior efficiency [1]. The state of charge (SOC) is a crucial performance metric for Li-ion batteries. It ensures the battery's safe functioning, prevents excessive charging or discharging, and extends its lifespan [2]. Therefore, improving the accuracy of SOC estimation is helpful to secure the safety of battery operation [3].

The estimation methods of state of charge (SOC) can be categorized into the following: Coulomb counting method (CCM) [4], data-driven methods and model-based method. The CCM is widely used in SOC estimation because of its easy implementation. Data-driven methods, such as support vector machine [5], neural network [6], use a substantial quantity of battery operating data for training and analysis in order to

K. Li et al. (Eds.): LSMS/ICSEE 2024, CCIS 2218, pp. 254–266, 2025.
https://doi.org/10.1007/978-981-96-0232-2_20

estimate SOC. The efficacy of such methods relies on the quality and amount of the data. Inadequate data might lead to the occurrence of over-fitting.

Among them, the model-based method is relatively popular, which can be attributed to its high accuracy and suitability to online applications [7]. There are two types of Li-ion battery models: the electrochemical model (EM) and equivalent circuit model (ECM). On the one hand, the EM has high accuracy as it can explain the chemical reaction inside the battery. However, due to its complexity and high order, it has limited practical utilization. On the other hand, the ECM simulates the dynamic response of the battery and establishes the relationship between current and voltage by using ideal electrical components like resistors, capacitors, and inductors. ECM model has been used extensively in SOC estimation because it makes a balance between accuracy and complexity. Unfortunately, since the ECM model does not describe the electrochemical reaction occurring inside the battery, it is impossible to precisely link electrical components to a specific physical phenomenon involving the battery [8].

In order to enhance the precision of the model, it is necessary to explicitly take into account the nonlinearity of the battery. Widanage et al. established a nonlinear equivalent circuit model (NLECM) composed of linear and nonlinear blocks by using the frequency domain method [9]. The NLECM model provides access to significant information that the conventional ECM model is unable to provide. The nonlinear block in the NLECM model explains the nonlinear dependency of the voltage on the current as a result of the dynamics of charge transfer, where the linear block represents the electrical characteristics of the internal ohmic resistance. Nevertheless, the problem of estimating SOC using the NLECM model still has to be resolved [10].

In [11], Li et al. uses the unscented Kalman Filter based on the Li-ion battery. The results show that this method may provide good computation efficiency and precise SOC estimate, and that it is suitable for embedded system applications. In addition, the extended Kalman filter [12], the cubature Kalman filter [13] and the particle filter [14] are also used for SOC estimation. These algorithms assume that the measurement and process noise covariance remain constant. Nevertheless, the noise covariance may vary under real operating conditions, hence affecting the accuracy of the SOC estimation.

The main contributions of this work can be attributed to two key components: (1) The nonlinear equivalent circuit model (NLECM) of lithium-ion battery, refer to [9], is established. The NLECM model consists of a linear ECM and a nonlinear overvoltage function. The linear block represents the behavior of lithium-ion batteries, whereas the nonlinear overvoltage function accounts for the non-linear relationship between voltage and current caused by diffusion limitation; (2) A window variational adaptive extended Kalman filter (WVAEKF) based on the NLECM model is proposed for estimating battery SOC. Based on the change of the error innovation sequence of the NLECM model, the length of the window is dynamically modified to estimate the innovation covariance matrix, thus enhancing the accuracy of the SOC estimation.

The rest of this article is organized as follows. In Sect. 2 describes the NLECM model and WVAEKF algorithm of Li-battery. Section 3 analyzes the results. Finally, Sect. 4 presents the conclusions of this work.

2 NLECM Model and SOC Estimation

2.1 Structure of NLECM Model

The NLECM model consists of three blocks: an open circuit voltage (OCV) block, a nonlinear function block, and a linear ECM as shown in Fig. 1(a). The identification of each block is summarized in the following subsections.

Multisine Signal Design

The multisine signal has been used as a characterization technique that can capture and separate a Li-ion battery's main linear response, nonlinearity, and noise distortion in comparison to conventional HPPC tests. In addition, the multisine signal exhibits a broader frequency range, greater magnitude, and a shorter duration compared to the typical HPPC test.

According to reference [9], the excitation signal in this article is designed as a pulse-multisine signal. The amplitude and bandwidth of the system can be stimulated based on a driving cycle, which is particularly important for practical electric vehicle (EV) applications. Consequently, the precision of the established ECM has improved in comparison to the HPPC data.

Construction of the ECM Model

The construction of the ECM is commonly performed by directly fitting or improving the model parameters using the measured current and voltage data sets. However, it needs to presuppose the structure of the ECM model. In this paper, the method of parameter identification in Ref. [14] is adopted. The impedance of the battery dynamics is estimated, and the suitable order and structure of the ECM model are determined in the form of a transfer function. This allows for additional fitting to acquire the model parameters.

The battery impedance was initially determined using a non-parametric approach, which involved representing it as a transfer function in Eq. (1). According to the method of determining the order and structure in Ref. [14], in this study, the model order is set to two.

$$Z_m(k) = \frac{b_{nb}(j\omega_k)^{nb} + \cdots + b_1 j\omega_k + b_0}{(j\omega_k)^{nb} + \cdots + a_1 j\omega_k + a_0} \tag{1}$$

where $Z_m(k)$ denotes the transfer function model, where $\omega_k = 2\pi f_s/N$ is the discrete angular frequency at harmonic k and nb is the model order.

Furthermore, as demonstrated by Eq. (2), In order to obtain the parameters of an ECM, the transfer function needs to be expanded using a partial fraction expansion.

$$Z_m(k) = -R_0 - \frac{R_{p1}}{\tau_1 j\omega_k + 1} - \frac{R_{p2}}{\tau_2 j\omega_k + 1} \tag{2}$$

where R_0 represents the internal resistance, R_{p1} and R_{p2} are the polarization re-sistances, τ_1 and τ_2 are the time constants.

Non-linear Function Block

The nonlinear function shown in Fig. 1(a) is inspired by the charge transfer reaction in electrode mechanics. As shown in Fig. 1(b), Li-ion batteries have porous solid-phase electrodes that are immersed in a liquid electrolyte, such that lithium ions can move between negative and positive electrode.

The electrons generated by the electrode reaction pass through the electrode-electrolyte interface to achieve charge balance. This interface prevents the flow of electrons and creates an overvoltage. Therefore, the charge transfer must overcome this overvoltage. Overvoltage is defined as:

$$\eta = \phi_s - \phi_e - U \tag{3}$$

where ϕ_s is the potential of the electrode; ϕ_e is potential of electrolyte. The Butler-Volmer equation is frequently employed in electrochemical models to depict charge transfer kinetics by establishing a relationship between the over-voltage of the electrode and the current density, as shown in Eq. (4).

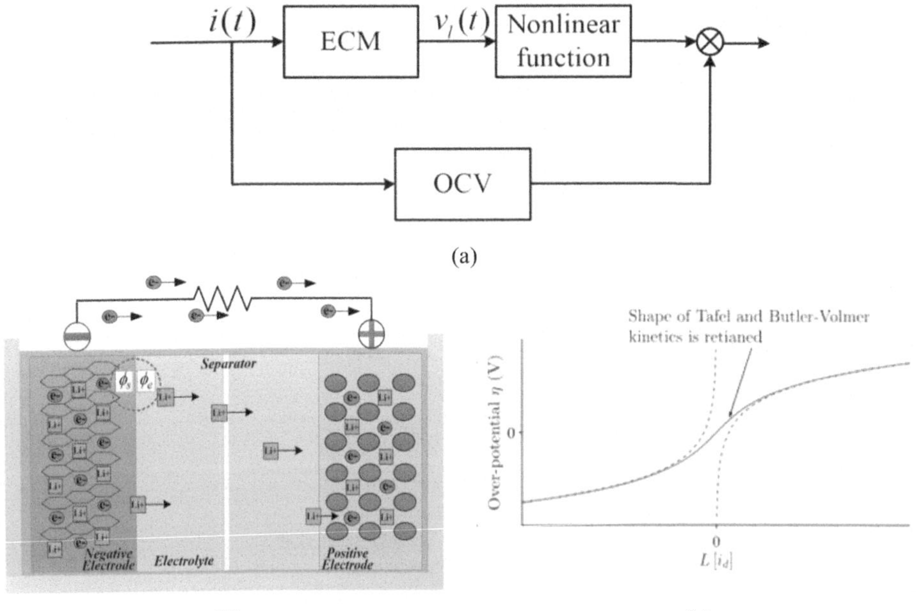

Fig. 1. Structure of NLECM model (a) Lithium battery NLECM model structure [14]. (b) A plot of the tafel and Butler-Volmer kinetic. (c) The principle of a lithium-ion battery [14].

$$\eta_{ct} = \frac{2RT}{F}\sinh^{-1}\left(\frac{I_{batt}(t)}{2i_0}\right) \tag{4}$$

Where T is the temperature, i_0 is the exchange current, F is Faraday's constant, and R is the molar gas constant.

Equation (4) represents the nonlinear relationship between current and overvoltage, which is rarely explained in the traditional battery ECM. By drawing a comparison diagram between the measured over-potential of each SOC and the output of the modeled ECM, the existence of any nonlinear deviation can be checked and captured by the nonlinear function.

Figure 1(c) shows the Tafel and Butler-Volmer kinetic Considering the S-type characteristics of the ideal Butler-Volmer dynamics, the S-type function is used to simulate the nonlinear characteristics, because its bounded and differentiable characteristics can guarantee the convergence of the function.

$$f(\overline{v}_l) = \frac{c_1 \overline{v}_l}{\sqrt{1 + c_2 \overline{v}_l^2}} \tag{5}$$

where c_l is the linear and c_2 is the non-linear sigmoid coefficients in Eq. (5). Two S-shaped coefficients can be estimated for each SOC and temperature.

The OCV Model Block

The battery undergoes a complete charging and discharging process during the battery modeling operation, utilizing an extremely low current (often C/25) that can potentially reduce any battery polarization. The SOC-OCV curve is created by averaging the observed charging and discharging voltage curves.

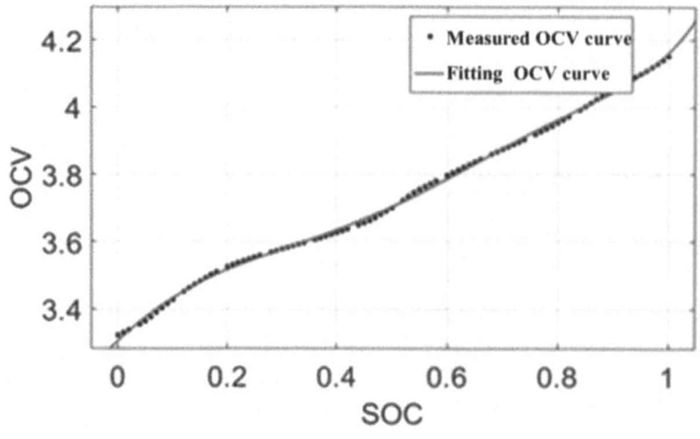

Fig. 2. The SOC-OCV curve.

As shown in Fig. 2, SOC and OCV are fitted. The function relationship is as follows:

$$\begin{aligned} OCV &= p_1 * SOC^6 + p_2 * SOC^5 + p_3 * SOC^4 \\ &+ p_4 * SOC + p_5 * SOC^2 + p_6 * SOC + p_7 \end{aligned} \tag{6}$$

2.2 The WVAEKF Algorithm Based on NLECM Model

As shown in Table 1, the standard AEKF is a fusion of the Extended Kalman Filter (EKF) plus an adaptive law. The EKF consists of two components: the model prediction step and the state correction phase. The states at step k are predicted in the prediction step using the information from step k-1. In the correction phase, the projected states are adjusted using the newly obtained measurements at time step k. The noise covariance is updated online by the adaptive law of the AEKF. Table 1 shows the estimation process of the conventional AEKF algorithm.

The state and measurement equations that are often utilized in the conventional AEKF are stated as follows.

$$\begin{cases} x_{k+1} = f(x_k, u_k) + \omega_k \\ y_k = g(x_k, u_k) + v_k \end{cases} \tag{7}$$

where, x_k, x_{k+1} represents the system state vectors at time k, k + 1, y_k is the measurement vector, w_k and v_k represent the Gaussian process and measurement noise, respectively.

In NLECM model, state vectors of state equation and observation equation are:

$$x(k) = \{SOC(k), U_1(k), U_2(k)\} \tag{8}$$

where *SOC(k)* represents the SOC at time k, $U_1(k)$, $U_2(k)$ are voltages at both ends of two RC branches. In NLECM model, the state equation is:

$$\begin{pmatrix} SOC(k+1) \\ U_1(k+1) \\ U_2(k+1) \end{pmatrix} = \begin{pmatrix} 1 & 0 & 0 \\ 0 & e^{-\frac{\Delta t}{R_1C_1}} & 0 \\ 0 & 0 & e^{-\frac{\Delta t}{R_2C_2}} \end{pmatrix} \times \begin{pmatrix} SOC(k) \\ U_1(k) \\ U_2(k) \end{pmatrix} + \begin{pmatrix} -\frac{\Delta_t}{Q} \\ R_1(1 - e^{-\frac{\Delta t}{R_1C_1}}) \\ R_2(1 - e^{-\frac{\Delta t}{R_2C_2}}) \end{pmatrix} i_k + w(k) \tag{9}$$

And the measurement equation is:

$$\begin{aligned} &U(k) = U_{OCV}(k) + U_{NLECM}(k) + v(k) \\ &(U_{NLECM}(k) = \frac{c_1 U_l}{\sqrt{1 + c_2 U_l^2}}; \; U_l = -R_0 I - U_1 - U_2) \end{aligned} \tag{10}$$

where $U_{OCV}(k)$ represents the value of OCV, U_l is the linear voltage output by ECM model.

Parameter matrix (A, B, C) are

$$A_k = \begin{pmatrix} 1 & 0 & 0 \\ 0 & e^{-\frac{\Delta t}{R_1C_1}} & 0 \\ 0 & 0 & e^{-\frac{\Delta t}{R_2C_2}} \end{pmatrix}, B_k = \begin{pmatrix} -\frac{\Delta t}{Q} \\ R_1(1 - e^{-\frac{\Delta t}{R_1C_1}}) \\ R_2(1 - e^{-\frac{\Delta t}{R_2C_2}}) \end{pmatrix} C_k = \left(\frac{dU_{ocv}}{dSOC} \; -\frac{\partial U}{\partial U_1} \; -\frac{\partial U}{\partial U_2} \right) \tag{11}$$

In the conventional adaptive law, the detection error innovation sequence has a set window of time, meaning that it only takes into account changes in the error innovation within a predetermined time frame. However, various operating situations and environments, as well as errors in the battery sensor, might alter the change of error in a practical application. Therefore, it is essential to determine how the error innovation sequence distribution has changed and adjust the size of the moving window according to the change of the error distribution.

Table 1. A summary of the conventional AEKF steps.

The conventional AEKF Algorithm Steps
1) State estimation
$\hat{x}_k^- = f(\hat{x}_{k-1}^+, u_{k-1})$
2) State estimation covariance update
$P_k^- = \hat{A}_k P_{k-1} \hat{A}_k^T + Q_{k-1}$
3) Error innovation sequence calculation
$e_k = y_k - g(\hat{x}_k^-, u_k)$
4) Kalman gain matrix
$L_k = P_k^- C_k^T [C_k P_k^- C_k^T + R_k]^{-1}$
5) State estimate update
$\hat{x}_k^+ = \hat{x}_k^- + L_k e_k$
6 Error covariance update
$P_k^+ = (I - L_k C_k) P_k^-$
7 Conventional adaptive law
$H_k = \frac{1}{M} \sum_{i=k-M+1}^{k} e_k e_k^T$
$R_k = H_k + C_k P_k^- C_k^T$
$Q_k = L_k H_k L_k^T$

According to Ref. [16–18], a window with a length of 2N is set in this paper. The error innovation sequence in the window is expressed as $[e_{k\text{-}2N}, e_{k\text{-}2N+1} \ldots\ldots e_{k\text{-}1}]$, ek ~ N(0,σ2), which is set up to detect the variation of the error innovation sequence.

If the error innovation sequence produced by the NLECM model remains unchanged within the first window, the duration of the identification window is extended, and a greater number of error innovation sequences are utilized to estimate the innovation covariance matrix. When there is a modification in the error innovation sequence, the updated error innovation sequence is chosen to estimate the covariance matrix of the innovation. In order to identify the change of error innovation sequence amplitude, the error sequence is squared:

$$M_k = e_k^2 \tag{12}$$

The sequence M_k is divided into two N-length regions. The length of the window is determined by identifying any changes in the distribution of the error innovation sequence inside the window. The probability density function of sequence M_k is:

$$f_M(m) = \prod_{n=k-2N}^{k-1} \frac{1}{\sqrt{2\pi\sigma^2 m_n}} e^{-\frac{m_n}{2\sigma^2}} \tag{13}$$

Next, the variance is calculate using the maximum likelihood function method. The logarithmic maximum likelihood function can be obtained by taking logarithms on both sides of the above formula. Then the partial derivative of the variance is 0, and the variance can be calculated.

$$A = \ln(f_M(m)) = -\frac{1}{2}\sum_{n=k-2N}^{k-1}\left[\ln\left(2\pi\sigma^2 m_n\right) + \frac{m_n}{\sigma^2}\right] \tag{14}$$

The variance σ^2 is

$$\sigma^2 = \frac{1}{2N}\sum_{n=k-2N}^{k-1} m_n \tag{15}$$

In the center of the detection window is a border detection point. Presently, the probability density function of Mk throughout the whole window is calculated by multiplying the probability density function of Mk in the two regions prior to and following the modification:

$$f_M(m) = \prod_{n=k-N}^{k-1}\frac{1}{\sqrt{2\pi\sigma_1^2 m_n}}e^{-\frac{m_n}{2\sigma_1^2}}\prod_{n=k-2N}^{k-N-1}\frac{1}{\sqrt{2\pi\sigma_2^2 m_n}}e^{-\frac{m_n}{2\sigma_2^2}} \tag{16}$$

Similarly, the maximum likelihood estimation is used to obtain:

$$B = \ln(f_M(m)) = -\frac{1}{2}\sum_{n=k-N}^{k-1}\left[\ln\left(2\pi\sigma_1^2 m_n\right) + \frac{m_n}{\sigma_1^2}\right] - \frac{1}{2}\sum_{n=k-2N}^{k-N-1}\left[\ln\left(2\pi\sigma_2^2 m_n\right) + \frac{m_n}{\sigma_2^2}\right] \tag{17}$$

$$\begin{aligned}\sigma_1^2 &= \frac{1}{N}\sum_{n=k-N}^{k-1} m_n\\ \sigma_2^2 &= \frac{1}{N}\sum_{n=k-2N}^{k-N-1} m_n\end{aligned} \tag{18}$$

Finally, it is derived:

$$B - A = N \cdot \ln\left(\frac{\sigma^2}{\sigma_1\sigma_2}\right) \tag{19}$$

where σ^2 is the variance of the whole window M sequence, σ_1 is the variance of the region before the change, σ_2 is the latter regional variance, B is he changing probability density function, A is the probability density function that does not change. The improved adaptive law is as follows:

$$H_k = \frac{1}{W_k}\sum_{i=k-W_k+1}^{k} e_k e_k^T\; W_k = \begin{cases} W_0\,(B-A) > Th \\ W_{k-1}{+}1(B-A) \le Th\ \&\ W_k \le W_{\max} \\ W_{max}\, W_k > W_{\max} \end{cases} \tag{20}$$

where, Th is threshold value (Usually is 1). W_{max} is the maximum window value. W_0 is the initial window length. ($W_k > W_{\max}$)

When $(B - A) \leq Th$ condition is satisfied, it indicates that the detected noise does not change significantly. The moving window will increase linearly with time, but the length of the window cannot exceed the maximum window length.

When $(B - A) \leq Th$ is not satisfied, it indicates that the error innovation sequence in the detection window has changed significantly. At this time, the window selects the changed sequence to update the innovation covariance matrix. The improved algorithm adaptively adjusts the size of the moving window based on changes in the error information sequence.

3 Experimental Results

3.1 Experimental Setup

The experimental setup for battery testing is described in Fig. 3 It consists of the experimental cell, a thermal chamber, a charge and discharge testing device (5V20A-NTFA), and a programmable upper computer. The BTS.8 hardware in the computer was used to monitor the cell and capture experimental data.

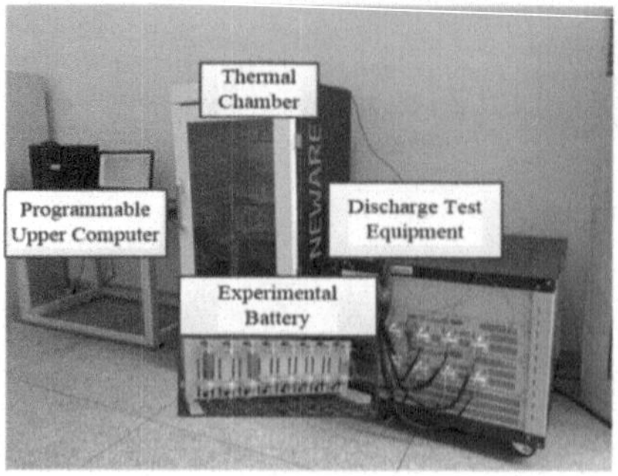

Fig. 3. The experimental setup.

3.2 The Results of WVAEKF SOC Estimation Algorithm

In this work, the current-voltage relationship of the battery was modeled using pulse-multisine current signals, and the signal was applied at five SOCs (10%, 20%, 50%, 80% and 95%) to parameterize the model. The remaining time parameter values are calculated by the n-D Lookup Table tool of MATLAB\Simulink. Figure 4 shows the results of parameter identification.

In this work, the NLECM model is established based on 18650 $LiNiCoAlO_2$ batteries. Figure 5 describes the calculation process of the WVAEKF algorithm. The error innovation sequence is found by comparing the NLECM output voltage with the real output voltage. On this basis, the battery SOC is estimated by the WVAEKF model. The WVAEKF algorithm selects different window lengths to correct the state estimation

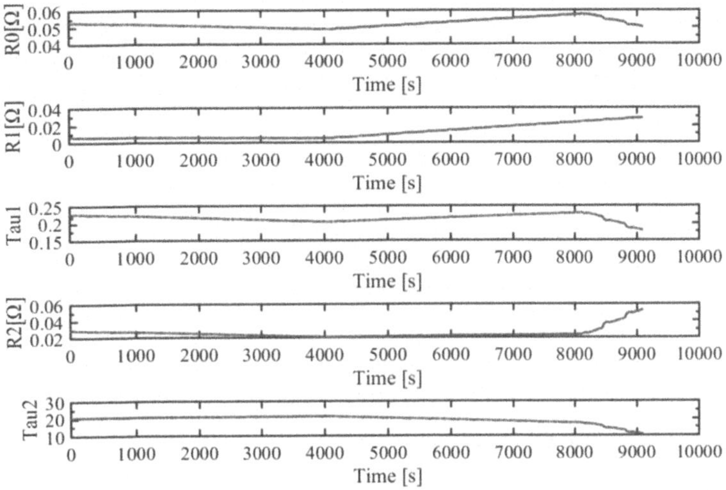

Fig. 4. Results of Parameter Identification using multisine signals.

according to the characteristics of the error sequence, which improves the accuracy of the SOC estimation. In this work, the initial window length is 2, the maximum window length is 6 and the change length of each window is 1.

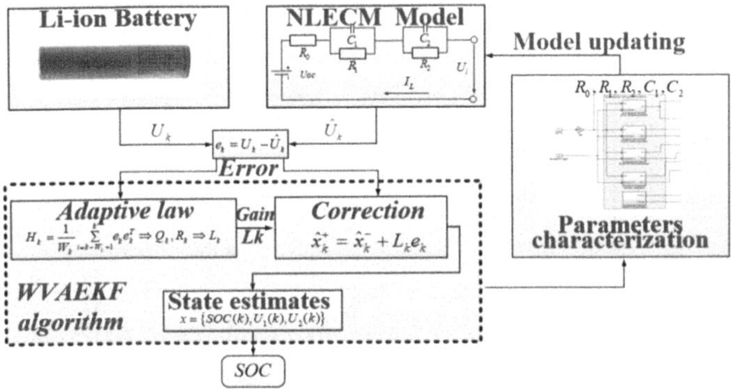

Fig. 5. The calculation process of WVAEKF algorithm.

Figure 6 shows the SOC results based on the WVAEKF, and the reference SOC is obtained by the experimental equipment. According to the research results, the prediction error of the algorithm for SOC is limited to 1%, and the RMSE of the algorithm is only 0.0054.

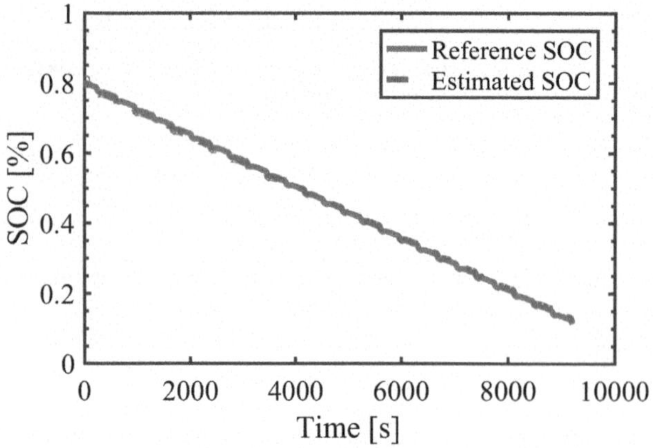

Fig. 6. The Reference SOC vs estimated SOC;

3.3 Comparison of Different SOC Methods

To quantitatively compare the performance of different SOC estimation methods based on the NLECM, the root mean square error of SOC estimation of these three methods is listed in Table 2 The investigation demonstrates that the WVAEKF suggested in this study reaches the maximum level of estimation accuracy. In comparison to the conventional AEKF method with a fixed window length of 6, the root mean square error of SOC estimation based on WVAEKF algorithm is reduced by 23% shown as Fig. 7.

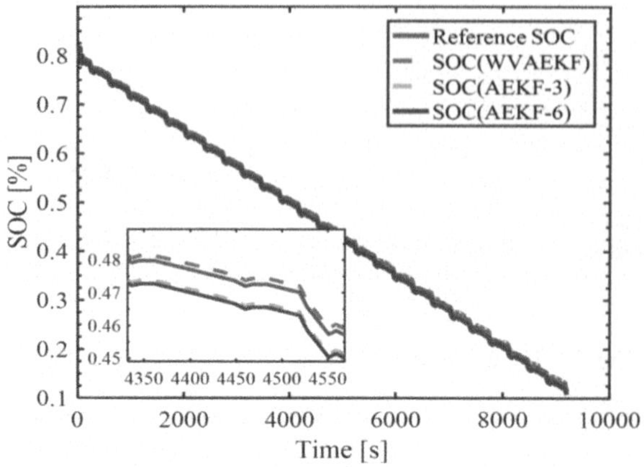

Fig. 7. The WVAEKF algorithm based on the NLECM model is compared with the traditional AEKF algorithm (fixed window: 3, 6) to estimate SOC.

Table 2. RMSE based on different SOC estimation methods.

Method	WVAEK	AEKF-3	AEKF-6
RMSE	0.0054	0.0063	0.0070

4 Conclusions

Based on the NLECM model, a WVAEKF algorithm is established to estimate the SOC of Li-ion batteries. A linear ECM, a nonlinear overvoltage function, and an OCV block make up the NLECM model. In order to parameterize the NLECM, a pulse-multisine is utilized, allowing parameterization over a wider bandwidth than the HPPC signal.

In this work, WVAEKF updates the noise covariance by identifying changes in the distribution of the error innovation sequence. After experimental validation, the SOC estimation error of the WVAEKF algorithm is limited to within 1%. The findings demonstrate the effective realization of the SOC estimation with high precision based on the NLECM model under the DST test.

In this work, the SOC calculation for a single battery cell has yielded exceptional results. Nevertheless, the battery pack often consists of numerous battery cells, therefore it is imperative to enhance the SOC estimation from an individual battery cell to the SOC estimation of the entire battery pack in the future.

Acknowledgments. The project was supported by the Scientific Research Foundation of Nanjing Institute of Technology (No. YKJ202212), the Jiangsu High-level Entrepreneurial and Innovative Talent Program (No. JSSCBS20230380), the Open Research Fund of Jiangsu Collaborative Innovation Center for Smart Distribution Network, Nanjing Institute of Technology (No. XTCX202307), the Postgraduate Research & Practice Innovation Program of Jiangsu Province (SJCX24_1303), and the Jiangsu Province University Student Innovation and Entrepreneurship Plan Project (202411276002Z).

Disclosure of Interests. The authors declare that they have no known competing financial interests or personal relationships that could have appeared to influence the work reported in this paper.

References

1. Wang, W., Wang, X., Xiang, C., Wei, C., Zhao, Y.: Unscented kalman filter-based battery soc estimation and peak power prediction method for power distribution of hybrid electric vehicles. IEEE Access **6**, 35957–35965 (2018). https://doi.org/10.1109/ACCESS.2018.2850743
2. Fan, C., Liu, K., Zhu, T., Peng, Q.: Understanding of lithium-ion battery degradation using multisine-based nonlinear characterization method. Energy **290**, 130230 (2024). https://doi.org/10.1016/j.energy.2024.130230
3. Xiong, R., Cao, J., Yu, Q., He, H., Sun, F.: Critical review on the battery state of charge estimation methods for electric vehicles. IEEE Access **6**, 1832–1843 (2018). https://doi.org/10.1109/Access.2017.2780258

4. Zhang, S., Guo, X., Dou, X., Zhang, X.: A data-driven coulomb counting method for state of charge calibration and estimation of lithium-ion battery. Sustain. Energy Technol. Assess **40**, 100752 (2020). https://doi.org/10.1016/j.seta.2020.100752
5. Michel, P.H., Heiries, V.: An adaptive sigma point kalman filter hybridized by support vector machine algorithm for battery SOC and SOH estimation. In: 2015 IEEE 81st Vehicular Technology Conference, pp. 1–7. IEEE, Glasgow, UK (2015). https://doi.org/10.1109/VTCSpring.2015.7145678
6. Tian, H., Ouyang, B.: Estimation of EV battery SOC based on KF dynamic neural network with GA. In: 2018 Chinese Control and Decision Conference, pp. 2720–2724. IEEE, Shenyang, China (2018). https://doi.org/10.1109/CCDC.2018.8407587
7. He, Z., Gao, M., Wang, C., Wang, L., Liu, Y.: Adaptive state of charge estimation for li-ion batteries based on an unscented kalman filter with an enhanced battery model. Energies **6**(8), 4134–4151 (2013). https://doi.org/10.3390/en6084134
8. Yokoi, Y., Ito, D., Yoshihara, Y., Ishigoori, K., Shitanda, I., Katayama, N.: A distributed constant equivalent circuit model and a method for estimating parameters for paper-based lactate biofuel cells. Electrochimica Acta **493**, 144384 (2024). https://doi.org/10.1016/j.electacta.2024.144384
9. Widanage, W.D., Barai, A., Chouchelamane, G.H., Uddin, K., McGordon, A., Marco, J., et al.: Design and use of multisine signals for Li-ion battery equivalent circuit modelling. part 2: model estimation. J. Power Sour. **324**, 61–69 (2016). https://doi.org/10.1016/j.jpowsour.2016.05.014
10. Fan, C., Liu, K., Ren, Y., Peng, Q.: Characterization and identification towards dynamic-based electrical modeling of lithium-ion batteries. J. Energy Chem. **92**, 738–758 (2024). https://doi.org/10.1016/j.jechem.2024.01.040
11. Zhang, L., Li, P., Zhang, Z.T., Yan, Z.: SOC estimation of li-ion battery based on unscented kalman filter. In: 2018 21st International Conference on Electrical Machines and Systems, pp. 2177–2182. IEEE, Jeju Korea (2018). https://doi.org/10.23919/ICEMS.2018.8548966
12. Zhu, C., Wang, S., Yu, C., Zhou, H., Fernandez, C., Guerrero, J.M.: An improved cauchy robust correction-sage Husa extended Kalman filtering algorithm for high-precision SOC estimation of Lithium-ion batteries in new energy vehicles. J. Energy Storage **88**, 111552 (2024). https://doi.org/10.1016/j.est.2024.111552
13. Shu, X., Li, G., Zhang, Y., Shen, S., Chen, Z., Liu, Y.: Stage of charge estimation of lithium-ion battery packs based on improved cubature kalman filter with long short-term memory model. IEEE Trans. Transp. Electrification **7**(3), 1271–1284 (2021). https://doi.org/10.1109/TTE.2020.3041757
14. Wang, Q., Sun, C., Gu, Y.: Research on SOC estimation method of hybrid electric vehicles battery based on the grey wolf optimized particle filter. Comput. Electr. Eng. **110**, 108907 (2023). https://doi.org/10.1016/j.egyr.2023.01.018
15. Fan, C., O'Regan, K., Li, L., Higgins, M.D., Kendrick, E., Widanage, W.D.: Data-driven identification of lithium-ion batteries: a nonlinear equivalent circuit model with diffusion dynamics. Appl. Energy **321**, 119336 (2022). https://doi.org/10.1016/j.apenergy.2022.119336
16. Emami, M., Taban, M.R.: A novel intelligent adaptive kalman filter for estimating the submarine's velocity: with experimental evaluation. Ocean Eng. **158**, 403–411 (2018). https://doi.org/10.1016/j.oceaneng.2018.04.021
17. Sun, D., Yu, X., Zhang, C., Wang, C., Huang, R.: State of charge estimation for lithium-ion battery based on an intelligent adaptive unscented Kalman filter. Int. J. Energy Res. **44**(14), 11199–11218 (2020). https://doi.org/info:doi/10.1002/er.5690
18. Fan, C., O'Regan, K., Li, L., Kendrick, E., Widanage, W.D.: Frequency domain non-linear characterization and analysis of lithium-ion battery electrodes. J. Energy Storage **36**(2), 102371 (2021). https://doi.org/10.1016/j.est.2021.102371

Battery State of Energy Estimation Method Based on Conditional Selection Neural Network Model

Kaituo Liu[1], Yihuan Li[1(✉)], and Qiao Peng[2]

[1] School of Control and Computer Engineering, North China Electric Power University, Beijing 102206, China
yihuan@ncepu.edu.cn

[2] Group of Information Technology, Analytics and Operations, Queen's University Belfast, Belfast BT9 5EE, UK

Abstract. There are differences in the discharge data of batteries at different life stages. Considering these differences, this paper proposes a conditional selection neural network model for estimating the SOE. It can flexibly select the appropriate network for prediction based on a set selection mechanism under different conditions, thereby improving the model's applicability and generalization ability. Firstly, the concept of Squeeze-and-Excitation is incorporated into the CNN and the resulting SE-CNN model is pretrained. Next, a selection mechanism is designed according to the degree of battery aging, thus the training data is divided into four parts based on the number of discharge cycles to further train the pre-trained network to form a "four-in-one" CS-SE-CNN model. During SOE estimation, the CS-SE-CNN model selects the prediction network based on the number of discharge cycles in the input data. The results demonstrate that the proposed method exhibits high accuracy and robustness in estimating the SOE of lithium-ion batteries.

Keywords: Conditional Selection · SE Blocks · Convolutional Neural Network · State of Energy

1 Introduction

With the widespread use of fossil fuels, global energy crisis and climate changes have become increasingly serious issues. Lithium-ion batteries, which can power electric vehicles and store energy on renewable grids, helping to reduce emissions from transportation and energy sectors, thus play an important role in accelerating the energy transition to clean energy [1, 2]. To ensure safe and reliable operation, a well-functioning Battery Management System (BMS) is crucial [3].

State of Energy (SOE) is an important monitoring parameter in BMS [4], and it is a crucial foundation for achieving load balancing, energy allocation, and power safety. Therefore, accurate SOE estimation is essential for lithium-ion batteries. SOE is defined

K. Li et al. (Eds.): LSMS/ICSEE 2024, CCIS 2218, pp. 267–278, 2025.
https://doi.org/10.1007/978-981-96-0232-2_21

as the ratio of the remaining available energy of the battery to the maximum available energy of the battery, as shown in Eq. (1).

$$SOE_t = SOE_0 - \frac{\int_0^t p(\xi)d\xi}{E_n} = SOE_0 - \frac{\int_0^t i(\xi)u(\xi)d\xi}{E_n} \tag{1}$$

where SOE_0 is the initial SOE value, SOE_t denotes the SOE value at time t, and E_n is the total available energy. The variables $p(\xi)$, $i(\xi)$ and $u(\xi)$ stand for power, current and voltage at time ξ, respectively.

Traditionally, the remaining energy in a battery is measured by State of Charge (SOC). SOC represents the remaining charge of the battery [5] and does not take into account the energy loss during discharge [6]. With the increasing widespread application of lithium-ion batteries, the functional requirements of battery management systems have become more sophisticated and complex. SOC is no longer suitable for predicting the remaining energy of power battery systems [7]. In contrast, the SOE metric focuses more on the energy that can be released by a lithium-ion battery during actual use [8]. SOE is directly related to energy consumption, and accurate SOE estimation provides a solid foundation for efficient battery use and the safety of BMS.

In recent years, research on SOE of lithium-ion batteries has been ongoing [9]. The power integration method is the basic method for SOE estimation. It is an open-loop estimation [10], which is susceptible to the accumulation of errors caused by noise, leading to a decrease in accuracy.

Lai et al. [11] applied particle filtering and extended Kalman filters to estimate SOE. Li et al. [12] combined battery and analytical models to predict battery terminal voltage characteristics and SOE. Long et al. [13] established a 2RC equivalent circuit model based on electrochemical laws. However, the complex chemical reactions within batteries make it difficult to establish accurate models. The significant differences among different battery models result in poor model portability, making them challenging to apply widely.

Data-driven methods do not require modeling the complex chemical reactions inside batteries, they only need a large amount of data to fit the results. Tian et al. [14] used Least Squares Support Vector Regression (LSSVR) for SOE estimation, avoiding complex parameter calculations in circuit models. Chen et al. [15] achieved joint estimation of SOE and SOC based on optimized bidirectional gated recurrent neural network. However, as the volume of data increases, the training time becomes longer, leading to an increase in cumulative errors [16]. Ma et al. [17] proposed an estimation method for SOC and SOE based on Long Short-Term Memory Neural Network (LSTMNN), reducing data volume and accelerating network training through memory gate circuits. To further reduce the impact of noise, Wu et al. [18] introduced an optimized Convolutional Neural Network-Bidirectional Gated Recurrent Unit (CNN-Bi-GRU) and an improved Kalman Bidirectional Smoothing algorithm, adding square root and reverse smoothing methods to the classic Square Root Extended Kalman Smoothing (SREKS) for filtering and denoising neural network outputs. However, there are differences in the discharge data of batteries at different life stages, which may cause the model to lose some features and affect accuracy.

To mitigate the negative impact of the differences in discharge data on the network, this paper proposes a conditional selection neural network model. The main contributions of this paper are as follows:

- Establish a convolutional neural network (CNN) with SE blocks, denoted as SE-CNN, where different weights are assigned to the network's hidden states to enhance the impact of important information on SOE.
- Give more attention to the differences in input data, divide the battery into four life stages based on the cycle discharge count, and accordingly divide the training data into four parts. Each part of the data is separately used to train SE-CNN, resulting in four networks with the same structure but different parameters. These four networks form a "four-in-one" CS-SE-CNN model. During the SOE forecasting period, the appropriate network is selected based on the cycle discharge count of the input data.

2 Method Background and Theoretical Analysis

2.1 Convolutional Neural Network

CNN is inspired by biological visual perception, and it possesses strong feature extraction capabilities, making it highly advantageous in handling image data [19].

The CNN mainly consists of convolutional layer, pooling layer, and fully-connected layer. Each kind of layer has a different function. In convolution layer, features are extracted from images through convolutional operations, where convolutional kernels slide over the image and perform dot product operations with local pixels to generate feature maps. To learn more nonlinear features, non-linear activation functions are added after the convolutional layer, allowing the network to approximate arbitrary complex functions and enhance its expressive power. The pooling layer compresses features, reduces the number of parameters in the network, and helps prevent overfitting issues. After passing through multiple convolutional layers and pooling layers, the feature maps are flattened and connected to the fully connected layer for tasks such as classification or regression.

The information transmission and training process of CNN are similar to the learning process of creatures. The activation function simulates the function of neural electrical signals passing among neurons of creatures, where information is transmitted only when the electrical signal exceeds a certain threshold. During CNN training, the loss function and optimizer enable CNN to learn specific content.

2.2 Squeeze-And-Excitation Blocks

Squeeze-and-Excitation Blocks [20] are channel attention modules. They can be easily added to CNN to improve performance. SE blocks add parameters to each channel of the convolutional block, allowing the network to assign different attention to each feature map during training. A basic structure of SE blocks is shown in Fig. 1.

First, $\mathbf{F}_{tr}: \mathbf{X} \rightarrow \mathbf{U}, \mathbf{X} \in \mathbb{R}^{H' \times W' \times C'}, \mathbf{U} \in \mathbb{R}^{H \times W \times C}$, is the convolution operation.

$$\mathbf{u}_c = \mathbf{v}_c * \mathbf{X} = \sum_{s=1}^{C'} \mathbf{v}_c^s * \mathbf{x}^s, \tag{2}$$

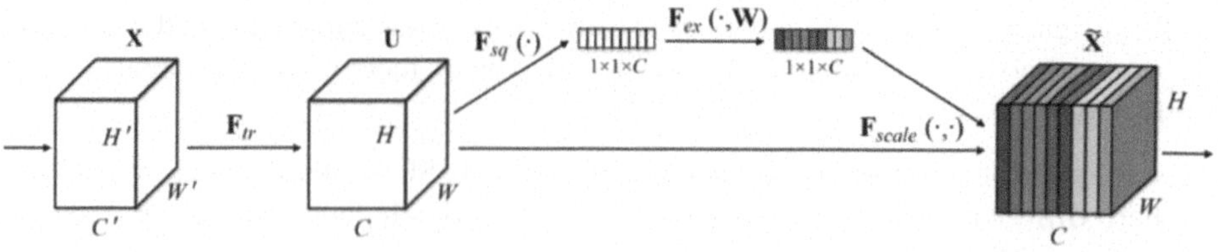

Fig. 1. Structure of SE Blocks

where $*$ represents convolution, $\mathbf{v}_c = [\mathbf{v}_c^1, \mathbf{v}_c^2, \ldots, \mathbf{v}_c^{C'}]$ and $\mathbf{X} = [\mathbf{x}^1, \mathbf{x}^2, \ldots, \mathbf{x}^{C'}]$ (the bias term is omitted for simplicity), while $\mathbf{v}_c^s$ is a two-dimensional spatial kernel that represents a single channel of $\mathbf{v}_c$, which acts on the corresponding channel of $\mathbf{X}$.

The output U from $\mathbf{F}_{tr}$ corresponds to the second 3D matrix on the left side in Fig. 1, or it can be referred to as C feature maps of size $H \times W$. The symbol $\mathbf{u}_c$ represents the c-th feature map in U.

Then comes the Squeeze operation, which compresses features along the spatial dimensions to transform each 2D feature map into a scalar, which to some extent captures the global receptive field [20]. Formally, a statistic $\mathbf{z} \in \mathbb{R}^C$ is generated by shrinking U through spatial dimensions $H \times W$,where the c-th element of z is calculated by:

$$z_c = \mathbf{F}_{sq}(\mathbf{u}_c) = \frac{1}{H \times W} \sum_{i=1}^{H} \sum_{j=1}^{W} u_c(i, j). \tag{3}$$

Excitation operation is similar to a gate mechanism in recurrent neural networks.

$$\mathbf{s} = \mathbf{F}_{ex}(\mathbf{z}, \mathbf{W}) = \sigma(g(\mathbf{z}, \mathbf{W})) = \sigma(\mathbf{W}_2 \delta(\mathbf{W}_1 \mathbf{z})), \tag{4}$$

where δ refers to the ReLU function, σ represents the sigmoid function, $\mathbf{W}_1 \in \mathbb{R}^{\frac{C}{r} \times C}$ and $\mathbf{W}_2 \in \mathbb{R}^{C \times \frac{C}{r}}$. We parameterize the gating mechanism by utilizing two fully connected (FC) layers, comprising a dimensionality-reduction layer with parameters W1 and reduction ratio r, followed by a ReLU layer, and finally a dimensionality-increasing layer with parameters W2 [20].

The final step is a Reweight operation, which treats the weights outputted by Excitation as the importance of each feature channel after feature selection. These weights are then multiplied element-wise and added to the original features, completing the recalibration of the original features in the channel dimension. The final output of the block is obtained by rescaling the transformation output U with the activations:

$$\tilde{\mathbf{x}}_c = \mathbf{F}_{scale}(\mathbf{u}_c, s_c) = s_c \cdot \mathbf{u}_c, \tag{5}$$

where $\widetilde{\mathbf{X}} = [\widetilde{\mathbf{x}}_1, \widetilde{\mathbf{x}}_2, \ldots, \widetilde{\mathbf{x}}_C]$ and $\mathbf{F}_{scale}(\mathbf{u}_c, s_c)$ represents channel-wise multiplication between the feature map $\mathbf{u}_c \in \mathbb{R}^{H \times W}$ and the scalar s_c.

2.3 Conditional Selection Neural Network

This paper proposes a prediction method based on conditional selection neural networks, which combines multiple neural networks. When input data is provided, the appropriate network is selected for prediction based on a predefined conditional selection mechanism.

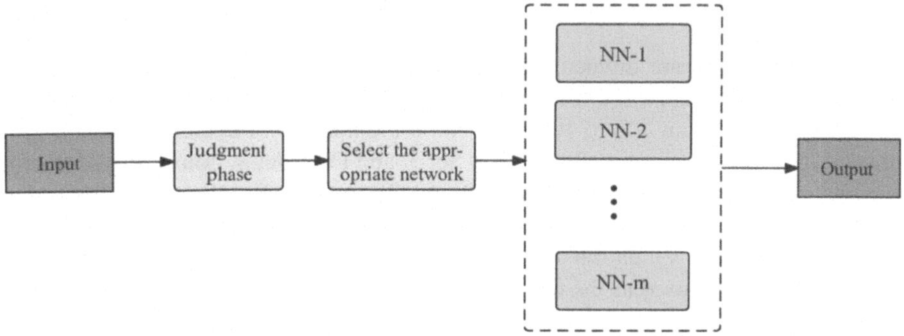

Fig. 2. Structure of Conditional Selective Neural Network

The structure of the conditional selection neural network is shown in Fig. 2. Firstly, m neural network models are established, which can be roughly obtained in two ways. The first way involves designing neural networks with different structures. These networks vary in their ability to extract and fuse data, as well as their focus on different data features. By designing corresponding selection mechanisms, the advantages of m networks with different structures can be integrated. The second way involves using only one network but training it with m different datasets to obtain m networks with the same structure but different parameters. The advantage of this method lies in its focus on the differences among input data. Based on the features of the input data, the appropriate network can be selected for prediction. This method not only enhances the model's generalization ability but also reduces the probability of learning incorrect features and improves prediction accuracy.

Afterwards, design an appropriate selection mechanism: During prediction, based on the features or conditions of the input data, a selection mechanism is designed to choose an appropriate network for prediction. The selection mechanism can be a hard selection (directly choosing a network based on specific conditions) or a soft selection (weighted fusion of predictions from different networks).

The advantage of a conditionally selected neural network lies in its ability to flexibly choose an appropriate model for prediction based on different conditions or features, thereby enhancing the model's applicability and generalization ability.

3 Data Preparation and Model Establishment

3.1 Data Preparation

Dataset. MIT dataset [21] is used in this work, where the tested lithium-ion phosphate (LFP)/graphite batteries were manufactured by A123 Systems (APR18650M1A) and were cycled to failure under forced convection temperature set to 30 °C indoors. The nominal capacity of these batteries is 1.1 Ah, and the nominal voltage is 3.3 V.

Since the estimation of SOE is the focus, this study only considers the discharge part. All cells were discharged with a CC-CV discharge at 4C to 2.0V with a current cutoff of C/50.

The dataset is divided into three "batches," each consists of approximately 48 batteries. The last 10 cells in batch 1 are used to validate the performance of the proposed method in this study.

These 10 batteries are divided into 2 groups. The first group includes 8 batteries, named as the training set, used for pre-training the network. The number of cycles for each battery ranges from 700 to 1000. Based on the number of cycles, the training set is divided into Data Set 1, Data Set 2, Data Set 3, and Data Set 4, consisting of the first 200 cycles, cycles 201–400, cycles 401–600, and the remaining cycles of discharge data, respectively. Samples generated from these four datasets are first shuffled and randomly split into training and validation sets in an 8:2 ratio. The second group includes the remaining 2 batteries, used as the test set.

Input Data Sample. To better utilize the feature extraction capabilities of convolution, the measured time series signals are converted into 3D images as input to the network [22].

Current, voltage, and temperature, have different scales, directly transforming these variables into 3D image inputs will significantly degrade the network performance. Therefore, data normalization should be applied first. Then, multiple data blocks are segmented from one discharge cycle, with each data block containing 36 time steps of data. Each segmentation of the data block slides 12 time steps, meaning that adjacent data blocks have 24 overlapping time steps. Each data block is a 36×3 matrix representing the current, voltage, and temperature of 36 adjacent time steps. Finally, each data block is transformed into a 3D image with dimensions of $6 \times 6 \times 3$, serving as an input, and the state of energy (SOE) of the last time step corresponding to the data block serves as the label, composing a data sample.

3.2 Model Construction

The network framework proposed in this paper is established through three consecutive steps. Firstly, based on the dimension of the input data, a suitable CNN model is constructed. Next, an SE module is added after each convolutional layer to improve the inter-channel dependencies, forming an SE-CNN model, which is pre-trained using the training set. Finally, datasets corresponding to different life stages are used to separately train the pre-trained SE-CNN model, resulting in a "four-in-one" model—the CS-SE-CNN model. The detailed process of each stage is elaborated below.

Step 1. CNN Model. The designed CNN architecture is shown in Fig. 3, which includes 3 convolutional layers, one Flatten layer, and 2 fully connected layers. As mentioned earlier, the input is a $6 \times 6 \times 3$ matrix, and the output size of each layer is indicated at the top of each layer. $w \times h \times c$ represents the size of the feature maps, where w, h, and c denote width, height, and channels, respectively. For each convolutional layer, $N@k$

$\times\ m \times d$ denotes the filter design, which means there are N filters of size $k \times m \times d$ convolving with the input of that layer. The depth of the input is d, and the number of output feature maps must be N.

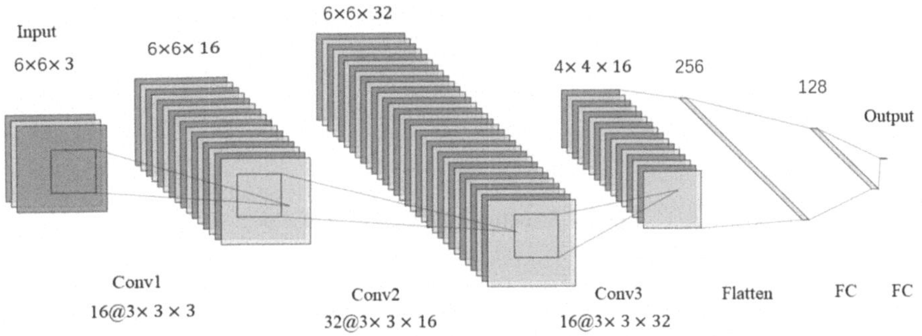

Fig. 3. Designed CNN

Step 2. SE-CNN Model. SE blocks add parameters to each channel of the convolutional block, allowing the network to assign different attention to each feature map during training. In this paper, SE Blocks are added after each convolutional layer, enabling the SE-CNN model to improve inter-channel dependencies with minimal increase in computational cost. The structure of the SE-CNN model is illustrated in Fig. 4.

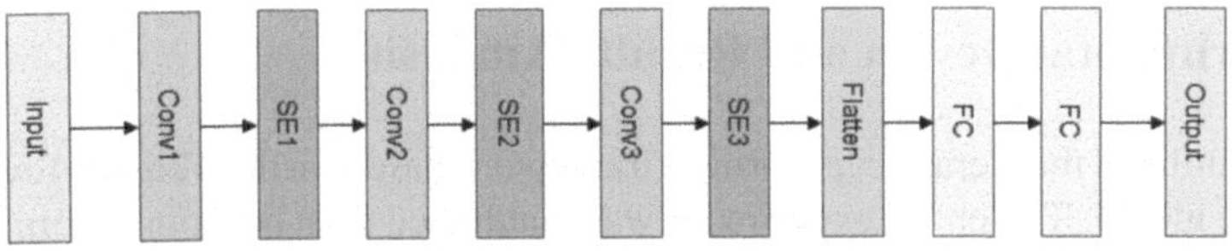

Fig. 4. Structure of SE-CNN

Pre-train the SE-CNN model using the training set.

Step 3. CS-SE-CNN Model. At different stages of a battery's life, as the battery ages, even if the battery is at the same SOE (State of Energy), there are differences in discharge data. Most data-driven methods overlook this difference in input data and directly use all data without distinguishing it for training the network. This not only leads to a decrease in the model's generalization ability, but may also mislead the model to learn incorrect features, resulting in erroneous predictions when processing these data, thereby reducing the accuracy of SOE predictions.

This paper pays more attention to the differences among input data and uses the number of discharge cycles as a criterion for distinguishing the life stage of the data and selecting the network. The battery is divided into four life stages based on the number of discharge cycles: early stage, mid-stage, late stage, and end stage. The training data is also divided into four corresponding parts.

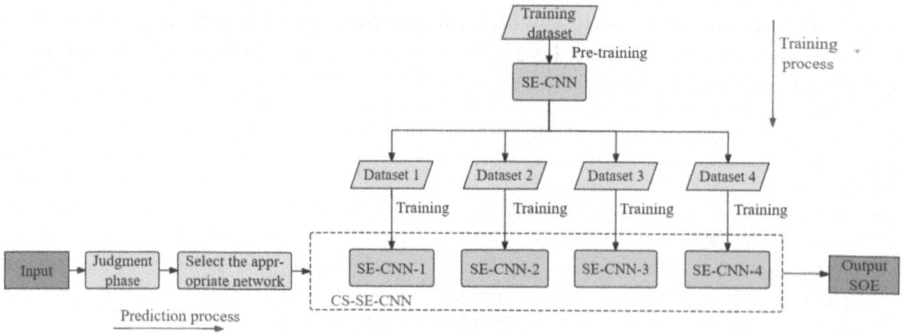

Fig. 5. Structure of CS-SE-CNN

The training and prediction process are shown in Fig. 5. First, pre-train the neural network using training set, and denote the network model as SE-CNN. Pre-trained model can accelerate the convergence speed of subsequent training. Next, train the pre-trained network using data from the early stage of battery life to obtain a network denoted as SE-CNN-1. Similarly, repeat this process for the mid-stage, late stage, and end stage of battery life to obtain corresponding network models: SE-CNN-2, SE-CNN-3, and SE-CNN-4. The combination of these four networks is referred to as CS-SE-CNN.

When estimating SOE, determine the life stage of the input data based on the number of discharge cycles and select the corresponding network for prediction, which outputs the SOE.

4 Experimental Design and Results Analysis

This study utilized the Keras deep learning framework based on the TensorFlow backend to build the model. The entire experiment was conducted on a computer equipped with an Intel(R) Core(TM) i5-9300H CPU and NVIDIA GeForce GTX 1650 GPU. The input to the network consists of current, voltage, and temperature data from 36 time steps, and the output is the predicted SOE value at the last time step.

4.1 Evaluation Metrics for SOE

To better reflect the predictive performance of the algorithm, MAE (Mean Absolute Error) and RMSE (Root Mean Square Error) are used for evaluation. MAE is a commonly used metric to assess the accuracy of prediction models, measuring the average absolute error between predicted values and actual values. RMSE is used to measure the average square root error between predicted values and actual values. The smaller the MAE and RMSE, the smaller the prediction error of the model, indicating a stronger predictive ability of the model.

$$MAE = \frac{1}{m}\sum_{i=1}^{m}|y_i - \hat{y}_i| \tag{6}$$

$$RMSE = \sqrt{\frac{1}{m}\sum_{i=1}^{m}(y_i - \hat{y}_i)^2} \tag{7}$$

Here, m represents the total number of input time series data points. y_i represents the actual SOE value, and $\hat{y}_i$ represents the predicted value by the model.

4.2 Experiment Results Analysis

The CS-SE-CNN model was trained using the prepared dataset, and its feasibility was tested using data from Battery 1 and Battery 2. The SOE estimation results and error comparisons for different models are displayed in Fig. 6, Fig. 7, and Fig. 8 respectively.

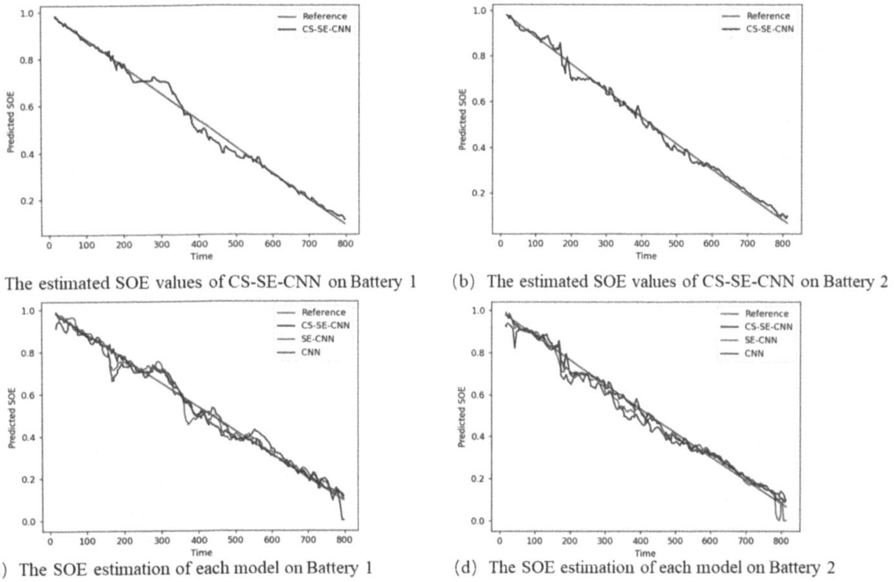

(a) The estimated SOE values of CS-SE-CNN on Battery 1

(b) The estimated SOE values of CS-SE-CNN on Battery 2

(c) The SOE estimation of each model on Battery 1

(d) The SOE estimation of each model on Battery 2

Fig. 6. The SOE estimation of each model on Battery 1 and Battery 2.

As shown in Fig. 6, throughout the discharge periods of both batteries, the SOE estimation results of CS-SE-CNN exhibit higher accuracy compared to CNN and SE-CNN. Comparing the results of SE-CNN and CNN also confirms the role of the SE module in model training, as it can adaptively adjust the weights of each feature map. Furthermore, comparing the results of CS-SE-CNN and SE-CNN, the addition of the selection mechanism has improved the accuracy of the model in SOE estimation. Particularly, at lower SOE levels, both CNN and SE-CNN models exhibit larger errors, while CS-SE-CNN performs well.

In Fig. 7, when estimating the SOE of Battery 1, the maximum error for CNN is 0.118, for SE-CNN is 0.126, and for CS-SE-CNN is 0.072. When estimating the SOE of

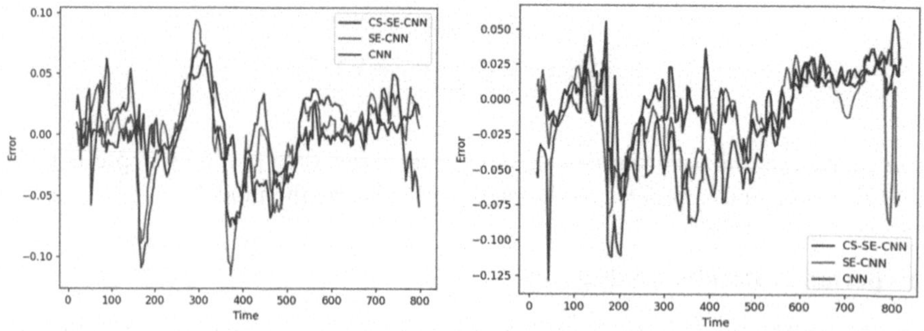

(a) The estimation error of each model on Battery 1's SOE (b) The estimation error of each model on Battery 2's SOE

Fig. 7. The SOE estimation errors of each model on Battery 1 and Battery 2.

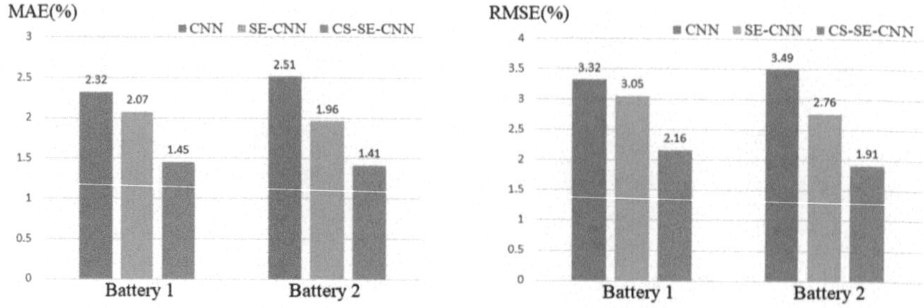

(a) The MAE of SOE estimation on Battery 1 and Battery 2 (b) The RMSE of SOE estimation on Battery 1 and Battery 2

Fig. 8. The MAE and RMSE of each model's SOE estimation on Battery 1 and Battery 2.

Battery 2, the maximum error for CNN is 0.132, for SE-CNN is 0.094, and for CS-SE-CNN is 0.074. Further, the MAE and RMSE of SOE estimation with different models are illustrated in Fig. 8. When estimating the SOE of Battery 1, the CS-SE-CNN model has an MAE of 1.45% and RMSE of 2.16%. When estimating the SOE of Battery 2, the CS-SE-CNN model has an MAE of 1.41% and RMSE of 1.91%. This indicates that CS-SE-CNN has higher estimation accuracy and good robustness.

5 Conclusions

This paper proposes a conditional selection neural network model for estimating the State of Energy (SOE) of batteries. To improve the accuracy of the model, multiple neural network models are established. When using the model, an appropriate network is selected for prediction based on a predefined selection mechanism. Firstly, the SE-CNN model is built and pre-trained using all available data. Then, using the number of discharge cycles as the selection criteria, the training data is divided into four parts based on the number of discharge cycles. Each part is used to train the pre-trained network, resulting in the "four-in-one" CS-SE-CNN model. During SOE estimation, the appropriate prediction network is selected from the CS-SE-CNN model based on the number of discharge cycles in the input data. When estimating SOE, the MAE for

CNN is 2.32%, and the MAE for CS-SE-CNN is 1.41%, resulting in a reduction of 0.91%. These results indicate that CS-SE-CNN can eliminate the negative impact of the differences in discharge data across different life stages of batteries on the network, thereby enhancing the accuracy and robustness of the model.

Acknowledgments. This work was supported by the National Natural Science Foundation of China (No. 52307242), and the Fundamental Research Funds for the Central Universities (2023JC001).

References

1. Zhang, S., Zhang, X.: A novel non-experiment-based reconstruction method for the relationship between open-circuit-voltage and state-of-charge/state-of-energy of lithium-ion battery. Electrochim. Acta **403**, 139637 (2022)
2. Desaulty, A.M., Monfort Climent, D., Lefebvre, G., et al.: Tracing the origin of lithium in Li-ion batteries using lithium isotopes. Nat. Commun. **13**(1), 1–10 (2022)
3. Fu, B., Wang, W., Li, Y., Peng, Q.: An improved neural network model for battery smarter state-of-charge estimation of energy-transportation system. Green Energy Intell. Transp. **2**(2), 100067 (2023)
4. Xi, Z., Wang, R., Fu, Y., Mi, C.: Accurate and reliable state of charge estimation of lithium ion batteries using time-delayed recurrent neural networks through the identification of overexcited neurons. Appl. Energy **305**, 117962 (2022)
5. Mei, P., Karimi, H.R., Chen, F., Yang, S., Huang, C., Qiu, S.: A learning-based vehicle-cloud collaboration approach for joint estimation of state-of-energy and state-of-health. Sensors **22**(23), 9474 (2022)
6. Cao, M., Zhang, T., Wang, J., Liu, Y.: A deep belief network approach to remaining capacity estimation for lithium-ion batteries based on charging process features. J. Energy Storage **48**, 103825 (2022)
7. Dong, G., Chen, Z., Wei, J., Zhang, C., Wang, P.: An online model-based method for state of energy estimation of lithium-ion batteries using dual filters. J. Power Sour. **301**, 277–286 (2016)
8. Mao, X., Song, S., Ding, F.: Optimal BP neural network algorithm for state of charge estimation of lithium-ion battery using PSO with Levy flight. J. Energy Storage **49**, 104139 (2022)
9. Laadjal, K., Marques Cardoso, A.J.: A review of supercapacitors modeling, SoH, and SoE estimation methods: Issues and challenges. Int. J. Energy Res. **45**(13), 18424–18440 (2021)
10. Qian, C., Xu, B., Chang, L., et al.: Convolutional neural network based capacity estimation using random segments of the charging curves for lithium-ion batteries. Energy **227**, 120333 (2021)
11. Lai, X., Huang, Y., Han, X., Gu, H., Zheng, Y.: A novel method for state of energy estimation of lithium-ion batteries using particle filter and extended Kalman filter. J. Energy Storage **43**, 103269 (2021)
12. Li, K., Wei, F., Tseng, K.J., Soong, B.-H.: A practical lithium-ion battery model for state of energy and voltage responses prediction incorporating temperature and ageing effects. IEEE Trans. Industr. Electron. **65**(8), 6696–6708 (2018)
13. Long, T., Wang, S., Cao, W., Zhou, H., Fernandez, C.: An improved variable forgetting factor recursive least square-double extend Kalman filtering based on global mean particle swarm optimization algorithm for collaborative state of energy and state of health estimation of lithium-ion batteries. Electrochim. Acta **450**, 142270 (2023)

14. Tian, H., Li, A., Li, X.: SOC estimation of lithium-ion batteries for electric vehicles based on multimode ensemble SVR. J. Power Electron. **21**(9), 1365–1373 (2021)
15. Chen, L., Song, Y., Lopes, A.M., Bao, X., Zhang, Z., Lin, Y.: Joint estimation of state of charge and state of energy of lithium-ion batteries based on optimized bidirectional gated recurrent neural network. IEEE Trans. Transp. Electrification **10**(1), 1605–1616 (2024)
16. Zhou, L., Lai, X., et al.: State estimation models of lithium-Ion batteries for battery management system: status, challenges, and future trends. Batteries **9**(2), 131 (2023)
17. Ma, L., Hu, C., Cheng, F.: State of charge and state of energy estimation for lithium-ion batteries based on a long short-term memory neural network. J. Energy Storage **37**, 102440 (2021)
18. Wu, F., Wang, S., Liu, D., Cao, W., Fernandez, C., Huang, Q.: An improved convolutional neural network-bidirectional gated recurrent unit algorithm for robust state of charge and state of energy estimation of new energy vehicles of lithium-ion batteries. J. Energy Storage **82**, 110574 (2024)
19. Li, Z., Liu, F., Yang, W., Peng, S., Zhou, J.: A survey of convolutional neural networks: analysis, applications, and prospects. IEEE Trans. Neural Netw. Learn. Syst. **33**(12), 6999–7019 (2022)
20. Hu, J., Shen, L., Sun, G.: Squeeze-and-excitation networks. In: Proceedings of the IEEE Conference on Computer Vision and Pattern Recognition, pp. 7132–7141 (2018)
21. Severson, K.A., Attia, P.M., Jin, N., et al.: Data-driven prediction of battery cycle life before capacity degradation. Nat. Energy **4**(5), 383–391 (2019)
22. Li, Y., Li, K., Liu, X., Wang, Y., Zhang, L.: Lithium-ion battery capacity estimation — a pruned convolutional neural network approach assisted with transfer learning. Appl. Energy **285**, 116410 (2021)

A Lithium-Ion Battery Health State Assessment Based on Bi-LSTM-Transformer Algorithm

Chong Li[1,5], Hui Dai[2], Jiaolong Ye[2], Dianan Liu[3], Weijie Lin[4], Guanhao Du[4], and Zhile Yang[5(✉)]

[1] Henan Institute of Science and Technology, Xinxiang 453000, China
[2] CRRC Zhuzhou Locomotive Co., Ltd., Zhuzhou 412000, China
[3] Shendianneng Technology Group Co., Ltd., Shenzhen 518131, China
[4] Shenzhen Hithium Energy Storage Technology Co., Ltd., Shenzhen 518000, China
[5] Shenzhen Institute of Advanced Technology, Chinese Academy of Sciences Shenzhen, Shenzhen 518055, China
zl.yang@siat.ac.cn

Abstract. In battery management systems, accurate prediction of the remaining useful life (RUL) and the state of health (SOH) of the battery is crucial for enhancing battery longevity. In this study, a novel prediction model combining bidirectional LSTM and Transformer is proposed for improving SOH prediction accuracy. First, to enhance the model's generalization ability and address the issue of gradient vanishing, the bidirectional LSTM algorithm was employed. Then to better utilize the limited data, the Transformer module was added to improve the feature extraction capability. Finally the University of Maryland (CALCE) lithium-ion battery dataset was used to validate this algorithm. In the battery life test, the Bi-LSTM-Transformer model has a residual error of 99.32% and the model training and testing time is less than that of RNN, LSTM, and Bi-LSTM, which provides an accurate prediction model for the state of health (SOH) of the battery as well as the remaining life prediction (RUL).

Keywords: Battery Management · State of Health · Remaining Life Prediction · Bi-LSTM-Transformer

1 Introduction

As an efficient electrochemical energy storage system, lithium-ion batteries are characterized by high energy density, no memory effect, low self-discharge rate, and long life. However, with the increase of charging and discharging times, the battery will show different levels of aging phenomena. And State of Health (SOH) and Remaining Useful Life (RUL) are quantitative indicators for evaluating the aging degree of lithium-ion batteries.

Battery prediction requires a large amount of data support, but because of the experimental conditions and other reasons, the quality of the resulting battery experimental data is poor, which can affect the accuracy of the prediction model. By means of approximate

K. Li et al. (Eds.): LSMS/ICSEE 2024, CCIS 2218, pp. 279–288, 2025.
https://doi.org/10.1007/978-981-96-0232-2_22

estimation. Hu used machine learning combined with data preprocessing to propose the FuSH and GPR algorithms, which have good combined estimation performance in terms of estimation accuracy and computational efficiency [1]. Gong proposed a machine algorithm based on Gaussian regression, which was tested in different battery datasets, and it was shown that the model achieves high prediction accuracy and exhibits strong generalization ability for SOH prediction of different types of lithium-ion batteries under different charging and discharging scenarios [2]. Hu proposed a SARIMA prediction model that enhances the safety and reliability of lithium-ion batteries by using maximum likelihood estimation and Akaike's information criterion for more accurate parameter filtering [3]. Chen established a quantitative prediction model for lithium-ion battery capacity uncertainty and proposed a 1dCNN-BiLSTMQR-KDE algorithm, which was tested on the NASA dataset with good results [4]. By expanding the capacity of the battery dataset, Tian proposes a deep neural network (DNN) to extend from the estimation of capacity degradation to the complete prediction of the constant-current charging curve, using a limited quantity of training data in different battery datasets to improve the estimation of the battery curve running in different scenarios [5]. Wang evaluated the performance of five deep learning models on 100 cells, summarizing the effectiveness of popular deep learning-based SOH estimation models from various viewpoints [6]. By approximating the size of the sample of words, Amogne proposed a Bi-LSTM-AT model using normalized capacity as the state of health. Multi-step prognostics using the sliding window method yielded significant SOH estimates [7]. By controlling for error, Lyu proposed a multi-step time prediction algorithm-Parallel State Fusion LSTM (PSF-LSTM), which utilizes similar samples to control the cumulative error and improves prediction accuracy and efficiency [8]. Dang introduced Gaussian regression based on the codec model to lessen the cumulative error from recursive predictions and to more precisely predict the RUL of zinc-ion batteries [9].

The above studies, while making significant progress in battery State of Health (SOH) prediction and Remaining Useful Life (RUL) estimation, But can't improve prediction accuracy better. In order to better acquire the data features, enhancing the model's accuracy and mitigate gradient vanishing in a limited dataset, this paper introduces a Bi-LSTM-Transformer algorithm. This approach has yielded excellent optimization results.

2 Method

2.1 Data-Driven

An approach to modelling based on collected data is called a data-driven approach, this method does not take into account any of the properties of the battery, but only the data characteristics of the object of study. Because of its simplicity and ease of use, it has become one of the popular methods for studying battery prediction at present. The data-driven approach can be divided into the following steps:

(1) The CALCE dataset used in this study is available from the Center for Advanced Life Cycle Engineering at the University of Maryland [10, 11], The dataset contains records from four lithium-ion batteries, each performing the three repeated operations of charging, discharging, and impedance measurement.

(2) Parameters such as voltage statistic, current statistic, test time and battery capacity statistic are extracted based on the dataset extracting charging data and incremental capacity curve.
(3) Experiments are trained directly on limited data, limiting the accuracy of deep learning predictions. Therefore the Transformer module was used to optimise the data in a rational way.
(4) Based on the features of the selected battery dataset, Prediction of battery capacity after charging time and prediction of battery health state using four deep learning networks, RNN, LSTM, BI-LSTM, and BI-LSTM-Transformer. And they are compared and analyzed with each other.

2.2 Definition of SOH

The State of Health (SOH) of a battery, which can be estimated by the battery checker and the number of battery cycles, it provides a quantitative assessment of the battery's aging process. Ct in Eq. 1 represents the actual battery capacity in the charging cycle, and C0 is the rated capacity. Characterizing the ability of the current battery to hold electrical energy as compared to a fresh battery, it refers to the comparative ratio of the full charge energy of the current battery to that of a battery in its fresh state.

$$\mathrm{SOH} = \frac{\mathrm{C_t}}{C_0} \times 100\% \tag{1}$$

2.3 RNN Model

The classical model of RNN model is shown in Fig. 1. A recurrent neural network is a neural network, it consists of an output unit (α), a hidden unit (β), and an input unit (γ) that is specialized for processing sequential data. The network has two output data, one of which is feedback to itself and the other is output to the neuron at the next moment. In general, recurrent neural networks have the ability to detect temporal dependencies in sequences of data and incorporate these dependencies into the model during the learning process. As shown in the Fig. 1, unfolding the left part enables to obtain a traditional recurrent neural network (RNN).

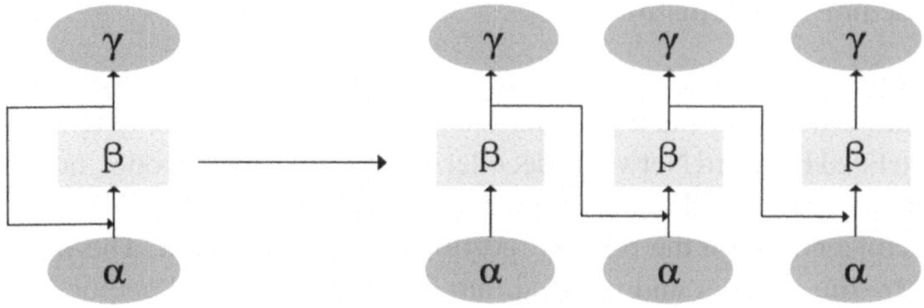

Fig. 1. RNN Model

2.4 LSTM Model

As shown in Fig. 2 is a LSTM Model, It is a special kind of recurrent neural network suitable for sequential data processing. In LSTM Model, the gradient explosion problem is solved by effectively controlling the flow of information within the cell through gate mechanisms such as forgetting gates, input gates and output gates. And can detect long-term dependencies present in sequences. The performance is better than traditional RNN, and the stability is also better.

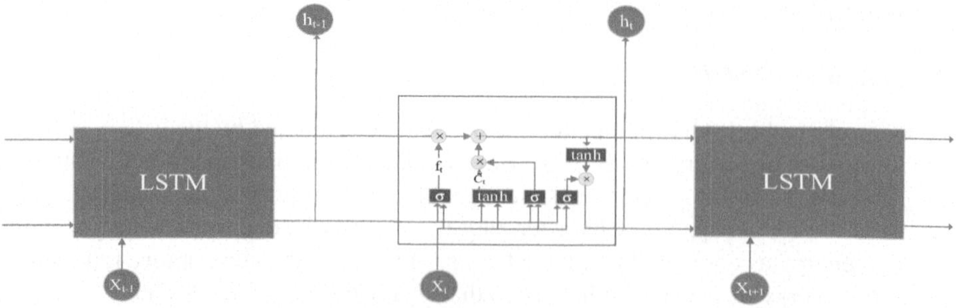

Fig. 2. LSTM Model

The formulas for the forgetting gate, input gate, and output gate are shown below, W_c for the weight parameters, b_f, b_c for the bias parameters, for the current input, and C_{t-1} for the state of the previous time step.These parameters represent the flow and update of information in the cell.

$$f_t = \sigma(W_c \cdot [h_{t-1}, x_t] + b_f) \tag{2}$$

$$\widehat{C}_t = \tanh(W_c \cdot [h_{t-1}, x_t] + b_c) \tag{3}$$

$$C_t = f_t \times C_{t-1} + i_t \times \widehat{C}_t \tag{4}$$

2.5 Transformer Mechanism

As shown in Fig. 3, the standard Transformer is a sequence-to-sequence framework with an encoder and a decoder. Each encoder has two sub-layers: Self-Attention layer and Position-wise Feed Forward Network; decoder compared to the encoder, not only has two layers, but also an additional attention layer (Encoder-Decoder Attention), which helps the decoder concentrate on the relevant portions of the input data. The input sequence is processed by the encoder and mapped into a high-dimensional vector, which is then used by the decoder to produce the output sequence.

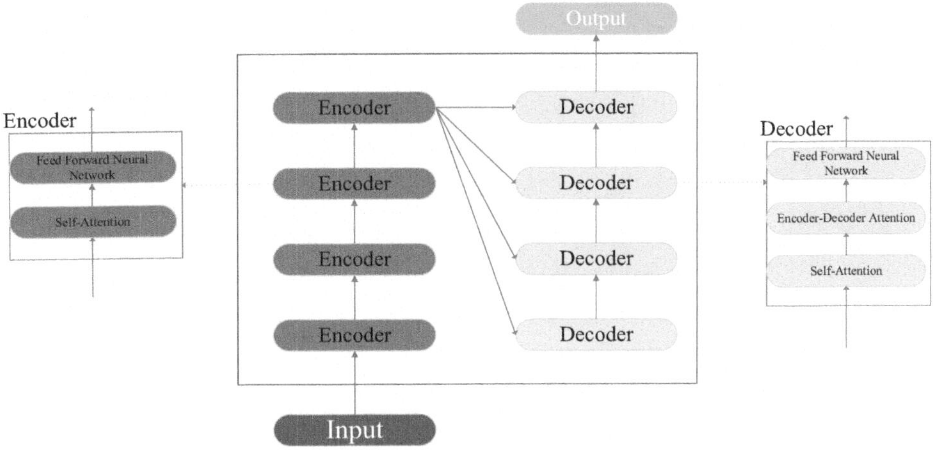

Fig. 3. Transformer Mechanism

3 Experiment

3.1 Data Acquisition

CALCE battery dataset was used in this experiment. The dataset was tested at a temperature of 1 °C. The charging mode was to charge the lithium-ion battery in constant current mode until the battery voltage reaches 4.2 V. Then the battery was charged in constant voltage mode until the charging current drops to 20 mA. The discharging mode was to discharge the battery in constant current mode until the battery voltage drops to 2.7 V. The termination condition For the battery to reach the end-of-life criteria, i.e., the rated capacity drops to 30% of it. In this study, because it is carried out at a constant temperature, the inputs to this paper are voltage and current, and the output is the capacity of the battery.

3.2 Remaining Battery Life Decay Predictions

A graph of the trend of the four types of lithium-ion battery capacity decay with the number of charge/discharge cycles and the results predicted by various algorithms are shown in the following figure. As we can see from the figure, with the change in battery chemistry and aging effects, All types of batteries exhibit a gradual decrease in capacity as the number of charges and discharges increases. Four main prediction models are included in the figure: traditional Recurrent Neural Network (RNN), simple Long Short-Term Memory Network (LSTM), Bidirectional Long Short-Term Memory Network (Bi-LSTM), and the Bi-LSTM-Transformer model proposed in this study. Each of these models attempts to predict the remaining capacity of the battery. Although all the prediction models can track the battery capacity decay trend to some extent, the prediction curve of the Bi-LSTM-Transformer model is closest to the curve of the actual battery capacity decay. This demonstrates the Bi-LSTM-Transformer algorithm's superior accuracy and performance in capturing the intricate dynamics of battery capacity decay found in this dataset (Figs. 4, 5, 6 and 7).

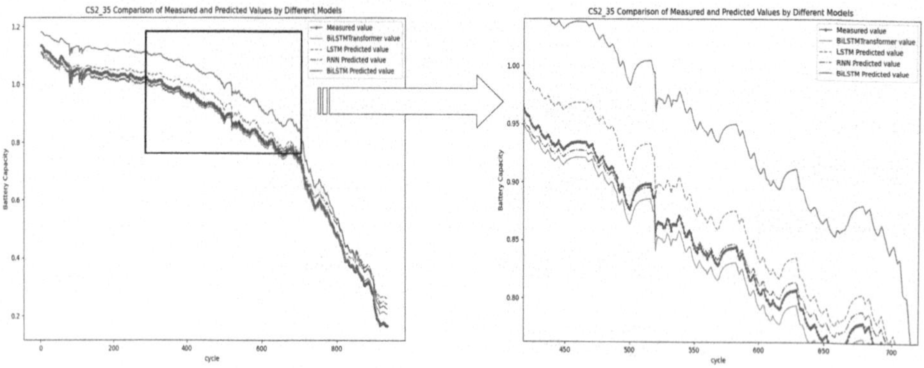

Fig. 4. CS2_35 RUL

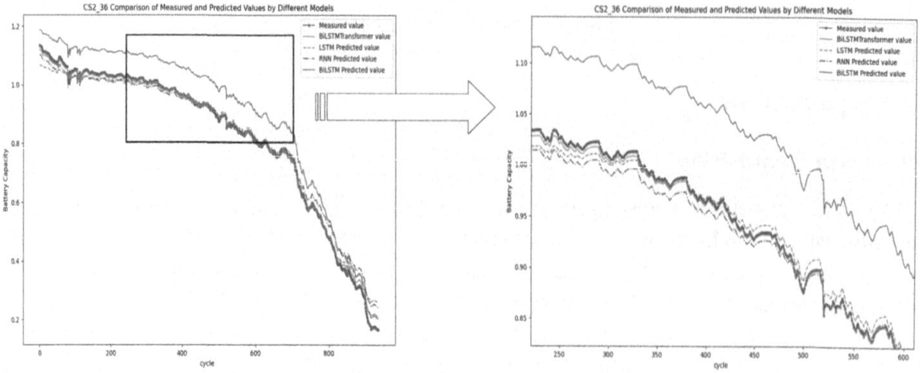

Fig. 5. CS2_36 RUL

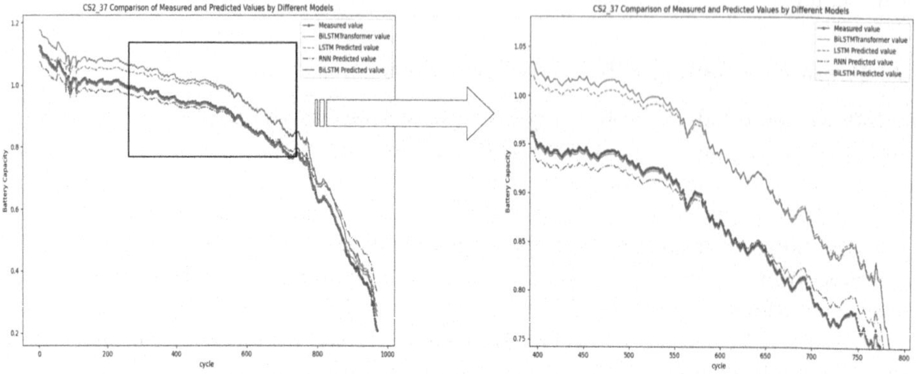

Fig. 6. CS2_37 RUL

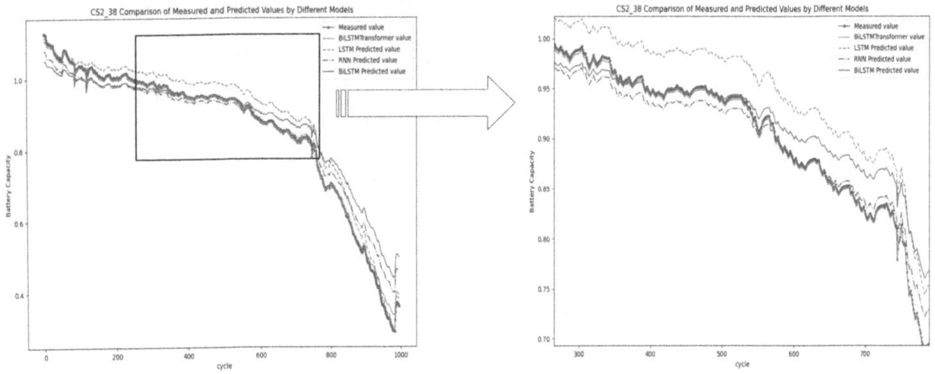

Fig. 7. CS2_38 RUL

3.3 Battery State of Health Estimation

The State of Health (SOH) of Li-ion batteries decays with the number of charge/discharge.As we can see from the graphs, although the performance advantage of the Bi-LSTM-Transformer algorithm is not as significant in some datasets as in some other scenarios, it performs much better in terms of stability. This means that the Bi-LSTM-Transformer model is able to provide more consistent and reliable prediction results under battery species. This enhanced stability is mainly due to the fact that Bi-LSTM-Transformer combines the advantages of bi-directional LSTM with the Transformer's self-attention mechanism. Bidirectional LSTM allows the model to access both history and prospects contextual details, which is especially critical for predicting battery performance, as the future state of the battery is not only influenced by previous usage history, but also potentially by future usage patterns. And the self-attention mechanism in the Transformer part can help the model to recognize the complex patterns and key dependencies in the battery data, which is important for understanding and predicting the complex dynamic characteristics of battery SOH (Figs. 8, 9, 10 and 11).

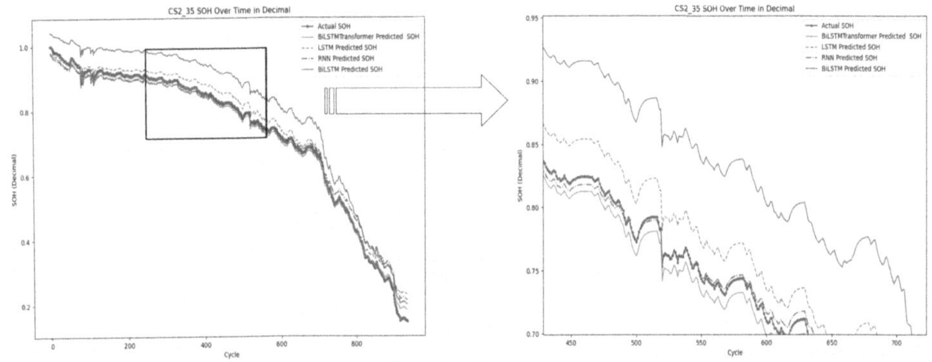

Fig. 8. CS2_35 SOH

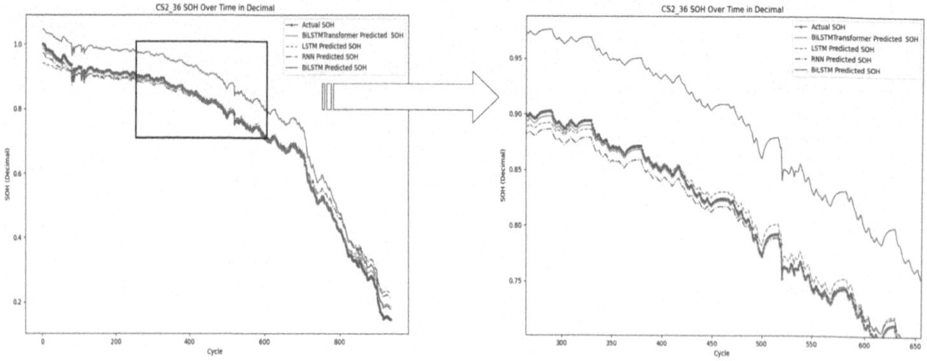

Fig. 9. CS2_36 SOH

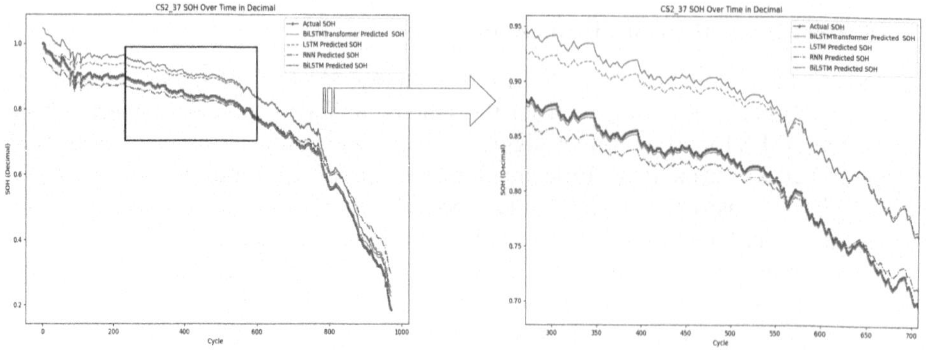

Fig. 10. CS2_37 SOH

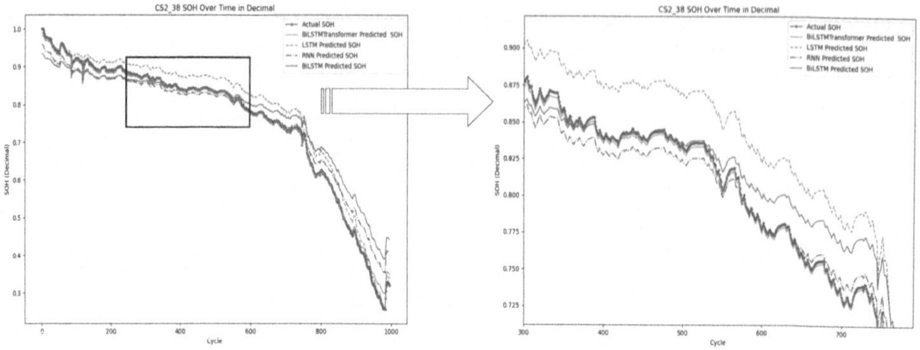

Fig. 11. CS2_38 SOH

3.4 Evaluation Indicators

First, the three metrics used to evaluate the prediction performance of RUL are: relative error (RE), mean absolute error (MAE), and RMSE. These three evaluation metrics are defined as follows:

$$RE = \frac{|RUL^{pred} - RUL^{true}|}{RUL^{true}} \tag{5}$$

$$RMSE = \sqrt{\frac{1}{n-T}\sum_{t=T+1}^{n}(x_t - \hat{x}_t)^2} \tag{6}$$

$$MAE = \frac{1}{n-T}\sum_{t=T+1}^{n}\|x_t - \hat{x}_t\| \tag{7}$$

Then, training each battery dataset, after five iterations, the average scores of all algorithms are shown in the table below. These three values indicate the deviation between the true value and the predicted value, with a deviation closer to zero indicating a more accurate prediction. By setting the learning rate to 0.01 and the number of training times to 2000, we can see from the Table 1 that the stability and the average value of the Bi-LSTM-Transformer algorithm are the best among the four algorithms, and the predicted value is also the closest to the true value; the prediction result of RNN is also closer to the true value, and the stability is good; the prediction results of LSTM and Bi-LSTM are not as good as those of the other two, and the stability is not as good as those of the other two. LSTM and Bi-LSTM are not as good as the other two and are not as fast as RNN.

Table 1. MAE, MSE, RE

Evaluation	RNN	LSTM	Bi-LSTM	Bi-LSTM-Transformer
MAE	0.0247	0.0369	0.0301	0.0153
MSE	0.0010	0.0016	0.0012	0.0005
RE	0.0343	0.0524	0.0465	0.0068

4 Conclusion

We propose a composite neural network model that consists of multiple neural networks for RUL prediction. First, we extract data from the dataset and pass it to the next network; then we use the Transformer module to learn the characteristics of the battery capacity decay, and the processed data is analyzed using a Bi-LSTM model. By comparing our algorithm with other algorithms, our algorithm is better and get lower RE, MAE and RMSE scores. The model effectively enhances feature extraction, improves model accuracy and mitigates problems such as gradient vanishing, good results have been achieved in battery remaining life prediction (RUL) as well as state of health.

Acknowledgements. This study was funded by the Science and Technology Project of Shenzhen (grant numbers GJHZ20210705141401005, CJGJZD20220517141405012 and KCXST20221021111402006), the "Nanling Team Project" of Shaoguan City, the Shenzhen Excellent Innovative Talents (grant number RCYX20221008093036022) and the Shenzhen-Hong Kong Joint Funding Project (A) (grant number SGDX20230116092053005).

References

1. Hu, X., Membe, S., Che, Y., Lin, X., Onori, S.: Battery health prediction using fusion-based feature selection and machine learning. IEEE Trans. Transp. Electrif. **7**(2), 382–398 (2021)
2. Gong, D., Gao, Y., Kou, Y., Wang, Y.: State of health estimation for lithium-ion battery based on energy features. Energy **257** (2022)
3. Hu, Y., Wang, S., Huang, J., Takyi-Aninakwa, P., Chen, X.: A novel seasonal autoregressive integrated moving average method for the accurate lithium-ion battery residual life prediction. Int. J. Electrochem. Sci. **17**(5) (2022)
4. Chen, C., Tao, G., Shi, J., Shen, M., Zhu, Z.H.: A lithium-ion battery degradation prediction model with uncertainty quantification for its predictive maintenance. IEEE Trans. Ind. Electron. **71**(4), 3650–3659 (2024)
5. Tian, J., Xiong, R., Shen, W., Lu, J., Yang, X.-G.: Deep neural network battery charging curve prediction using 30 points collected in 10 min. Joule **5**(6), 1521–1534 (2021)
6. Wang, F., Zhai, Z., Liu, B., Zheng, S., Zhao, Z., Chen, X.: Open access dataset, code library and benchmarking deep learning approaches for state-of-health estimation of lithium-ion batteries. J. Energy Storage **77** (2024)
7. Wang, F.-K., Amogne, Z.E., Chou, J.-H., Tseng, C.: Online remaining useful life prediction of lithium-ion batteries using bidirectional long short-term memory with attention mechanism. Energy **254** (2022)
8. Lin, P., Wang, S., Jin, P., Yuan, H., Ma, Z., Di, Y.: Parameter adaptive joint estimation of state of charge and available capacity based on multi-innovation-state estimator fusion. J. Energy Storage **83** (2024)
9. Dang, W., Liao, S., Yang, B., Yin, Z., Liu, M., Yin, L., et al.: An encoder-decoder fusion battery life prediction method based on Gaussian process regression and improvement. J. Energy Storage **59** (2023)
10. He, W., Williard, N., Osterman, M., Pecht, M.: Prognostics of lithium-ion batteries based on Dempster-Shafer theory and the Bayesian Monte Carlo method. J. Power Sour. **196**(23), 10314–10321 (2011)
11. Xing, Y., Ma, E.W.M., Tsui, K.-L., Pecht, M.: An ensemble model for predicting the remaining useful performance of lithium-ion batteries. Microelectron. Reliab. **53**(6), 811–820 (2013)

Degradation Prediction of PEMFC Using Long Short-Term Memory Network Based on Quantum-Behaved Particle Swarm Optimization

Yaolin Dong[1,2], Wei Wang[1,2](✉), Yuan Cao[1,2], and Huaiqi Xie[3]

[1] School of Automation, Central South University, Changsha, China
wangwei729@csu.edu.cn
[2] National Engineering Research Centre of Advanced Energy Storage Materials, Shenzhen, China
[3] Digicert Inc., Sunnyvale, CA 94085, US

Abstract. The commercialization of Proton Exchange Membrane Fuel Cells (PEMFC) is constrained by high costs and short lifespans. Conducting degradation prediction research will be beneficial for enhancing the durability and extending the lifespan of PEMFC. In this article, a novel long short-term memory (LSTM) model combined with the quantum-behaved particle Swarm optimization (QPSO) is proposed to achieve high accuracy of PEMFC degradation prediction. LSTM is utilized to describe the voltage degradation of PEMFCs, and QPSO is utilized to enhance the prediction performance by optimizing the hyperparameters within LSTM. The effectiveness of the presented prediction method was validated utilizing the durability test data of a fuel cell stack under constant current load. Experimental results demonstrate that presented method shows great prediction accuracy of PEMFC aging, and its prediction accuracy significantly surpasses conventional methods.

Keywords: Proton Exchange Membrane Fuel Cell · Degradation Prognosis · Quantum-Behaved Particle Swarm Optimization · Long Short-term Memory

1 Introduction

Climate change is increasingly garnering widespread attention globally, with energy shortages, environmental pollution, and global warming posing formidable challenges [1, 2]. Faced with the detrimental effects of traditional fossil fuels on the environment, there is an urgent need to expedite the development of a clean, low-carbon, safe, and efficient new energy system [3]. Proton exchange membrane fuel cells (PEMFCs), as highly efficient, zero-emission, rapidly responsive energy conversion devices capable of operating at room temperature, have already found extensive applications in transportation, energy storage, and power generation [4]. Currently, manufacturers like Toyota, Honda, and Mercedes-Benz are vigorously promoting the commercialization of fuel cells [5], although the high cost and short lifespan of the stacks remain significant obstacles to large-scale commercial deployment [6].

K. Li et al. (Eds.): LSMS/ICSEE 2024, CCIS 2218, pp. 289–299, 2025.
https://doi.org/10.1007/978-981-96-0232-2_23

As the operational years extend, fuel cells undergo a gradual aging process, exhibiting irreversible performance decline. When the performance of the stack declines to a certain extent, its lifespan can be declared terminated, i.e. end of life (EOL). PEMFC aging is primarily caused by degradation and aging of components such as the membrane, electrodes, catalyst, and gas diffusion layer. This process is a complex, multi-physics, multi-phase, and multi-scale coupled phenomenon, influenced by factors including improper thermal management, water management, improper gas management, and frequent load fluctuations [7].

To expedite the commercial application of fuel cells, extending their lifespan is crucial. Establishing an accurate model for predicting fuel cell aging can aid in assessing the State of Health (SOH) of PEMFCs and estimating the Remaining Useful Life (RUL). This, in turn, facilitates targeted system maintenance and prolongs the service life. To address this, prognostics and health management (PHM) is proposed for predicting the aging process and remaining lifespan of fuel cells [8]. Existing prediction methods can be roughly classified into three categories: model-driven, data-driven, and hybrid methods.

Model-driven methods predict the PEMFC Degradation trend by establishing mechanistic models or empirical models. Futter et al. [9] conducted research on electrochemical hydrogen peroxide formation, transport, electrochemical reactions of iron ions, and radical attack on the polymer, leading to the development of a comprehensive 2D along-the-channel PEMFC model, which revealed the impact of chemical membrane degradation on fuel cell performance. Ao et al. [10] proposed a method utilizing Frequency-Domain Kalman Filtering (FDKF) and four empirical models to predict the degradation and RUL of PEMFCs. Compared to the Extended Kalman Filter (EKF) approach, their method is more precise and takes shorter computation time. However, due to the unclear understanding of PEMFC degradation mechanisms, models established by such methods often struggle to accurately reflect the degradation process.

Data-driven methods, requiring no in-depth understanding of PEMFC's internal mechanisms, infer the degradation trends of PEMFCs by collecting historical operational data and employ artificial intelligence techniques to predict RUL. Such data-based approaches offer high computational efficiency and are well-suited for online prediction [11]. Chen et al. [12] utilized the Grey Neural Network (GNN) to establish a degradation model for PEMFC. They optimized the hyperparameters of GNN using particle swarm optimization algorithm. The predictive accuracy of this model was validated under real operating conditions of a fuel cell electric vehicle. Zuo et al. [13] applied an attention mechanism into Recurrent Neural Network (RNN), comparing the predictive performance of two RNN architectures, LSTM and GRU, in long-term durability tests of fuel cells. It was demonstrated that the attention-based LSTM achieved the highest accuracy. However, data-driven methods rely on a vast amount of experimental data, which often necessitating destructive testing of PEMFCs for thousands of hours to acquire.

Hybrid methods represent the fusion of model-driven and data-driven approaches. Hu et al. [14] proposed a hybrid prediction framework based on the unscented Kalman filter and transformer network. Zhou et al. [15] integrated nonlinear autoregressive neural network (NARNN) with Particle Filter (PF) algorithm, where NARNN can effectively captures the local nonlinear characteristics of degradation, and PF can predict long-term aging trends. The hybrid method combines the strengths of both approaches, but it typically has much higher complexity than a single model, requiring significant plenty of computational resources.

From the above review, it is evident that model-based prediction methods and hybrid methods require the establishment of physical models or empirical models based on the performance degradation trend of PEMFC, facing difficulties in modeling and limited applicability. In contrast, data-driven methods typically offer higher computational efficiency and broader applicability. In this article, a model-driven method based on Long Short-term Memory (LSTM) Network combined with the Quantum-behaved Particle Swarm Optimization (QPSO) was proposed to predict the degradation of PEMFC. The hyperparameters of the LSTM network, such as the number of hidden layers and learning rate, are optimized using the QPSO algorithm to enhance prediction accuracy and convergence speed. The effectiveness of the predictions is validated through durability testing data of PEMFCs under various operating conditions.

The structure of this article is as follows: Sect. 2 provides a detailed explanation of the data sources and experimental parameters. Section 3 presents the framework of the degradation prediction algorithm of PEMFC based on QPSO-LSTM. Section 4 comprises experimental validation. Finally, in Sect. 5, conclusions are drawn.

2 Experiment and Data Processing

The test data from the IEEE PHM 2014 Data Challenge is utilized in this article to train and validate the proposed QPSO-LSTM prediction model. This dataset is provided by the FCLAB Research Federation [16].

2.1 PEMFC Aging Test

FCLAB conducted two sets of experiments on two identical fuel cell stacks (FC1 and FC2), respectively, under constant load and quasi-dynamic load conditions. The experimental setup is shown in Fig. 1, and this platform is suitable for testing fuel cells below 1 kW. As shown in Fig. 2, a fuel cell stack consisting of 5 individual cells called FC1 operated under a constant current of 70 A for approximately 1154 h, while a ripple current of 70 A with 7 A oscillations at a frequency of 5 kHz was applied to a same fuel cell stack called FC2 for approximately 1020 h. The experimental parameters of FC1 and FC2 are listed in Table 1. In this article, only the experimental data from FC1 are temporarily selected for the training and validation of the proposed model.

Health monitoring data like voltage, current density, temperature, and gas flow rate were recorded through the test facilities during experiment. As shown in Fig. 2, it is evident that the voltages of both FC1 and FC2 decrease with the increase in testing time, indicating that stack voltage can be used to characterize the aging process of PEMFC. Therefore, voltage is selected as an indicator to measure the degradation level of PEMFC.

During the experiments, polarization curves and electrochemical impedance spectroscopy measurements were conducted approximately every week. Following each measurement session, a notable recovery phenomenon in voltage was observed, resulting in the voltage decay curve exhibiting strong non-linear characteristics over time. Additionally, due to measurement errors and random disturbances, the curve exhibited significant fluctuations, requiring data preprocessing before model training.

Fig. 1. PEMFC degradation experiment platform in FCLAB [16].

Table 1. Parameters of FC1 and FC2 during Experiment.

Parameter	Value
Number of cells	5
Active area	100 cm^2
Stack rated current	70 A
Temperature	54 °C
Hydrogen pressure	1.3 bar
Relative humidity	50%

2.2 Data Processing

The quantum of the original voltage data is extremely vast (A total of 143,862 data was recorded for FC1 during a 1154 h durability test, while 127,370 data was recorded for FC2 during a 1020 h test), and there are partial abnormal data and a large amount of random noise, which will significantly affect computational speed and prediction accuracy. By removing outliers from the raw data and calculating the average of the stack voltages per half hour, reconstructed data was obtained, reducing the data volume to 2309 for FC1 and 2042 for FC2.

As shown in Fig. 2, there are still some level of fluctuation exhibits in reconstructed data, so further smoothing was applied using Savitzky-Golay filter. The Savitzky-Golay filter is a specialized type of low-pass filter, renowned for its ability to effectively remove high-frequency noise while preserving the time-varying information of the signal. Its core concept involves applying weighted filtering to the data within a moving window. These weights are obtained through least-squares fitting of a specified order polynomial [17]. For the present study, the window length was set as 41, and the order of the polynomial

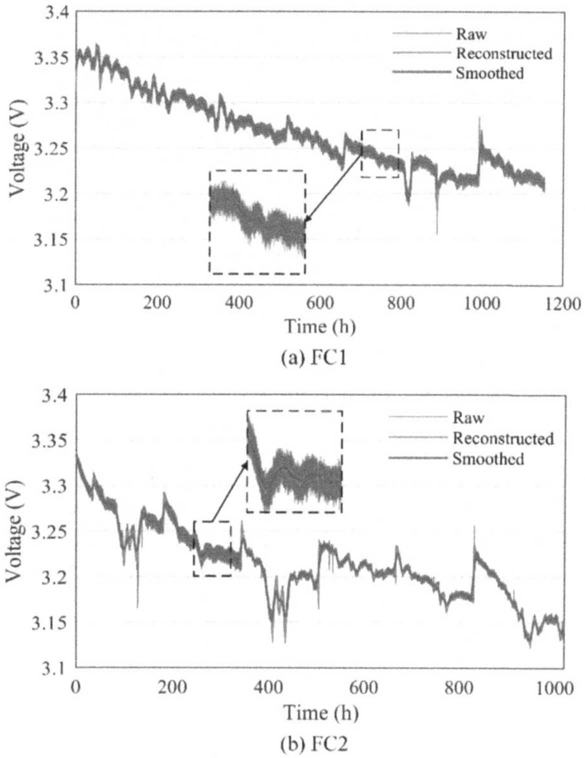

Fig. 2. Raw data, reconstructed data and smoothed data of FC1 and FC2.

was specified as 1, thereby achieving a balance between computational speed, noise reduction capability, and waveform preservation.

The raw data, reconstructed data, and smoothed data are depicted in Fig. 2. Even after preprocessing, the reconstructed data and smoothed data retains the underlying trends of the original voltage data. Subsequently, only the smoothed data are utilized for model training and validation.

3 Degradation Prediction Method for PEMFC

3.1 Long Short-Term Memory Network

Long Short-Term Memory (LSTM) represents an enhanced variant within the realm of recurrent neural network architectures. By incorporating gate mechanisms, it adeptly tackles the challenges of gradient vanishing and exploding that plague conventional RNN. The architectural of LSTM cells, as depicted in Fig. 3, primarily encompasses three crucial components: the forget gate, the input gate, and the output gate. X_t, h_t, C_t respectively represent the input quantity, the output quantity, and the internal state quantity of the LSTM cell at time t.

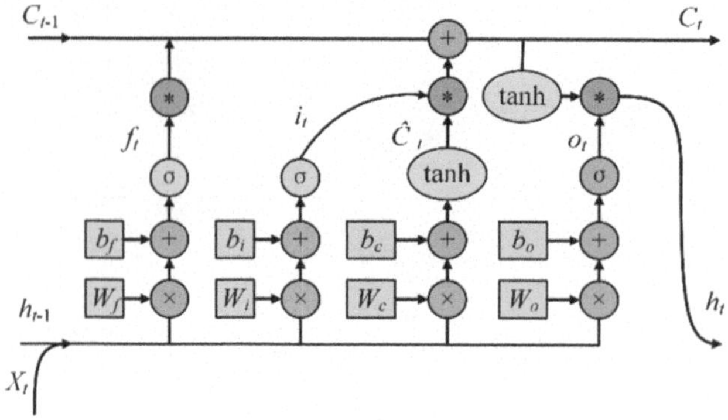

Fig. 3. Architecture of LSTM cells.

LSTM employs the forget gate f_t to determine the extent to which the internal state from the previous time step is retained, while utilized the input gate to extracts pertinent information from X_t and h_{t-1}, accomplish to update the cell state values.

$$f_t = \sigma(W_f \cdot [h_{t-1}, X_t] + b_f) \tag{1}$$

$$i_t = \sigma(W_i \cdot [h_{t-1}, X_t] + b_i) \tag{2}$$

$$\widehat{C}_t = tanh(W_C \cdot [h_{t-1}, X_t] + b_C) \tag{3}$$

$$C_t = f_t * C_{t-1} + i_t * \widehat{C}_t \tag{4}$$

Finally, by employing the output gate, LSTM extracts information from the cell state to obtain the output quantity.

$$o_t = \sigma(W_o \cdot [h_{t-1}, X_t] + b_o) \tag{5}$$

$$h_t = o_t * tanh(C_t) \tag{6}$$

where W_f, W_i, W_C, W_o are the weight matrices, b_f, b_i, b_C, b_o are the corresponding bias coefficients.

3.2 Improvement of LSTM for PEMFC Based on QPSO

LSTM has two types of parameters. One type is automatically updated during the model training process through gradient descent, including the weight matrices and bias coefficients of each gate. The other type is set to fixed values during network initialization, known as hyperparameters, which include the learning rate, number of iterations, number of neurons in the hidden layer, and so on. As the complexity of the model increases, the influence of hyperparameters on prediction accuracy gradually becomes more and more significant. For deep reservoir networks with two or more layers, optimizing hyperparameters through intelligent optimization algorithms can significantly improve prediction accuracy and convergence speed compared to manually adjusting network parameters. Therefore, an improved version of PSO was selected, which called QPSO, to optimize the hyperparameters of the LSTM network.

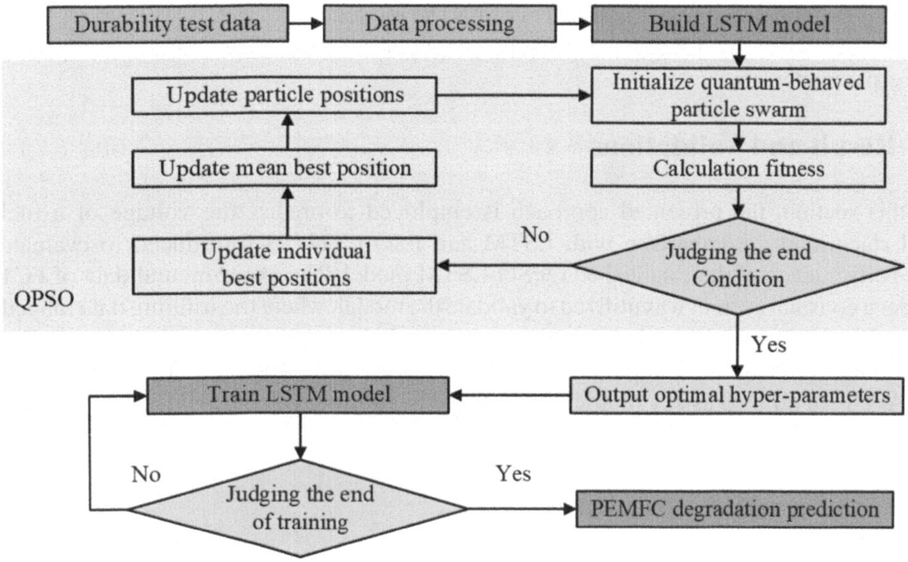

Fig. 4. Schematic of LSTM based on the QPSO.

Quantum Particle Swarm Optimization (QPSO) incorporates principles from quantum theory into particle behavior. It describes particle motion states using wave functions and calculates probability densities using the Schrödinger equation. Through Monte Carlo random simulation, it updates particle positions, significantly enhancing the randomness of particle movement, thus significantly reduces the possibility of falling into local optima. Compared to conventional PSO, QPSO has fewer control parameters and exhibits stronger global search capabilities.

As shown in Fig. 4, the major implementation steps of the presented LSTM for PEMFC prediction based on QPSO are described as follows.

step1. Build LSTM model for PEMFC prediction, and concatenate the hyperparameters that need to be optimized into a multi-dimensional vector.
Step2. Initialize quantum-behaved particle swarm and train the LSTM model using parameters obtained from the particle positions.
Step3. Utilize the mean squared error of the model's predicted values as the fitness function, then calculate the fitness of each particle.

$$f = \frac{1}{T}\sum\nolimits_{i=1}^{T}(y_i - \hat{y}_i)^2 \tag{7}$$

in which T denotes the number of sample points, y_i and $\hat{y}_i$ are the real values and predicted values respectively.

Step4. Judge the termination condition, if reaching the termination condition, stop the iteration and switch to *step 6*; otherwise, switch to *step 5*.
Step5. Updating individual best positions and mean best position of each particle according to current fitness values, then updating particle positions and switch to *step 2*.
Step6. Output optimal hyperparameters of LSTM model.
Step7. Establish the LSTM for PEMFC degradation prediction based on optimal hyperparameters.

4 Result and Validations

In this section, the presented approach is employed to predict the voltage of a fuel cell stack, and a comparison with LSTM and PSO-LSTM is introduced to evaluate the performance of the established QPSO-LSTM model. The experimental data of FC1 under a constant current was utilized to validate the model, where the training data ranged from 0 to 550 h, and the testing data spanned from 551 to 1154 h. To evaluate predictive performance, Root Mean Square Error (RMSE) and Mean Absolute Percentage Error (MAPE) are utilized as key metrics, formulated as follows.

$$\mathrm{RMSE} = \sqrt{\frac{1}{T}\sum\nolimits_{t=1}^{T} (y_i - \hat{y}_i)^2} \tag{8}$$

$$\mathrm{MAPE} = \frac{1}{T}\sum\nolimits_{t=1}^{T} \left| \frac{y_i - \hat{y}_i}{y_i} \right| \tag{9}$$

in which T denotes the number of sample points, y_i and $\hat{y}_i$ are the real values and predicted values respectively.

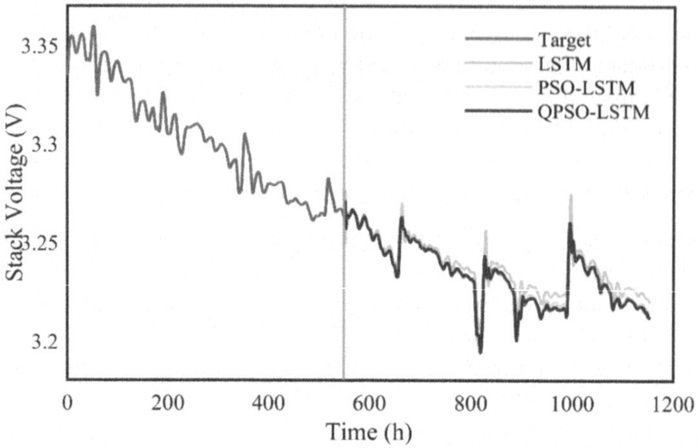

Fig. 5. Prognostic results of QPSO-LSTM, PSO-LSTM, and LSTM under FC1.

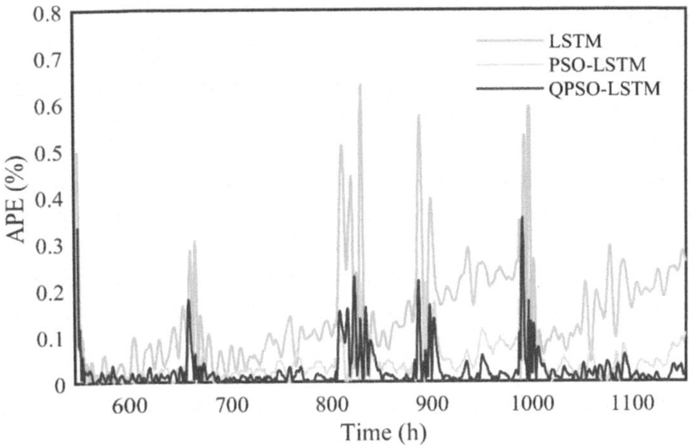

Fig. 6. The APE of degradation prediction under different prognostic methods.

Figure 5 and Table 2 demonstrates the prediction results of the three methods. It is observed that the RMSE of the QPSO-LSTM model is 1.5355e−3 and the MAPE is 0.02619%, which are the least value among the three models, indicating that the proposed method achieves the most accurate prediction of PEMFC degradation. All three models are able to capture the time dependence features in the data and achieve tracking of the actual voltage curve. However, through comparison of the three models, it can be noted that employing intelligent optimization algorithms to optimize hyperparameters can significantly improve the prediction performance of the LSTM. Furthermore, it is evident that the effectiveness of the proposed QPSO-LSTM method surpasses that of conventional PSO-LSTM method.

Table 2. Comparison of Prognostic Results for FC1.

Methods	Evaluation metric	
	RMSE	MAPE
LSTM	5.8162e−3	0.14516%
PSO-LSTM	1.9206e−3	0.04389%
QPSO-LSTM	1.5355e−3	0.02619%

Figure 6 illustrates the absolute percent error (APE) under different prognostic methods, which further indicates that the QPSO-LSTM model has achieved the highest prediction accuracy. Another notable phenomenon is the sudden increase in prediction errors around certain moments, coinciding with the time points of FC1 characterization tests ($t = 658, 823, 991$ h), which deserves further investigation.

5 Conclusion

In this article, a LSTM model based on QPSO is presented to achieve accurate degradation prediction of PEMFCs. As an improved structure of standard RNN, LSTM exhibits strong capabilities in capturing time dependence and nonlinear processing in PEMFC degradation prediction. By employing the novel QPSO algorithm to optimize the hyperparameters of each hidden layer in LSTM, the prognostic performance of the model is further enhanced. The effectiveness of the presented QPSO-LSTM approach has been validated under constant current load conditions and compared with standardized methods. Experimental results indicate that the QPSO-LSTM model achieves the highest prediction accuracy among all the methods and the predictive curve closely aligns with the actual data, demonstrating its superior prediction ability.

To further expand the research presented in this paper, the presented method will be applied to prediction under quasi-dynamic operating conditions. Additionally, accurately predicting remaining useful life based on voltage prediction results will become a priority issue of concern.

Acknowledgments. This work was supported by National Natural Science Foundation (NNSF) of China under Grant 62103443 and 61973322, by Hunan Natural Science Foundation under Grant 2022JJ40630.

Disclosure of Interests. The authors have no competing interests to declare that are relevant to the content of this article.

References

1. Zhao, F., et al.: Hydrogen fuel cell vehicle development in China: an industry chain perspective. Energy Technol. **8**, 2000179 (2020). https://doi.org/10.1002/ente.202000179
2. Ferriday, T.B., Middleton, P.H.: Alkaline fuel cell technology - a review. Int. J. Hydrogen Energy **46**, 18489–18510 (2021). https://doi.org/10.1016/j.ijhydene.2021.02.203
3. Mohammadi, A., Cirrincione, G., Djerdir, A., Khaburi, D.: A novel approach for modeling the internal behavior of a PEMFC by using electrical circuits. Int. J. Hydrogen Energy **43**, 11539–11549 (2018). https://doi.org/10.1016/j.ijhydene.2017.08.151
4. Ramezanizadeh, M., Alhuyi Nazari, M., Hossein Ahmadi, M., Chen, L.: A review on the approaches applied for cooling fuel cells. Int. J. Heat Mass Transf. **139**, 517–525 (2019). https://doi.org/10.1016/j.ijheatmasstransfer.2019.05.032
5. Tanç, B., Arat, H.T., Baltacıoğlu, E., Aydın, K.: Overview of the next quarter century vision of hydrogen fuel cell electric vehicles. Int. J. Hydrogen Energy **44**, 10120–10128 (2019). https://doi.org/10.1016/j.ijhydene.2018.10.112
6. Benaggoune, K., Yue, M., Jemei, S., Zerhouni, N.: A data-driven method for multi-step-ahead prediction and long-term prognostics of proton exchange membrane fuel cell. Appl. Energy **313**, 118835 (2022). https://doi.org/10.1016/j.apenergy.2022.118835
7. Wang, Y., et al.: A data-driven approach to lifespan prediction for vehicle fuel cell systems. IEEE Trans. Transp. Electrif. **9**, 5049–5060 (2023). https://doi.org/10.1109/TTE.2023.3237219

8. Jouin, M., Gouriveau, R., Hissel, D., Péra, M.-C., Zerhouni, N.: Prognostics and health management of PEMFC – state of the art and remaining challenges. Int. J. Hydrogen Energy **38**, 15307–15317 (2013). https://doi.org/10.1016/j.ijhydene.2013.09.051
9. Futter, G.A., Latz, A., Jahnke, T.: Physical modeling of chemical membrane degradation in polymer electrolyte membrane fuel cells: influence of pressure, relative humidity and cell voltage. J. Power Sour. **410–411**, 78–90 (2019). https://doi.org/10.1016/j.jpowsour.2018.10.085
10. Ao, Y., Laghrouche, S., Depernet, D., Chen, K.: Proton exchange membrane fuel cell prognosis based on frequency-domain Kalman filter. IEEE Trans. Transp. Electrif. **7**, 2332–2343 (2021). https://doi.org/10.1109/TTE.2021.3077506
11. Sun, X., Xie, M., Fu, J., Zhou, F., Liu, J.: An improved neural network model for predicting the remaining useful life of proton exchange membrane fuel cells. Int. J. Hydrogen Energy **48**, 25499–25511 (2023). https://doi.org/10.1016/j.ijhydene.2023.03.219
12. Chen, K., Laghrouche, S., Djerdir, A.: Degradation prediction of proton exchange membrane fuel cell based on grey neural network model and particle swarm optimization. Energy Convers. Manag. **195**, 810–818 (2019). https://doi.org/10.1016/j.enconman.2019.05.045
13. Zuo, J., et al.: Deep learning based prognostic framework towards proton exchange membrane fuel cell for automotive application. Appl. Energy **281**, 115937 (2021). https://doi.org/10.1016/j.apenergy.2020.115937
14. Hu, Y., Zhang, L., Jiang, Y., Peng, K., Jin, Z.: A hybrid method for performance degradation probability prediction of proton exchange membrane fuel cell. Membranes **13**, 426 (2023). https://doi.org/10.3390/membranes13040426
15. Zhou, D., Gao, F., Breaz, E., Ravey, A., Miraoui, A.: Degradation prediction of PEM fuel cell using a moving window based hybrid prognostic approach. Energy **138**, 1175–1186 (2017). https://doi.org/10.1016/j.energy.2017.07.096
16. IEEE PHM 2014 data challenge: Outline, experiments, scoring of results, winners. http://eng.fclab.fr/wpcontent/uploads/2014/04/IEEE-Details-After.pdf
17. Sadeghi, M., Behnia, F., Amiri, R.: Window selection of the Savitzky-Golay filters for signal recovery from noisy measurements. IEEE Trans. Instrum. Meas. **69**, 5418–5427 (2020). https://doi.org/10.1109/TIM.2020.2966310

An Extended Single-Particle Model Based on Physics-Informed Neural Network for SOC State Estimation of Lithium-Ion Batteries

Aina Tian[1], Luyao He[1], Kailang Dong[1], Tao Ding[1], Yang Gao[2], Jiuchun Jiang[1(✉)], and Xiaoguang Yang[3]

[1] Hubei Key Laboratory for High-Efficiency Utilization of Solar Energy and Operation Control of Energy Storage System, Hubei University of Technology, Wuhan 430068, China
jcjiang@hbut.edu.cn

[2] School of Electric Power, South China University of Technology, Guangzhou 510640, China

[3] Beijing Institute of Technology Shenzhen Automobile Research Institute, Shenzhen, China

Abstract. Because of its great accuracy, the electrochemical model is frequently utilized in the algorithm design process for lithium-ion batteries. Sadly, the electrochemical model requires a lot of time to solve since it is made up of many nonlinear partial differential equations. In order to solve an extended single particle model (ESPM) fast, a neural network based on physical information (PINN) is examined in this paper. The PINN-ESPM structure can not only estimate the state of charge, but also quickly and accurately estimate the lithium-ion concentration and potential under various application currents, which has stronger adaptability and scalability. In the process of neural network learning, different from the traditional neural network that needs to be trained by labeled data, the loss function is designed only based on the physical constraints brought by equations, boundary conditions and initial values, which makes it an unsupervised learning method. Finally, by comparing the PINN-ESPM proposed in this paper with the data obtained by the P2D model under various current conditions and the experiment battery voltage, the maximum relative error is maintained at 4%. The error of SOC based on the model is less than 4%. While under the same computing resources, PINN-ESPM is 500 times faster than the traditional numerical method.

Keywords: Lithium-ion battery · Extended single-particle model · Physics-informed neural network

1 Introduction

The development of electric vehicles (EVs) is widely recognized as an effective means to address the challenges of excessive carbon emissions and fossil oil depletion. Energy storage system is an indispensable part of electric vehicles, which largely determines its performance and commercial availability. Lithium-ion battery (LIB) plays a leading role in the energy storage of electric vehicles due to its high energy density and long service life [1]. In order to better manage the battery system, it is very important to establish the battery model and study the battery algorithm [2].

K. Li et al. (Eds.): LSMS/ICSEE 2024, CCIS 2218, pp. 300–316, 2025.
https://doi.org/10.1007/978-981-96-0232-2_24

Nevertheless, as LIB is a very intricate and nonlinear electrochemical system [3], only few LIB models are able to simultaneously attain great computational efficiency and high precision. For instance, the equivalent circuit model (ECM) is frequently used as the LIB model for algorithm design in battery management systems (BMS) [4–6]. Although ECMs have a straightforward design and a very quick calculation speed, under some operating conditions, their accuracy needs to be increased to match the demands of sophisticated battery algorithms [7]. Another empirical methodology that trains machine learning to discover associations is the data-driven model. Data-driven models are currently being utilized to forecast many battery metrics, including capacity, SOC, and SOH [8]. Data-driven models are difficult to explain the fundamental physics of battery cells. They need to be retrained if cell chemistry or operating circumstances change, in addition to the substantial amount of training data and initial training costs. On the other hand, this issue may be resolved by using the physics-based modeling method known as the electrochemical model (EM). As a result, EM-based BMS [9] has drawn increasing attention and has emerged as one of the research hotspots in recent years. The pseudo two-dimensional (P2D) model, which is also called the Doyle-Fuller-Newmann model [11], is one of the most often utilized EM models in both academia and industry [10]. The microscale dynamics of lithium-ion insertion and extraction are well described by the pseudo two-dimensional (P2D) model of the LIB process. It provides a mechanism for the occurrence of ohmic polarization and concentration polarization in LIB. Unfortunately, the mechanism model of lithium-ion batteries is difficult to use in car BMS since it is made up of nonlinear partial differential equations (PDEs) and frequently needs dozens of electrochemical parameters to represent. Consequently, PDAEs have been converted into ordinary differential equations (ODEs) by the application of model reduction techniques including the Padé approximation [12], balanced truncation [13], and correct orthogonal decomposition [14], leading to a number of reduced EMs.

Currently, computational techniques like the finite element method (FEM) and finite volume method (FVM) [15] are primarily used to solve the P2D model. Numerical techniques for iteratively solving each sub-equation in a certain order [17] and the explicit-implicit Runge-Kutta-Chebyshev method [16] are two more sophisticated yet computationally efficient approaches. Numerous strategies are employed in these researches to increase P2D models' computational efficiency. Unfortunately, the significant computational overhead of the original P2D model restricts its use for online state estimation in embedded BMS. As a result, these approaches are only viable when time constraints are reasonable and resources are somewhat ample [18].

Suppose that the current density in the electrolyte is uniform [19]. This supposition significantly increases the model's computation speed and qualifies it for use with car BMS. But under high C-rate settings, SPM assumes a uniform current density. Lithium-ions cannot flow freely due to the electrolyte's low diffusion coefficient, which causes a large departure from the P2D model. Numerous academics have embraced the approximation electrolyte lithium-ion concentration approach as a solution to this issue.

The development of physical information neural networks (PINNs) [20, 21] served as a catalyst for the creation of LIB models that balance computational efficiency and accuracy. Raissi et al. introduced PINN for the first time in 2018 [20]. In order to overcome the physical difficulties that PDEs represent, this research investigates the use of neural networks' capacity to fit nonlinear functions. These equations frequently give extra physical limitations that neural networks need to take into account while handling PDEs. Because it takes into account both kinds of data, PINN surpasses conventional neural networks in this regard. Numerous real-world issues have been effectively solved with PINN [22, 23]. These difficult partial differential equations may be quickly solved thanks to PINN's exceptional capacity to learn from physical past information. PINN was utilized by Misyris et al. [23] to forecast the power system's operational conditions, including rotor angle and frequency. In order to confine the neural network's output using physical knowledge, WaveY-Net only predicts the magnetic field's distribution and calculates the electric field using Maxwell's equations. WaveY-Net calculates two to four orders of magnitude quicker than conventional techniques.

Related investigations in the battery industry have also used PINN [24–27]. Li et al. [24] constructed state observers for important P2D model parameters using 2d-LSTM. Voltage, current, temperature, and other time series data from battery operations are sent into the network. Lithium-ion concentration is one of the five main parameters of the battery model that are represented by the outputs. In order to estimate the heat generation rate (HGR) of the battery under particular operating circumstances, Pang et al. [25] employed PINN to construct a bidirectional LSTM (Bi-LSTM) network model based on the Bayesian Optimization Algorithm (BOA). But the reason these researchers only took inspiration from PINN is that, in building the neural network's loss function, they neglected to incorporate the extra physical limitations that the electrochemical model introduced. These research' neural network training methods mostly rely on fitting simulated data, which restricts the capacity of the networks to pick up on the workings of lithium-ion batteries. A physically directed neural network was employed in Reference [27] to solve the PDE of lithium-ion concentration diffusion in electrode particles. Next, the battery's state of charge (SOC) and state of health (SOH) are estimated using the established battery model. To train the PINN network, the electrode particle concentration distribution under a certain applied current is approximated using a fully connected network (FCN) as the PINN architecture. In contrast to earlier papers [24, 25] that describe the development of loss functions, it takes into account the boundary conditions and physical limitations of differential equations. The sole inputs to the basic network structure are time T and the particle's radial coordinate R. Because the LIB model can only fully understand the lithium-ion diffusion process under certain applied current and electrochemical parameter conditions, this significantly restricts the application of the model; the network must be retrained whenever the battery's applied current varies. The network outlined in Reference [28] may output the battery's terminal voltage and accept any current sequence as input. Nevertheless, this approach disregards the battery's mechanism and handles the battery as a black box. Consequently, the model can only be age-independent.

Our approach creates the LIB model based on ESPM and leverages Fully Connected Network (FCN) as the PINN architecture to tackle these issues. Coordinate x, particle radius r, battery capacity SOC, and applied current I are the inputs of this network. The network, known as ESPM-Net, is employed to solve diffusion equations in electrolytes and electrode particles. Compared with the network [27, 28] described in the literature, this network allows fast and accurate calculation of various electrochemical parameters and dynamic driving conditions under various applied current conditions, such as terminal voltage, lithium ion concentration, electrolyte potential, lithium and concentration on the particle surface, and interface current density, with greater adaptability and scalability. The main contributions of this article are as follows:

(1) A ESPM based on electrolyte kinetics, called PINN-ESPM, is established. With this device, SPM accuracy at high magnification is significantly increased. It reduces the SPM error by approximating the lithium-ion distribution on the surface of the electrolyte and electrode particles using PINN, and then calculating the terminal voltage.
(2) Create a neural network ESPM-Net containing physical information, which is the core part of PINN-ESPM. This network can quickly solve the pseudo-two-dimensional diffusion equation of the LIB model. Under specific battery parameters, the network can approximate the lithium-ion concentration distribution on the electrode particle surface and the lithium-ion concentration distribution in the electrolyte under various application currents.
(3) A SOC estimation method based on PINN-ESPM is proposed. Using the physical constraints provided by PDE to design the loss function, PINN-ESPM can approximate SOC more accurately under dynamic conditions.

2 ESPM Model Based on PINN

The solid and electrolyte dynamics of lithium-ion batteries in the cathode, separator, and anode are described by Doyle et al.'s P2D model. However, during high-rate charge and discharge, the accuracy of the single-particle model declines due to the model's complexity and sluggish calculation speed. To solve the modified single-particle model, a neural network based on physical knowledge is therefore suggested. The cathode and anode may be thought of as single particles in RSPM since it is expected that the interface current density inside the cathode and anode areas is equally distributed at all times (see Fig. 1). With the use of RSPM as the objective function, PINN may use both geometric and physical data as input, adopt a fully connected neural network structure (see Fig. 2), and produce an accurate solution.

2.1 ESPM Model

Figure 1 shows the schematic diagram of the battery, from left to right is the current collector, anode, diaphragm, cathode and current collector. The explanations of all physical parameters in the following equations and expressions are shown in Table 1. Through the above assumptions, the solid-liquid phase diffusion equation of lithium battery can be expressed as:

$$\frac{\partial c_{s,i}(r,t)}{\partial t} = \frac{D_{s,i}}{r^2}\frac{\partial}{\partial r}\left(r^2\frac{\partial c_{s,i}(r,t)}{\partial r}\right) \tag{1}$$

$$\frac{\partial c_{e,i}(x,t)}{\partial t} = \frac{\partial}{\partial x}\left(D_{e,i}^{eff}\frac{\partial c_{e,i}(x,t)}{\partial x}\right) + \frac{(1-t_+^0)a_{s,i}}{\varepsilon_e} i_{loc,i}(x,t) \tag{2}$$

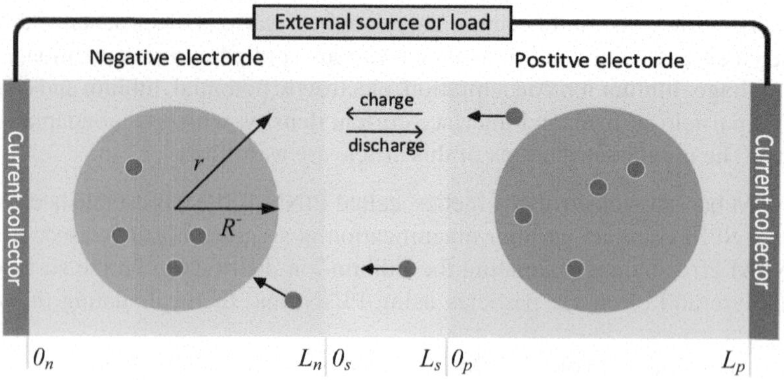

Fig. 1. Modified single particle model

where i is the area of the battery cell, $i = n$ is the negative electrode area, $i = s$ is the separator area (Formula 1 is not applicable to the separator area), $i = p$ is the positive electrode area, $c_{s,i}(r, t)$ represents the solid phase lithium ion concentration, $r \in (0, R_i)$ is the radial direction coordinate, $D_{s,i}$ is the solid phase lithium ion diffusion coefficient, $D_{e,i}^{eff}$ is the effective liquid phase diffusion coefficient, can be expressed by $D_{e,i}^{eff} = D_{e,i}\varepsilon_e^{brugg}$, $D_{e,i}$ is the liquid phase diffusion coefficient, ε_e is the electrolyte volume fraction, *brugg* is Bruggmann's constant, t_+^0 represents the lithium ion migration number, $a_{s,i}$ is the active surface area per unit electrode volume, $a_{s,i} = 3\varepsilon_{s,i}/R_i$, $i_{loc,i}$ is the interface current density. Along the case of ESPM, we presume that the interface current density is uniformly distributed along the anode's thickness direction or along the cathode. With this assumption, the interface current density may be obtained in an analytical manner.

$$i_{loc,i,ESPM} = \frac{i_{app}}{a_{s,i} F L_i} \tag{3}$$

where i_{app} is the applied current density, $i_{app} = I/A$, I is the charge-discharge current ($I > 0$ is the discharge), and A is the cross-sectional area of the electrode. The ohmic potential drop caused by the electrolyte conductivity and the electrolyte concentration overpotential [29], the liquid phase potential difference can be obtained.

$$\Delta\varphi_e = \left(\frac{L_n}{2\sigma_{e,n}^{eff}} + \frac{L_s}{\sigma_{e,s}^{eff}} + \frac{L_p}{2\sigma_{e,p}^{eff}}\right) \cdot I + \frac{2RT}{F}\left(1 - t_+^0\right)\left(\ln c_{e,L_p} - \ln c_{e,0_n}\right) \tag{4}$$

Where $\sigma_{e,i}^{eff}$ represents the effective electrolyte conductivity, $\sigma_{e,i}^{eff} = \sigma_{e,i}\varepsilon_e^{brugg}$, $\sigma_{e,i}$ is the electrolyte conductivity.

In order to analytically solve the solid potential of the anode and cathode in ESPM, we use the Bulter-Volmer equation.

$$i_{loc,i,ESPM} = i_{0,i,ESPM}\left[\exp\left(\frac{\alpha F \eta_{i,ESPM}}{RT}\right) - \exp\left(-\frac{(1-\alpha) F \eta_{i,ESPM}}{RT}\right)\right] \tag{5}$$

where

$$i_{0,i,ESPM} = F\kappa_i\left(c_{s,\max,i} - c_{s,surf,i}\right)^{\alpha}\left(c_{s,surf,i}\right)^{1-\alpha}\left(c_{e,i,ave}\right)^{\alpha} \tag{6}$$

$$\eta_{i,ESPM} = \phi_{s,i} - \phi_{e,i,ave} - U_{eq,i} \tag{7}$$

In the above expression, $c_{e,i,ave}$ represents the given average lithium-ion concentration along the thickness direction of the positive or negative electrode, while $\phi_{e,i,ave}$ represents the given average electrolyte potential along the thickness direction of the anode or cathode, and the open circuit potential of the $U_{eq,i}$ electrode. It is important to note that in order to obtain a straightforward analytical statement, the average lithium-ion concentration and the average electrolyte potential are used in the above. The single particle hypothesis, which holds that one particle represents the anode's or cathode's interfacial current density, is the basis for this treatment. In calculating the interfacial current density, we assume the average electrolyte concentration and potential of a particle in the environment.

Further definition.

$$m_n = i_{app}/F\kappa_n L_n\left(c_{s,\max,n} - c_{s,surf,n}\right)^{\alpha}\left(c_{s,surf,n}\right)^{\alpha}\left(c_{e,n,ave}\right)^{\alpha} \tag{8}$$

$$m_p = i_{app}/F\kappa_p L_p\left(c_{s,\max,p} - c_{s,surf,p}\right)^{\alpha}\left(c_{s,surf,p}\right)^{\alpha}\left(c_{e,p,ave}\right)^{\alpha} \tag{9}$$

Substituting Eq. (3) into equation Eq. (5), we can get

$$m_n = \exp\left(\frac{\alpha F\eta_{n,ESPM}}{RT}\right) - \exp\left(-\frac{(1-\alpha)F\eta_{n,ESPM}}{RT}\right) \tag{10}$$

$$m_p = -\exp\left(\frac{\alpha F\eta_{p,ESPM}}{RT}\right) + \exp\left(-\frac{(1-\alpha)F\eta_{p,ESPM}}{RT}\right) \tag{11}$$

Assuming that $\alpha = 0.5$, we can solve the overpotential of anode and cathode by Eqs. (8–11).

$$\eta_{n,ESPM} = \frac{2RT}{F}\ln\left(\frac{\sqrt{m_n^2+4}+m_n}{2}\right) \tag{12}$$

$$\eta_{p,ESPM} = -\frac{2RT}{F}\ln\left(\frac{\sqrt{m_p^2+4}+m_p}{2}\right) \tag{13}$$

The battery cell terminal voltage based on ESPM is

$$V_{ESPM} = U_{eq,p} - U_{eq,n} - \Delta\varphi_e - \frac{2RT}{F}\ln\left(\frac{\sqrt{m_p^2+4}+m_p}{2}\right) - \frac{2RT}{F}\ln\left(\frac{\sqrt{m_n^2+4}+m_n}{2}\right) \tag{14}$$

Table 1. Lithium-ion battery parameters

Parameter	Unit	Description	Anode	Separator	Cathode
L	m	Length of domain	7.7×10^{-5}	8×10^{-5}	7.4×10^{-5}
R_p	m	Particle radius	1.7×10^{-5}	–	6×10^{-8}
ε_s	–	Electrode volume fraction	0.6	–	0.3477
ε_e	–	Electrolyte volume fraction	0.4	0.65	0.5
D_s	$m^2 \cdot s^{-1}$	Diffusion coefficient in solid phase	7×10^{-13}	–	1×10^{-17}
D_e	$m^2 \cdot s^{-1}$	Diffusion coefficient in the electrolyte	1×10^{-9}		
σ_e	$S \cdot m^{-1}$	Electrode conductivity	159.6	–	35.1
κ	$m^{2.5} \cdot s^{-1}$	Reaction rate constant	3.99×10^{-9}	–	4×10^{-9}
$c_{s,\max}$	$mol \cdot m^{-3}$	maximum solid phase	29000	–	30000
$c_{e,0}$	$mol \cdot m^{-3}$	initial electrolyte concentration	1000		
t_+^0	–	Lithium-ion transference number	0.5		
S	m^2	Current collector area	1.5		
$brugg$	–	Bruggeman coefficient	1.5		
α	–	Charge transfer coefficients for electrodes	0.5		
T	K	temperature	293		
F	$C \cdot mol^{-1}$	Faraday's constant	96487		
R	$J \cdot mol \cdot m^{-3}$	Universal gas constant	8.314		

2.2 PINN

The neural network based on physical information is used to solve the ESPM model. Unlike the traditional neural network, it needs to mark the data to train, but only relies on the equation to train, making it an unsupervised learning method. The trained model can only be output by a given input. The operation speed has been greatly improved, and the lithium-ion concentration distribution, terminal voltage, SOC curve can be obtained, and physical parameters such as SOC and current can be used as input, as shown in the figure. In this way, the SOC and terminal voltage changes of any battery capacity under any charge and discharge conditions can be obtained.

The input of the PINN network is position information x and time t. Different from previous studies, the input here adds the charge-discharge current $I_{ch/dis}$ and the initial lithium-ion concentration $c_{s0,surf}$ on the particle surface. By adding these two inputs, the

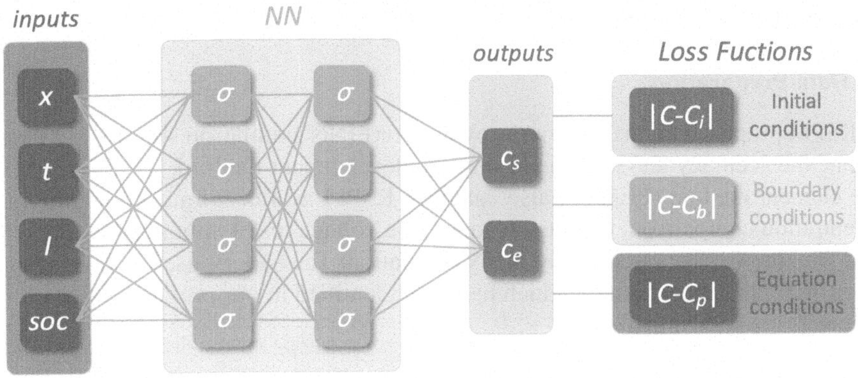

Fig. 2. PINN network structure

distribution of lithium-ion concentration on the particle surface and electrolyte lithium-ion concentration under various application current conditions and various battery capacity conditions can be solved, which has stronger adaptability and scalability. The two outputs obtained by the neural network are substituted into the loss function. By continuously optimizing the weight of the neural network, the value of the obtained output into the loss function is getting smaller and smaller, so that the two outputs are getting closer and closer to the real value. At the same time, tanh is chosen as the activation function in this study. The loss function here can be expressed as follows:

$$Loss = MSE_e + MSE_b + MSE_i \tag{15}$$

$$\begin{aligned} MSE_e &= \frac{1}{N_{e1}} \sum_{i=1}^{N_{e1}} \left\| \frac{\partial c_{s,i}(r,t)}{\partial t} - \frac{D_{s,i}}{r^2}\frac{\partial}{\partial r}\left(r^2 \frac{\partial c_{s,i}(r,t)}{\partial r}\right) \right\|^2 \\ &+ \frac{1}{N_{e2}} \sum_{i=1}^{N_{e2}} \left\| \frac{\partial c_{e,i}(x,t)}{\partial t} - \frac{\partial}{\partial x}\left(D_{e,i}^{eff} \frac{\partial c_{e,i}(x,t)}{\partial x}\right) - \frac{(1-t_+^0)a_{s,i}}{\varepsilon_e} i_{loc,i}(x,t) \right\|^2 \end{aligned} \tag{16}$$

$$\begin{aligned} MSE_b &= \frac{1}{N_{b1}} \sum_{i=1}^{N_{b1}} \left\| \left.\frac{\partial c_{s,i}(r,t)}{\partial t}\right|_{r=0} \right\|^2 + \frac{1}{N_{b2}} \sum_{i=1}^{N_{b2}} \left\| \left.\frac{\partial c_{s,i}(r,t)}{\partial t}\right|_{r=r_p} \right\|^2 \\ &+ \frac{1}{N_{b3}} \sum_{i=1}^{N_{b3}} \left\| c_{e,n}(L_n,t) - c_{e,s}(0_s,t) \right\|^2 + \frac{1}{N_{b4}} \sum_{i=1}^{N_{b4}} \left\| c_{e,s}(L_s,t) - c_{e,p}(0_p,t) \right\|^2 \\ &+ \frac{1}{N_{b5}} \sum_{i=1}^{N_{b5}} \left\| \left. D_{e,s}^{eff}\frac{\partial c_{e,s}(x,t)}{\partial x}\right|_{x=L_n} - \left. D_{e,s}^{eff}\frac{\partial c_{e,s}(x,t)}{\partial x}\right|_{x=0_s} \right\|^2 \\ &+ \frac{1}{N_{b6}} \sum_{i=1}^{N_{b6}} \left\| \left. D_{e,s}^{eff}\frac{\partial c_{e,s}(x,t)}{\partial x}\right|_{x=L_s} - \left. D_{e,s}^{eff}\frac{\partial c_{e,s}(x,t)}{\partial x}\right|_{x=0_p} \right\|^2 \end{aligned} \tag{17}$$

$$MSE_i = \frac{1}{N_{i1}} \sum_{i=1}^{N_{i1}} \left\| c_{s,i}(r,0) - c_{s0} \right\|^2 + \frac{1}{N_{i2}} \sum_{i=1}^{N_{i2}} \left\| c_{e,i}(x,0) - c_{e0} \right\|^2 \tag{18}$$

3 Research Method

3.1 Data Preparation

The RSPM proposed in this paper uses python for PINN simulation solution. In order to evaluate the accuracy and performance of the development method, we use COMSOL finite element method (FEM) to also calculate RSPM as a benchmark for lithium-ion concentration comparison, and then compare the voltage results obtained by the experiment with the voltage obtained by PINN simulation. The battery used in this paper is a lithium-iron phosphate battery, which has been identified in our previous research. The parameters are shown in Table 2.

Table 2. Lithium iron phosphate battery specification

Type	Pouch battery
Cathode	$LiFePO_4$
Anode	Graphite
Nominal Capacity	25 Ah
Nominal Voltage	3.3 V
Charge Voltage	3.6 V
Discharge Voltage	2.0 V
Size (H/W/T)	238 mm/140 mm/7.5 mm

The overpotential of the electrode can also be expressed as $\eta_i = \phi_{s,i} - U_i(\theta_i)$, where U_i is the open circuit potential of the electrode. In our previous study, the open circuit potentials of the cathode and anode of lithium iron phosphate battery have been obtained.

$$\begin{aligned} U_n = {} & 0.1379 + 0.7526e^{(-35.61\theta_n)} - 0.0153\tanh\left(\frac{\theta_n - 0.6142}{0.0156}\right) \\ & - 0.1312\tanh\left(\frac{\theta_n - 0.3173}{0.0721}\right) - 0.1212\tanh\left(\frac{\theta_n - 0.2120}{0.0940}\right) \\ & - 0.1291\tanh\left(\frac{\theta_n - 0.4524}{0.1584}\right) - 0.1099\tanh\left(\frac{\theta_n - 0.3976}{0.1596}\right) \\ & - 0.1083\tanh\left(\frac{\theta_n - 0.4246}{0.1539}\right) - 0.1543\tanh\left(\frac{\theta_n - 0.4003}{0.0985}\right) \\ & - 0.7192\tanh\left(\frac{\theta_n - 0.3684}{0.1573}\right) - 0.025 \end{aligned} \tag{19}$$

$$\begin{aligned} U_p = {} & 4.751 - 10^7 \times 4.839e^{-8.562\theta_p^{-8.562}} \\ & + 0.001929\tanh\left(\frac{\theta_p - 0.5344}{0.9648}\right) - 0.7222\tanh\left(\frac{\theta_p - 0.9216}{0.02259}\right) \\ & - 1.641\tanh\left(\frac{\theta_p + 0.0009297}{0.006564}\right) - 0.4082\tanh\left(\frac{\theta_p + 0.7821}{0.2462}\right) \end{aligned} \tag{20}$$

Here $\theta_i = \frac{c_{s,i,surf}}{c_{s,i,\max}}$ represents the degree of lithiation of the electrode, which can also be expressed as SOL. The stoichiometric parameters, $\theta_{0\%}^+, \theta_{100\%}^+, \theta_{0\%}^-, \theta_{100\%}^-$,, characterize the lithium insertion of cathode and anode under 0% and 100% SOC, respectively.

3.2 Training Method

In order to improve the training speed and accuracy of PINN, the physical parameters are converted into dimensionless values here. The total lateral length of the battery cell is selected as the reference length $L_N = L_n + L_s + L_p$, the initial lithium ion concentration of the solid phase is the solid phase reference concentration $c_{sN,i} = c_{s,i}(r, 0)$, the end time of charge and discharge is the reference time $t_N = t_{ch/dis}$, and the charge and discharge current is the reference current $I_N = I_{ch/dis}$. Then $R_n^* = 1$, $c_{s,n}^*(r, 0) = 1$ and $T^* = 1$ the dimensionless diffusion coefficient is expressed as:

$$D_{s,i}^* = \frac{D_{s,i} t_N}{L_N} \tag{21}$$

$$D_{e,i}^* = \frac{D_{e,i} t_N}{L_N} \tag{22}$$

The particle surface diffusion flux in the equation is also dimensionless, expressed as:

$$i_{loc,i}^* = \frac{i_{loc,i} t_N}{c_{sN,i} L_N} \tag{23}$$

The number of iterations *epoch* = 5000 and the learning rate are set to 1×10^{-3} throughout the training process. After several tries, the learning rate schedule is established to guarantee the network's quick and seamless convergence. In Sect. 4, the final training results are examined and discussed with Adam serving as the optimizer.

3.3 SOC State Estimation

As the average lithium-ion concentration of the positive and negative particles in the upper battery mechanism model, the SOC of the battery mostly relates to the quantity of power accessible inside the battery. The significant computational complexity of the battery mechanism model is the primary cause of its difficulties for battery SOC calculation. The PINN approach may be used to quickly estimate the state of charge of a battery based on the simplified mechanism model that was described in Sect. 2. The simplified mechanism model, which differs from the conventional equivalent circuit empirical model for estimating state of charge (SOC), must calculate the average solid-phase lithium-ion concentration of the battery's positive and negative electrodes.

At time t, the SOC of the battery is recorded as $SOC(t)$, the average lithium-ion concentration of the positive solid particles is $c_{s,p,mean}(t)$, and the average lithium-ion concentration of the negative solid particles is $c_{s,n,mean}(t)$. It is easy to know that the three are in accordance with the following relationship:

$$SOC(t) = \frac{c_{s,p,mean}(t) - c_{s,p,SOC=0\%}}{c_{s,p,SOC=100\%} - c_{s,p,SOC=0\%}} \tag{24}$$

$$SOC(t) = \frac{c_{s,n,mean}(t) - c_{s,n,SOC=0\%}}{c_{s,n,SOC=100\%} - c_{s,n,SOC=0\%}} \tag{25}$$

Among them, $c_{s,p,SOC=0\%}$ and $c_{s,p,SOC=100\%}$ represent the lithium ion concentration of the positive electrode particles of the battery when the SOC is 0% and 100% respectively, while $c_{s,n,SOC=0\%}$ and $c_{s,n,SOC=100\%}$ represent the corresponding lithium ion concentration of the negative electrode particles. Taking the battery shown in Table 2 as an example, the corresponding parameter values conform to Table 3.

Table 3. Lithium-ion concentration parameters in particles corresponding to different SOC (mol/m^3)

	Negative particles	Positive particles
$c_{s,SOC=100\%}$	15953	373
$c_{s,SOC=0\%}$	918	27336

Using the physical information neural network, based on the improved ESP model of the battery, the SOC of the battery can be estimated.

4 Results and Discussion

4.1 Model Verification

In order to evaluate the accuracy and performance of the developed method, we use the RSPM model solved by PINN and the same model solved by the finite element method (FEM) in COMSOL as a benchmark to compare the lithium-ion concentration, and then compare it with the voltage of the lithium-ion phosphate battery under the experiment.

Figure 3 shows the comparison of lithium-ion concentration and terminal voltage on the surface of positive and negative electrode particles under different discharge capacities. When the SOC of the battery is 1, Fig. 3a and Fig. 3c show that the lithium-ion concentration on the surface of the positive and negative electrodes calculated by COMSOL and PINN is in good agreement at a lower rate. At the same time, Fig. 3e shows that the terminal voltage at a lower rate is also very consistent, and the absolute error of the voltage is below 30 mV. In contrast, at a higher rate, the lithium-ion concentration on the surface of the positive and negative electrodes calculated by the experiment and PINN has a large deviation at the beginning of the discharge, and the corresponding terminal voltage deviation also becomes larger, reaching below 60mV, but it maintains a good agreement in the subsequent discharge process.

When the battery SOC is 0.5, Fig. 3b and Fig. 3d show that The concentration of lithium ions on the positive and negative electrode surfaces as determined by PINN agrees well with the concentration of lithium ions on the same surfaces as determined by COMSOL at the start and throughout the discharge. In contrast, at the end of the discharge, the lithium-ion concentration on the surface of the positive and negative electrodes solved by PINN has a large deviation from the lithium-ion concentration on the surface of the positive and negative electrodes solved by COMSOL. Figure 3f shows that the terminal voltage calculated by PINN is in good agreement with the experimental terminal voltage curve at the beginning and process of discharge, and the error is kept at 10 mV. At the end of discharge, the voltage deviation becomes larger, but it is also kept below 15 mV.

Figure 4 shows the comparison of lithium-ion concentration and terminal voltage on the surface of positive and negative electrode particles under different capacity charging states. When the SOC of the battery is 0, Fig. 4a and Fig. 4c show that the lithium-ion concentration on the surface of the positive and negative electrodes calculated by COMSOL and PINN is in good agreement at higher magnification. At the same time, Fig. 4e shows that the terminal voltage at higher magnification is also in good agreement, and the absolute error of the voltage is below 5mV. In contrast, at a lower rate, the lithium-ion concentration on the surface of the positive and negative electrodes calculated by COMSOL and PINN has a large deviation at the beginning of charging, and the corresponding terminal voltage deviation also becomes larger, but it also remains below 15 mV, and maintains a good agreement in the subsequent charging process.

When the SOC of the battery is 0.5, Fig. 4b and Fig. 4d show that Both at the start of the charging process and throughout, surface concentration of positive and negative particles solved by PINN and the lithium-ion concentration on the surface of the electrodes solved by COMSOL coincide well. However, at the end of charging, the COMSOL solution for the concentration of lithium ions on the positive and negative electrode surfaces and the lithium-ion concentration on those surfaces, as solved by PINN, disagree significantly. Figure 3f shows that the terminal voltage calculated by PINN is in good agreement with the experimental terminal voltage curve at the beginning and process of charging, and the error is kept at 10 mV. At the end of charging, the voltage deviation becomes larger, but it is also kept below 15 mV.

Figure 5 displays the battery terminal voltage change during the battery dynamic condition experiment and the simulation voltage of the model used in this work under NEDC conditions. As can be observed, there is a good match between the measured voltage and the calculated voltage. In the stage of 13,000 s before charging and discharging, the error is kept below 50 mV, and the error becomes larger at the end of charging and discharging, but it is also kept within 200 mV.

4.2 The Validity of SOC is Verified Based on PINN-EPSM

Similar to this, under the NEDC scenario, Fig. 6 illustrates the comparison between the battery's estimated state of charge as determined by the ampere-hour integration technique and the state of charge determined by the physical information neural network based on the ESPM model. It is evident that the maximum error of the SOC estimate result is less than 4% when there is a specific error between the ESPM model and the P2D model.

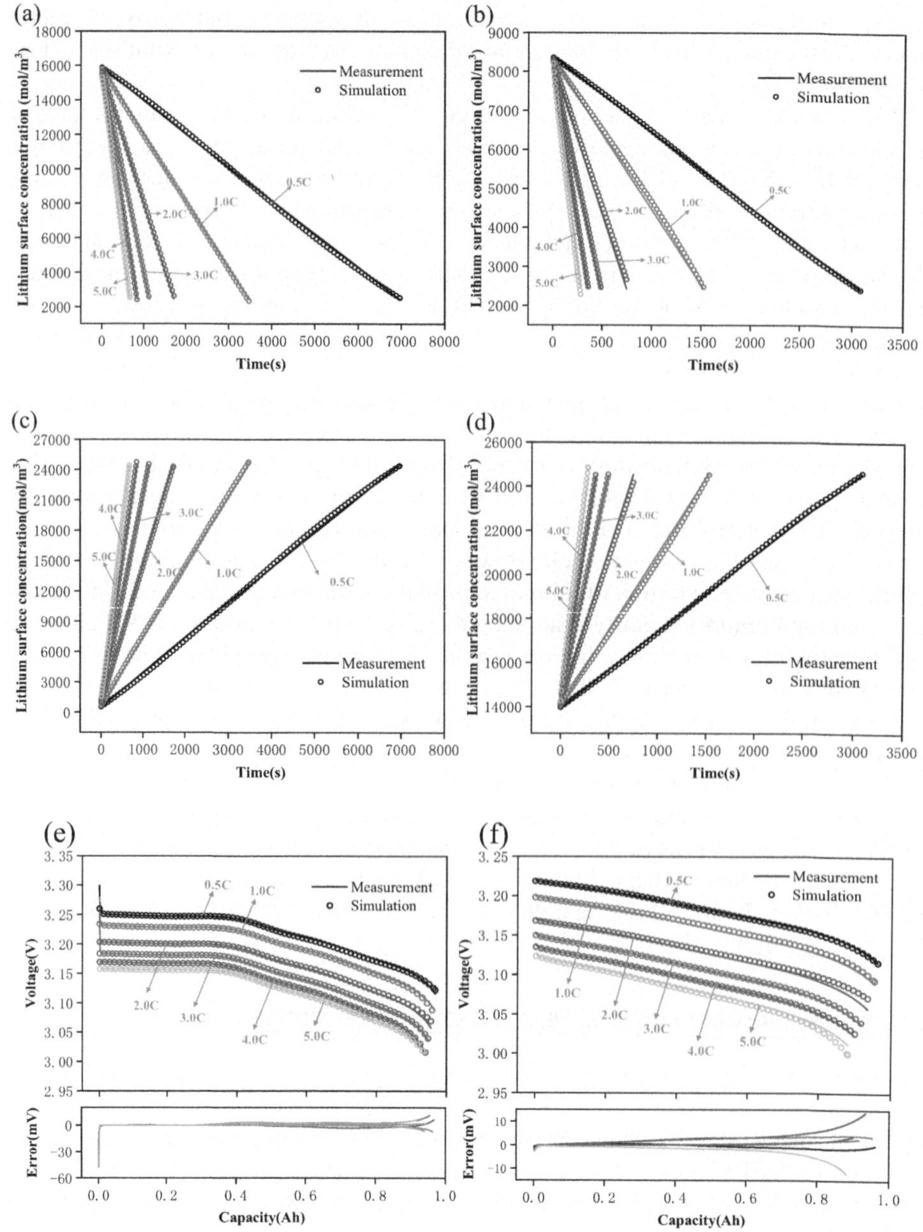

Fig. 3. The lithium-ion concentration on the surface of the anode particles of the same model was compared at different rates (a) SOC = 1 and (b) SOC = 0.5 under COMSOL and PINN solutions. The lithium-ion concentration on the surface of the positive electrode particles discharged at different rates (c) SOC = 1 and (d) SOC = 0.5, the terminal voltage curve and voltage error curve at different rates (e) SOC = 1 and (f) SOC = 0.5. (0.5 C is shown by curves and black dots. 1.0 C is represented by the red curves and dots. 2.0 C is represented by the blue dots and curves. 3.0 C is represented by curves and green dots. 4.0 C is shown by curves and purple dots. 5.0 C is represented by the orange curves and dots.)

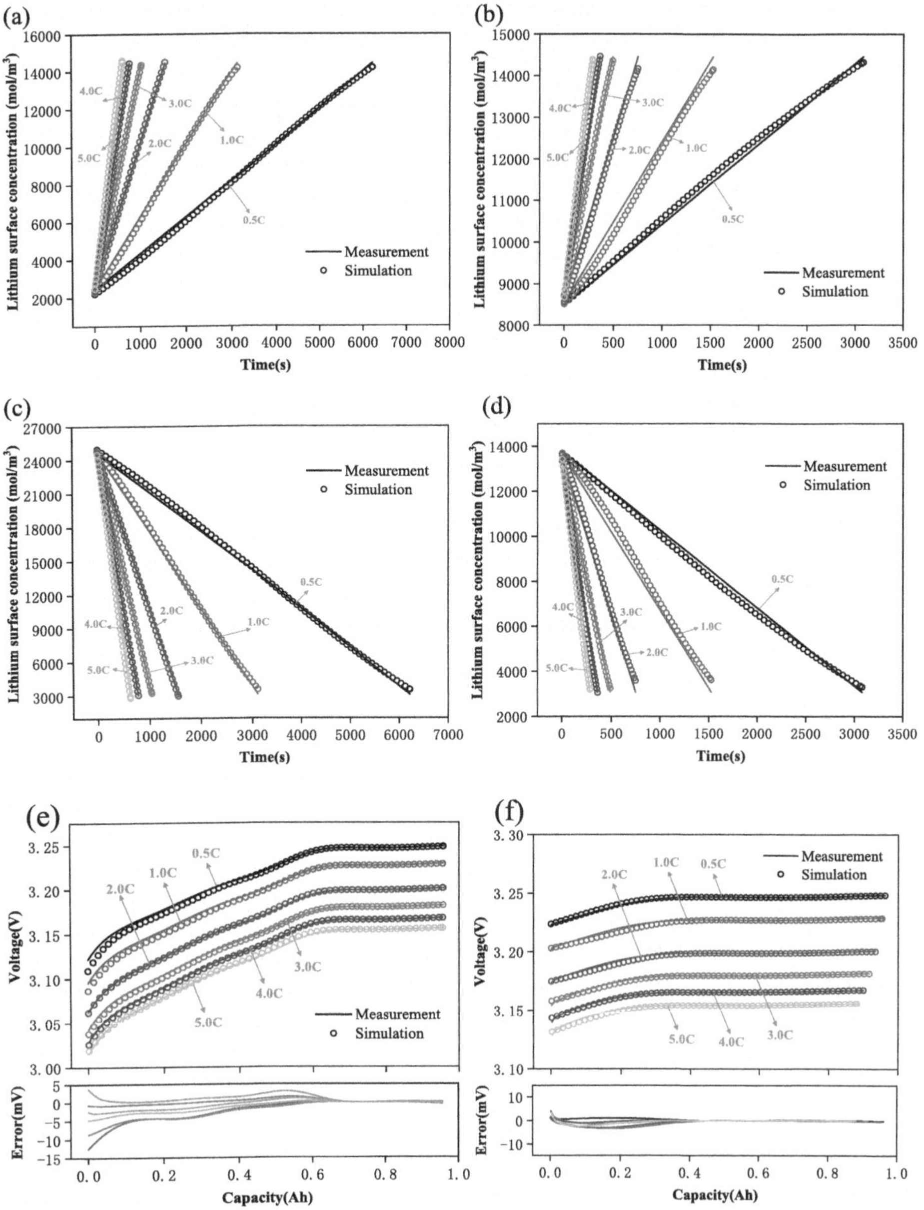

Fig. 4. The lithium-ion concentration on the surface of the anode particles charged at different rates (a) SOC = 1 and (b) SOC = 0.5 under the same model solved by COMSOL and PINN was compared. The lithium-ion concentration on the surface of the positive electrode particles charged at different rates (c) SOC = 1 and (d) SOC = 0.5, the terminal voltage curve and voltage error curve charged at different rates (e) SOC = 1 and (f) SOC = 0.5. (0.5 C is shown by curves and black dots. 1.0 C is represented by the red curves and dots. 2.0 C is represented by the blue dots and curves. 3.0 C is represented by curves and green dots. 4.0 C is shown by curves and purple dots. 5.0 C is represented by the orange curves and dots.)

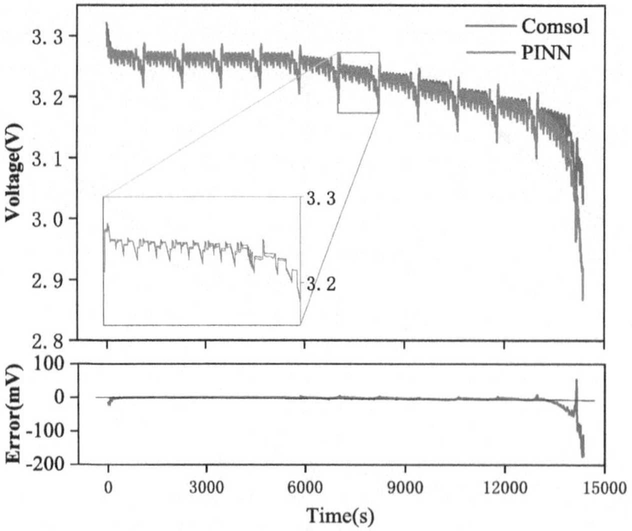

Fig. 5. NEDC dynamic operating condition voltage situation

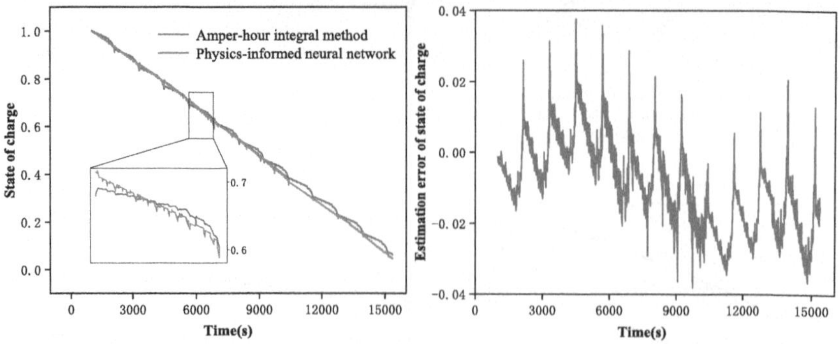

Fig. 6. Based on the ESPM, the SOC of the battery is estimated by using the physical information neural network.

5 Conclusions

In this paper, an ESPM based on electrolyte dynamics, called PINN-ESPM, is proposed. Compared with the traditional ESPM model, it has greater advantages. Firstly, by solving the diffusion equation, the lithium-ion concentration distribution under different applied currents can be obtained. Secondly, a neural network ESPM-Net containing physical information is designed. It no longer needs traditional labeled data to train the neural network, but only relies on the physical constraints brought by the model to design the loss functions, which is called an unsupervised learning method. ESPM-Net is used to solve the diffusion equation. Under static conditions, the solution speed of the battery model is 500 times faster than that of the traditional method, and the maximum relative error is maintained at 2%. Under dynamic conditions, the maximum relative error is also maintained at 4%.

However, the paradigm put out in this work is not without flaws. For example, under the condition of highly complex model, the training time will be too long, and when solving PDEs, it will lead to unstable solutions. Therefore, in future studies, we plan to further improve the ESPM-Net network structure. In addition, in the process of model establishment, the influence of temperature on the model is not considered, and the influence of thermal runaway and aging on the battery is ignored. Therefore, in the future research, the temperature index will also be put into the ESPM-Net structure as an input.

Acknowledgments. This work is supported by the National Natural Science Foundation of China (grant number 52207233, 52177212), Supported by the Science and Technology Research Program of the Hubei Provincial Department of Education (grant number T2021005, Q20212401), Supported by Natural Science Foundation of Hubei Province (grant number 2023AFA033), Supported by Guiding Fund of Government's Science and Technology (grant number 2022BGE230).

References

1. Kim, T., Song, W., Son, D.Y., et al.: Lithium-ion batteries: outlook on present, future, and hybridized technologies. J. Mater. Chem. A **7**(7), 2942–2964 (2019)
2. Chen, S., Wei, X., Zhang, G., et al.: All-temperature area battery application mechanism, performance, and strategies. Innovation **4**(4), 100465 (2023)
3. Horiba, T.: Lithium-ion battery systems. Proc. IEEE **102**(6), 939–950 (2014)
4. Zhang, X., Lu, J.L., Yuan, S.F., et al.: A novel method for identification of lithium-ion battery equivalent circuit model parameters considering electrochemical properties. J. Power Sour. **345**(1), 21–29 (2017)
5. Chen, W., Liang, J., Yang, Z., et al.: A review of lithium-ion battery for electric vehicle applications and beyond. Innov. Solut. Energy Transit. **158**(1), 4363–4368 (2019)
6. He, H., Xiong, R., Guo, H., et al.: Comparison study on the battery models used for the energy management of batteries in electric vehicles. Energy Convers. Manag. **64**(1), 113–121 (2012)
7. Jiang, B., Zhu, Y., Zhu, J., et al.: An adaptive capacity estimation approach for lithium-ion battery using 10-min relaxation voltage within high state of charge range. Energy **263**(PC) (2022)
8. Gao, T., Lu, W., et al.: Machine learning toward advanced energy storage devices and systems. IScience **24**(1), (2021)
9. Chaturvedi, N.A., Klein, R., Christensen, J., et al.: Algorithms for advanced battery-management systems. IEEE Control. Syst. Mag. **30**(3), 49–68 (2010)
10. Marc, D., Yuris, F.: Computer simulations of a lithium-ion polymer battery and implications for higher capacity next-generation battery designs. J. Electrochem. Soc. **150**(6) (2003)
11. Doyle, M., Fuller, T.F., Newman, J.: Modeling of galvanostatic charge and discharge of the lithium/polymer/insertion cell. J. Electrochem. Soc. **140**(6), 1526 (2019)
12. Marcicki, J., Canova, M., Conlisk, A.T., et al.: Design and parametrization analysis of a reduced-order electrochemical model of graphite/LiFePO4 cells for SOC/SOH estimation. J. Power Sour. **237**(1), 310–324 (2013)
13. Mayhew, C., He, W., Kroener, C., et al.: Investigation of projection-based model-reduction techniques for solid-phase diffusion in Li-ion batteries, pp. 123–128. IEEE (2014)

14. Cai, L., White, R.E.: An efficient electrochemical-thermal model for a lithium-ion cell by using the proper orthogonal decomposition method. J. Electrochem. Soc. **157**(11), 1188–1195 (2010)
15. Marcello, T., Lalo, M., Bhushan Gopaluni, R., et al.: LIONSIMBA: a matlab framework based on a finite volume model suitable for li-ion battery design, simulation, and control. J. Electrochem. Soc. **163**(7), 1192–1205 (2016)
16. Bermejo, R., Sastre, P.G.: An implicit-explicit Runge-Kutta-Chebyshev finite element method for the nonlinear lithium-ion battery equations. Appl. Math. Comput. **361**(1), 398–420 (2019)
17. Han, S., Tang, Y., Kjaleghi, R.S.: A numerically efficient method of solving the full-order pseudo-2-dimensional (P2D) Li-ion cell model. J. Power Sour. **490**(1) (2021)
18. Wang, X., Li, J., Chen, S., et al.: Online detection of lithium plating onset for lithium-ion batteries based on impedance changing trend identification during charging processes. IEEE Tran. Transp. Electrif. **9**(2), 3487–3497 (2022)
19. Ning, G., Popov, B.N.: Cycle life modeling of lithium-ion batteries. J. Electrochem. Soc. **151**(10), 1584–1591 (2004)
20. Raissi, M., Perdikaris, P., Karniadakis, G.E.: Physics-informed neural networks: a deep learning framework for solving forward and inverse problems involving nonlinear partial differential equations. J. Comput. Phys. **378**(1), 686–707 (2018)
21. Raissi, M., Karniadakis, G.E.: Hidden physics models: machine learning of nonlinear partial differential equations. J. Comput. Phys. **357**(1), 125–141 (2017)
22. Chen, M., Lupoiu, R., Mao, C., et al.: High speed simulation and freeform optimization of nanophotonic devices with physics-augmented deep learning. ACS Photonics, 3110–3123 (2022)
23. George, S.: Physics-informed neural networks for power systems. arXiv-CS-systems and control (2019)
24. Li, W., Zhang, J., Kringbeck, F.: Physics-informed neural networks for electrode-level state estimation in lithium-ion batteries. J. Power Sour. **506**(1), 45–58 (2021)
25. Pang, H., Wu, L., Liu, J., et al.: Physics-informed neural network approach for heat generation rate estimation of lithium-ion battery under various driving conditions. J. Energy Chem. **78**(1), 1–12 (2023)
26. Renato, G., Nascimento, M.C., Chetan, S., Kulkarni., et al.: Hybrid physics-informed neural networks for lithium-ion battery modeling and prognosis. J. Power Sour. **513**(1), 526–540 (2021)
27. Singh, S., Ebongue, Y.E., Rezaei, S., et al.: Hybrid modeling of lithium-ion battery: physics-informed neural network for battery state estimation. Batteries **9**(6), 301–312 (2023)
28. Schmitt, J., Horstkötter, I., Bäker, B.: Electrical lithium-ion battery models based on recurrent neural networks: a holistic approach. J. Energy Storage **58**(1), 61–78 (2023)
29. Gao, T., Lu, W.: Reduced-order electrochemical models with shape functions for fast, accurate prediction of lithium-ion batteries under high C-rates. Appl. Energy **35**(PA) (2024)
30. Li, W., Fan, Y., Ringbeck, F., et al.: Electrochemical model-based state estimation for lithium-ion batteries with adaptive unscented Kalman filter. J. Power Sour. **476**(1), 34–52 (2020)

Health Prognostic for Lithium-Ion Battery via Time-Series Transfer Learning

Yehui Zhu[1], Hanjing Cheng[1(✉)], Xiaoliang Wang[1,2], Zhongpeng Li[1], Ke-cai Cao[1], Hongjie Wu[1], and Juping Gu[1]

[1] School of Electronic and Information Engineering, Suzhou University of Science and Technology, Suzhou, China
chj@usts.edu.cn

[2] Jiangsu Industrial Intelligent and Low-carbon Technology Engineering Center, Nanjing, China

Abstract. Lithium-ion batteries have wide applications in the energy sector, and effective health prediction is crucial for ensuring their safety and performance. Most machine learning-based State of Health (SOH) estimation methods operate under the assumption that battery data follows the same distribution. However, the statistical characteristics of time-series data may change over time, affecting the accuracy of predictions. To address this issue, an adaptive time-series transfer learning method based on knee point is proposed in this paper. For different battery datasets, a knee method is employed to determine segmentation points of the time series, dividing the battery degradation curve into two stages. Then, healthy features are extracted from battery charge-discharge data for input into the transfer learning model, and the effectiveness of these features is validated. Finally, time-series transfer learning is used to train the health features, exploiting the shared information learned from different domains for SOH prediction. Experimental results demonstrate that the proposed method could achieve good performance across multiple battery data.

Keywords: Lithium-ion battery · State-of-health · Knee method · Time-series transfer learning

1 Introduction

Due to their low self-discharge rate and high energy density, lithium-ion batteries are widely used in consumer electronics and electric vehicles. [1] For lithium-ion batteries, the battery management system (BMS) is essential to extend battery life and ensure its safe operation. [2] One of the key functions of a BMS is to provide accurate state of health (SOH) estimates, which is key to ensuring the EV driving experience and safety. SOH is a metric that reflects the degree of aging and reliability of a battery, and is usually defined as the ratio of actual capacity to rated capacity. Since the capacity cannot be directly measured, predicting

K. Li et al. (Eds.): LSMS/ICSEE 2024, CCIS 2218, pp. 317–330, 2025.
https://doi.org/10.1007/978-981-96-0232-2_25

the SOH of a battery has become a research hotspot. Currently, SOH estimation methods are divided into two main categories: model-based methods and data-driven methods.

Model-based methods include the electrochemical models, the equivalent circuit models, the empirical models, the stochastic degradation model, and some filter algorithms or observer methods [3–5]. Although model-based methods can provide reliable estimates of SOH, these approaches typically require accurate battery models, which are always time consuming and rely heavily on prior knowledge. In addition, prediction accuracy tends to decrease with increasing number of charge-discharge cycles and changes in battery operating conditions. Therefore, it is a challenge for model-based methods to maintain a balance between SOH estimation accuracy and model complexity [6].

On the other hand, data-driven methods show superior capabilities in dealing with nonlinear mapping problems and have been increasingly applied in SOH estimation. Data-driven methods do not require an understanding of the specific differences between batteries to analyze and extract information from battery data. Instead, they use machine learning techniques such as neural networks (NN) [7], support vector machines (SVM) [8], Gaussian process regression (GPR) [9] and others. Compared to the above methods, neural networks are more suitable for battery SOH estimation due to their inherent non-linear mapping capability and self-learning ability. Xu et al. [10] improved the accuracy of neural networks by removing irrelevant features during data preparation and introducing skip connections in a Convolutional Neural Network-Long Short-Term Memory (CNN-LSTM) model to achieve accurate estimation of battery SOH. Ma et al. [11] considered the inconsistent distribution of label space, and introduced a new domain adaptation strategy to reduce cross-domain distribution discrepancy. In addition,RNN methods such as Bi-directional long short-term memory networks [12] and gated neural units [13] have achieved good performance in SOH prediction.

However, most machine learning methods assumes that the data follows the same distribution. As batteries age, different data distributions may emerge, which can lead to a decrease in the accuracy of training results from models trained on single-label data. Hence, existing methods still face the following challenges: (a) most methods assume that training and test data sets follow the same distribution, which does not match the nature of battery degradation; (b) the degradation point is difficult to determine for lithium-ion battery data.

To address the aforementioned problems, an adaptive time-based transfer learning (ATTL) method is proposed in this paper. In ATTL, the time series data can be adaptively split into two parts by a knee method. Then, features learned from both the capacity fade and performance fade stages are transformed into a unified feature space through time-series transfer learning.

The main contributions of this paper can be summarized as follows:

(1) In order to identify the different stages of the capacity degradation curve of lithium-ion batteries, a knee method was proposed to search for degradation points and determine the decision boundary.

(2) Time-series transfer learning method is introduced to address the non-uniform distribution of battery degradation over time and unifies the learned features into a single feature space for SOH estimation.
(3) The proposed ATTL method is conducted on publicly available lithium-ion battery datasets to evaluate its effectiveness.

The rest of this paper is organized as follows. Section 2 provides a detailed description of the proposed method. Section 3 introduces simulation results and analysis. Section 4 summarizes the paper.

2 Methodology

The overall framework of ATTL is shown in Fig. 1. The detailed description of each module within the model is as follows:

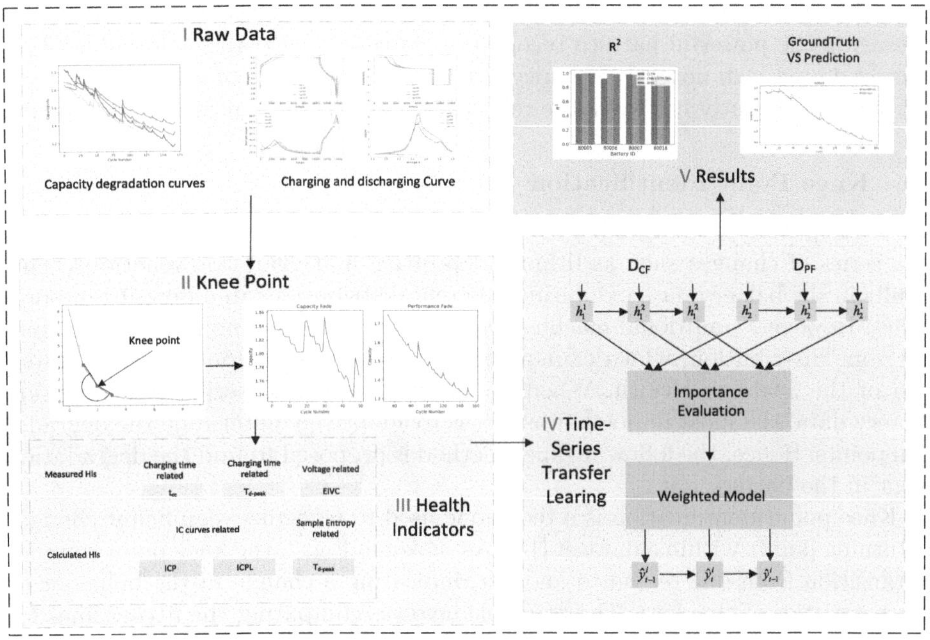

Fig. 1. The framework of the ATTL method.

Module I - Raw Data: The model first imports the original dataset and analyzes the changes in different data as the battery is charged and discharged. It confirms that the characteristics show certain trends with the number of cycles.

Module II - Knee Point: The Knee Point algorithm is utilized to identify the degradation points in the battery capacity degradation curve. The Knee Point is typically observed as a sudden change or a change in slope within the data points, indicating a significant change or a critical point in the dataset.

Module III - Health Indicators: Six features are selected for input into the prediction model, including three measurement features and three calculated features. In the calculated features, the incremental capacity (IC) curve and sample entropy are included to extract hidden information contained within the data.

Module IV - Time-Deries Transfer Learning: Transfer learning involves simultaneously learning shared knowledge within two domains, and then transforming the learned features into a unified feature space for SOH prediction.

Module V - Results: After processing through transfer learning, metrics such as RMSE and R^2 are employed to reflect the accuracy of predictions. It compares the results across different datasets to achieve precise predictions of SOH.

The entire framework, from initial data input to final prediction results, aims to provide a systematic and accurate method for SOH prediction by precisely partitioning the time series, emphasizing the extraction of relevant features, and leveraging the powerful pattern recognition capability of transfer learning. This improved approach not only enhances the model's handling of different datasets but also significantly improves the reliability and accuracy of prediction results.

2.1 Knee Point Identification

For battery data, as the number of charge and discharge cycles increases, it leads to a series of changes such as lithium deposition and oxidation reactions. This results in the battery data exhibiting different distributions in different temporal cycles. However, traditional machine learning methods assume that data follows the same distribution, which cannot meet the requirement for accurate estimation of the State of Health. When partitioning the time series of lithium-ion battery data, the most important aspect is to adaptively determine the degradation points. Hence, the following knee method is proposed to find the degradation point in the battery data.

Knee point identification is a technique used to recognize significant changes or turning points within a dataset [14]. As shown in Fig. 2, the knee point signifies a transition from one region to another, indicating a change in the behavior or characteristics of the data. The principle involves computing the fitting lines for each point and its left and right sides.

For a given set of data points (x_i, y_i), the aim is to identify a knee point k that divides the dataset into two parts. Each part can then be approximated using a linear regression model $y = mx + b$. The selection of the knee point is based on minimizing the total error E of these two parts. The error E can be the sum of absolute deviations or the square root of the sum of squared deviations. The total error E is defined as:

$$E = \sum_{i=1}^{k} |y_i - (m_{fwd,k}x_i + b_{fwd,k})| + \sum_{j=k+1}^{n} |y_j - (m_{bck,k}x_j + b_{bck,k})| \quad (1)$$

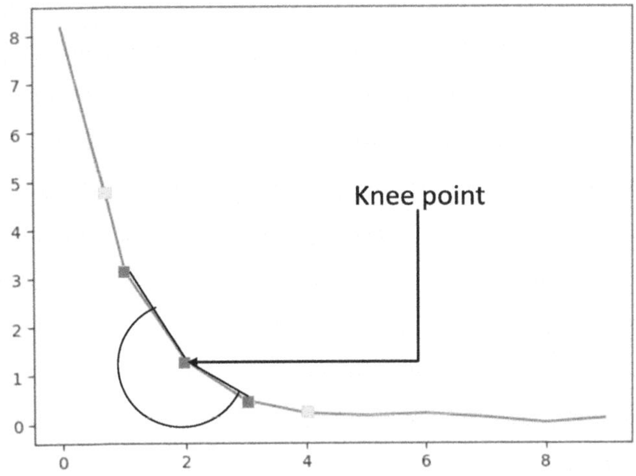

Fig. 2. A simple example with a knee point.

where n is the total number of data points, $m_{fwd,k}$ and $b_{fwd,k}$ are the parameters of the linear regression model calculated based on the data points (x_1, y_1) ... (x_k, y_k), and $m_{fwd,k}$ and $b_{fwd,k}$ are the parameters of the linear regression model calculated based on the data points (x_{k+1}, y_{k+1}) ... (x_n, y_n).

2.2 Extraction of Health Indicators

Since the capacity of a battery is difficult to measure directly in practical applications, it is necessary to extract health indicators from the charging and discharging process to describe the health status of the battery. Follow in [15], six health indicators including measured HIs and calculated HIs are selected in this paper.

The changes in voltage, current, temperature, and incremental capacity of the battery during charging and discharging are shown in Fig. 6. The HIs that can be directly measured include the charging time under constant current state (t_{cc}), the variation of charging voltage at equal time intervals (ETCV), and the temperature peak during discharge (TP). The calculated HIs include the peak value (ICP) of the incremental capacity (IC) curve, the position of the peak value (ICPL), and the sample entropy (SampEn). IC analysis is a commonly used method for predicting the health status of batteries. SampEn is an information entropy used to measure the uncertainty or disorder of data samples.

$$\frac{dQ}{dV} = \frac{I \cdot dt}{dV} = I \cdot \frac{dt}{dV} \tag{2}$$

2.3 Time-Series Transfer Learning

Inspired by the adaRNN [16], the Temporal Distribution Matching (TDM) module is used to learn shared knowledge between different states. The TDM loss function can be represented as:

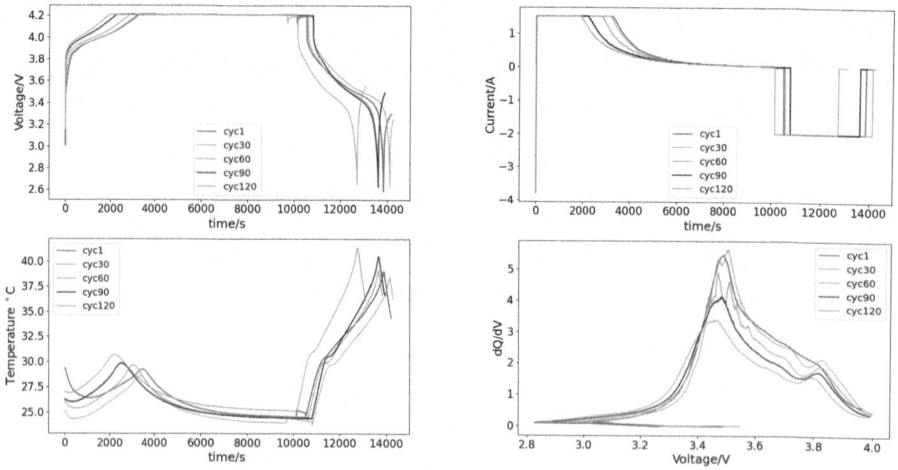

Fig. 3. Battery charging and discharging characteristic curve.

$$L(\theta, \alpha) = L_{pred}(\theta) + \lambda L_{tdm}(D_{CF}, D_{PF}; \theta, \alpha) \tag{3}$$

where θ denotes the learnable model parameters, α denotes the importance rating factor of each pair of data segments, and λ is a trade-off parameter. D_{CF} represents the capacity fade state and D_{PF} represents the performance fade state (Fig. 3).

In this paper, the L_{pred} can be computed by

$$\mathcal{L}_{pred}(\theta) = \frac{1}{|\mathcal{D}_{CF}|} \sum_{i=1}^{|\mathcal{D}_{CF}|} \ell(\mathbf{y}_i^{CF}, \mathcal{M}(\mathbf{x}_i^{CF}; \theta)) + \frac{1}{|\mathcal{D}_{PF}|} \sum_{i=1}^{|\mathcal{D}_{PF}|} \ell(\mathbf{y}_i^{PF}, \mathcal{M}(\mathbf{x}_i^{PF}; \theta)) \tag{4}$$

where (x_i, y_i) represents the i-th token segment starting at period D_{CF} and D_{PF}. $\ell(\cdot, \cdot)$ denotes the MSE loss, and θ denotes the learnable model parameters. M is the learned model.

The L_{tdm} can be represented as:

$$\mathcal{L}_{tdm}(\mathcal{D}_{CF}, \mathcal{D}_{PF}; \theta) = \sum_{t=1}^{V} \alpha_{CF,PF}^{t} d(\mathbf{h}_{CF}^{t}, \mathbf{h}_{PF}^{t}; \theta) \tag{5}$$

where $\alpha_{CF,PF}^{t}$ denotes the distribution importance between the periods D_{CF} and D_{PF} at state t. h_{CF}^{t} and h_{PF}^{t} represent the hidden states of D_{CF} and D_{PF}.

According to the standard Gate Recurrent Unit (GRU) calculation method, all hidden states of the GRU can be calculated easily. $\delta(\cdot)$ represents update rules for hidden elements The state computation can be expressed as

$$\mathbf{h}_i^t = \delta(\mathbf{x}_i^t, \mathbf{h}_i^{t-1}) \tag{6}$$

The update function for parameter α is as follows:

$$\alpha_{CF,PF}^{t,(n+1)} = \begin{cases} \alpha_{CF,PF}^{t,(n)} \times (1 + \sigma(d_{CF,PF}^{t,(n)} - d_{CF,PF}^{t,(n-1)})) & d_{CF,PF}^{t,(n)} \geq d_{CF,PF}^{t,(n-1)} \\ \alpha_{CF,PF}^{t,(n)} & otherwise \end{cases} \tag{7}$$

where $\alpha_{CF,PF}^{t,(n+1)}$ and $\alpha_{CF,PF}^{t,(n)}$ represents the α of the epoch n+1 and epoch n. $\sigma(\cdot)$ is the sigmoid function. where $\alpha_{CF,PF}^{t,(n+1)}$ and $\alpha_{CF,PF}^{t,(n)}$ represents the α of the epoch n+1 and epoch n. $\sigma(\cdot)$ is the sigmoid function.

3 Simulation and Verification

In this section, the effectiveness of ATTL is validated by assessing the necessity of knee point selection, the correlation of selected HIS, and the effectiveness of the time-series transfer learning model.

3.1 Experiment Setting

The hardware and software configurations used were: Intel(R) Core(TM) i5-9300H, 16GB memory, NVIDIA GeForce GTX 1050, Python 3.6, PyTorch 1.10.2, CUDA 11.8.

The dataset provided by the NASA Ames Prognostics Center of Excellence [17] is utilized in this paper. The detailed information is described in Fig. 4 and the capacity degradation curves of the NASA battery are shown in Fig. 5. The hyperparameters are: input size of 6, batch size of 20, hidden size of 100, epoch of 250, sequence length of 10, number of layers of 1, and learning rate of 0.0005. Note that we use the default hyperparameters in CNN-LSTM-Skip [10] for the comparison.

3.2 Evaluation Criteria

State of Health (SOH) is an indicator of battery performance degradation, reflecting the extent of battery aging. In this paper, SOH is defined based on battery capacity, representing the percentage of the current capacity relative to the original capacity:

$$SOH = \frac{C_{Now}}{C_{Rated}} \times 100\% \tag{8}$$

where C_{Now} represents the current battery capacity, C_{Rated} denotes the rated capacity of the batteries.

The following statistical indicators are used to evaluate the estimation performance of this method. Root Mean Square Error (RMSE) is a typical indicator of regression models, representing the deviation between estimated and actual values. For larger errors, the weight is higher. Mean Absolute Error (MAE) is used to represent the average absolute error between predicted and true values. When the value approaches 0, the performance of the model will improve The

Battery ID	B0005、B0006、B0007、B0018
Measured operating conditions：charge, discharge, impedance	
Room temperature：24 degrees Celsius	
Charging: Charging was carried out in a constant current (CC) mode at 1.5A until the battery voltage reached 4.2V and then continued in a constant voltage (CV) mode until the charge current dropped to 20mA. **Discharging:** Discharge was carried out at a constant current (CC) level of 2A until the battery voltage fell to 2.7V, 2.5V, 2.2V and 2.5V for batteries 5 6 7 and 18 respectively. **Impedance:** Impedance measurement was carried out through an electrochemical impedance spectroscopy (EIS) frequency sweep from 0.1Hz to 5kHz.	
Experimental termination conditions: The experiments were stopped when the batteries reached end-of-life (EOL) criteria, which was a 30% fade in rated capacity (from 2Ahr to 1.4Ahr).	

Fig. 4. The detailed information of dataset.

determination coefficient R^2 is used to evaluate goodness of fit and regression. The closer the value is to 1, the better the fitting effect. The calculation formulas for RMSE, MAE, and R^2 are as follows:

$$RMSE = \sqrt{\frac{1}{N}\sum_{i=1}^{N}\left(S\hat{O}H_i - SOH_i\right)^2} \tag{9}$$

$$MAE = \frac{1}{N}\sum_{i=1}^{N}\left|S\hat{O}H_i - SOH_i\right| \tag{10}$$

$$R^2 = 1 - \frac{\sum_i \left(S\hat{O}H_i - SOH_i\right)^2}{\sum_i \left(S\bar{O}H - SOH_i\right)^2} \tag{11}$$

where $SOH_i, S\hat{O}H_i$ and $S\bar{O}H$ are the reference value, estimation value, and mean value, respectively. N is the number of sample data of the testing set.

3.3 Simulation Results and Analysis

HIs Analysis. After extracting the battery HIs, it is necessary to quantitatively analyze the validity. The use of HIs with low validity as inputs to the model

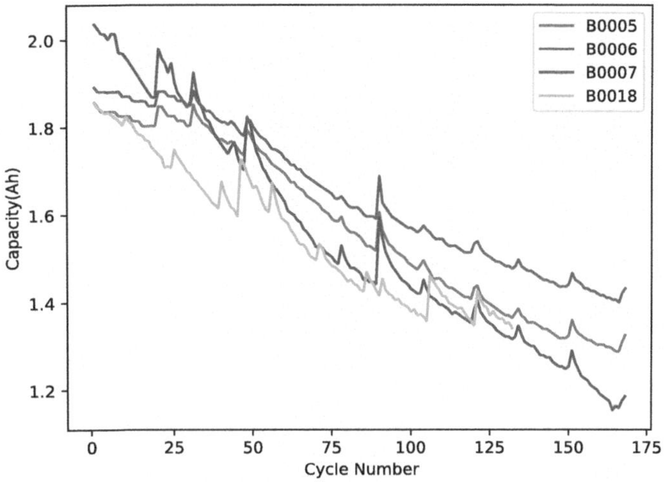

Fig. 5. The capacity degradation curves of NASA battery.

will negatively affect the accuracy of the estimation. The Pearson correlation coefficient is used to measure the strength and direction of the linear relationship between two continuous variables. Its calculation formula is as follows:

$$PCC = \frac{\sum_{i=1}^{l}(z_i - \bar{z})(q_i - \bar{q})}{\sqrt{\sum_{i=1}^{l}(z_i - \bar{z})^2}\sqrt{\sum_{i=1}^{l}(q_i - \bar{q})^2}} \tag{12}$$

where z and q are reference values for HI and capacity, respectively. l is the number of samples. The value of PCC is $[-1,1]$, where 1 and -1 represent a complete linear correlation and 0 represents the uncorrelation of the two variables.

As shown in Fig. 6, the absolute correlation values of most features are above 0.8, indicating that the selected HIs are effective.

Experiment on Segmentation Point. As shown in Fig. 7, we divide the degradation of lithium battery data into two stages through the proposed knee method: the capacity fade (CF) stage and the performance fade (PF) stage.

Different segmentation points including the manually selected temporal segmentation points, the point determined by the coral loss function in [16] and the point determined by the knee method are compared. The results are presented in Table 1. It can be seen that the segmentation points selected by the knee method achieved the best performance, with an RMSE reduction of 1.4% compared to the

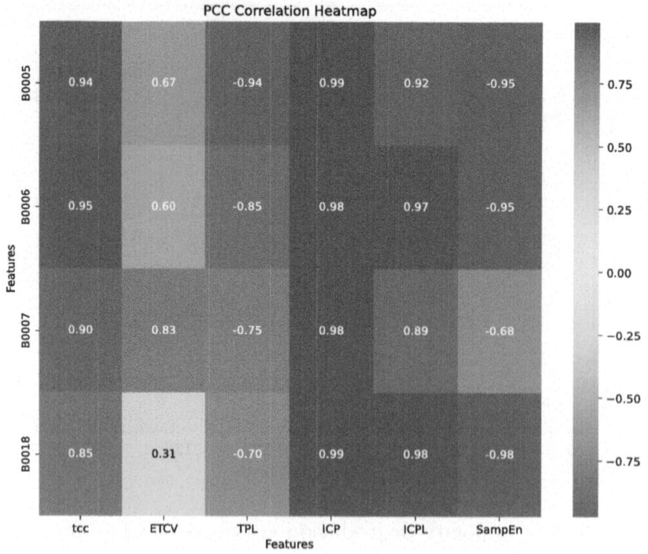

Fig. 6. PCC values for each battery.

other segmentation points. In addition, the R-squared coefficient also increases by 0.1%, which indicates that the knee method can adaptively determine the most appropriate segmentation point for lithium-ion battery time-series data.

Table 1. Prediction results of different segmentation points.

Segmentation Point	Manually-45	Manually-55	CORAL	Knee
RMSE ↓	0.0152	0.0149	0.0142	**0.0130**
R^2 ↑	0.9973	0.9975	0.9977	**0.9983**

Experiment on Various Batteries. The detailed simulation results of different batteries are shown in Table 2. It can be seen that ATTL achieves a coefficient of determination (R^2) above 0.96 for B0005, B0006, B0007 and B0018, with some reaching up to 0.99, and the RMSE values are all within 5%. In terms of other evaluation metrics, ATTL outperforms the LSTM network, with performance improvements of over 10%. These results indicate that the time series transfer learning model performs well in SOH prediction tasks, maintaining high accuracy across different scenarios.

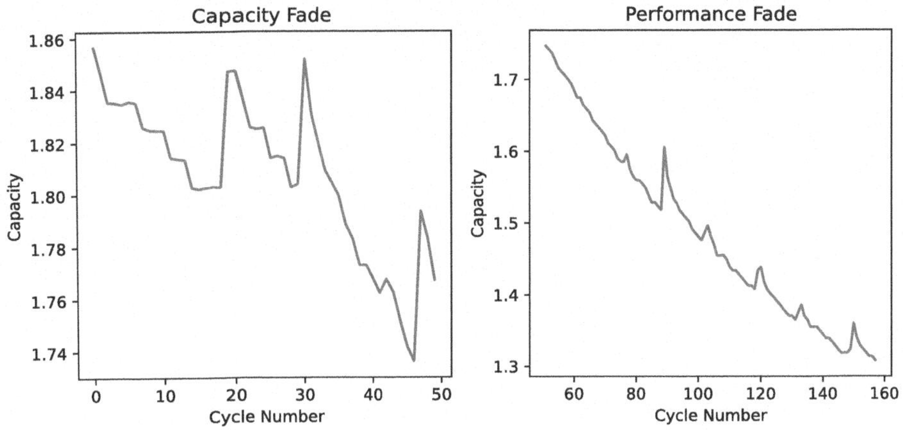

Fig. 7. The CF stage and the PF stage of the lithium battery degradation.

Table 2. Prediction results on various batteries.

Method	battery ID	RMSE ↓	MAE ↓	R^2 ↑
LSTM	B0005	0.0274	0.0203	0.9927
	B0006	0.0603	0.0413	0.9234
	B0007	0.0396	0.0282	0.9832
	B0018	0.0472	0.0395	0.9619
ATTL	B0005	0.0130	0.0097	0.9983
	B0006	0.0314	0.0252	0.9845
	B0007	0.0153	0.0119	0.9976
	B0018	0.0472	0.0340	0.9637

A comparative analysis of the prediction results of battery B0005 is carried out. Figure 8 shows the predicted and actual values of SOH for both methods. It can be seen that the curves fitted by ATTL for SOH estimation capture the fluctuations more accurately in the battery degradation curve. These peaks represent periods when the battery performance is at a relatively high level. In addition, ATTL can also maintain consistency with the trend of gradual capacity degradation, indicating its ability to fit the entire battery degradation curve well.

Model Comparison. In this section, the proposed ATTL method will be compared with the traditional LSTM and CNN-LSTM-Skip. Figure 9 compares the R^2 values of three models trained on different datasets. From Fig. 9, it can be observed that for various battery datasets, the accuracy of all three models exceeds 90%. For batteries B0005 and B0007, ATTL exhibits the highest accuracy.

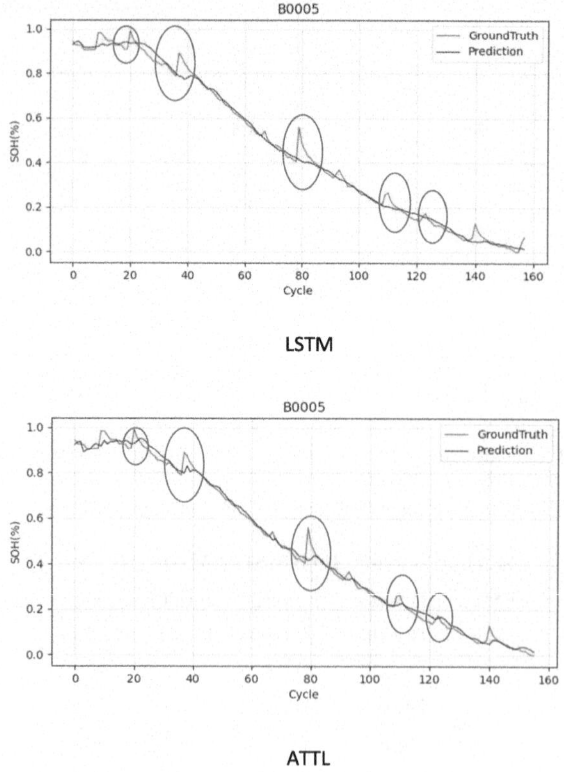

Fig. 8. Prediction vs. ground truth.

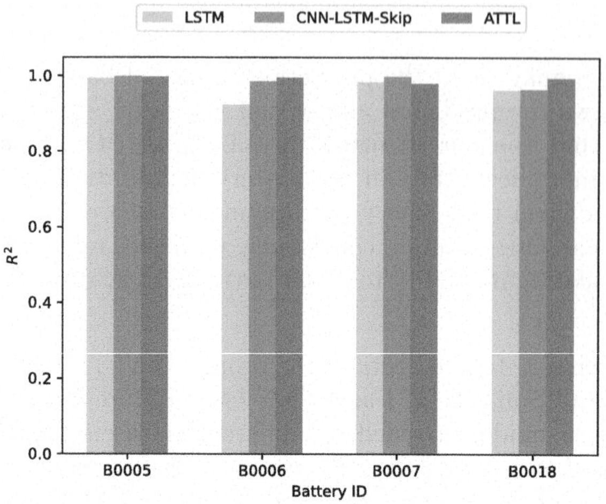

Fig. 9. The SOH estimation comparison results with other methods.

4 Conclusion

Considering that there are different stages of battery degradation and the feature distribution of different stages is also different, an adaptive time series transfer learning model is used to predict battery degradation. The knee points method is used to divide various battery datasets the battery capacity degradation curve into two stages, capacity fade and performance fade. In addition, extracting features with high Pearson correlation coefficients as health factors to improve the accuracy of predicting the health status (SOH) of lithium-ion batteries.

Finally, a time-series transfer learning model is proposed for predicting SOH. This model extracts implicit information from the data and is suitable for time series modeling processes with different data distributions. The evaluation metrics used in this study for SOH, including MAE, and RMSE both achieve estimation accuracy below 5%, which are lower than those of other methods. Comparative studies have shown that our proposed method achieves higher accuracy across different battery datasets. There is still room for improvement in the method of determining the timing split point, and it will be further studied in the future.

References

1. Ruan, J., Song, Q., Yang, W.: The application of hybrid energy storage system with electrified continuously variable transmission in battery electric vehicle. Energy **183**, 315–330 (2019)
2. Li, X., Ma, Y., Zhu, J.: An online dual filters RUL prediction method of lithium-ion battery based on unscented particle filter and least squares support vector machine. Measurement **184**, 109935 (2021)
3. Sadabadi, K.K., Jin, X., Rizzoni, G.: Prediction of remaining useful life for a composite electrode lithium ion battery cell using an electrochemical model to estimate the state of health. J. Pow. Sources **481**, 228861 (2021)
4. Bian, X., Wei, Z., Li, W., Pou, J., Sauer, D.U., Liu, L.: State-of-health estimation of lithium-ion batteries by fusing an open circuit voltage model and incremental capacity analysis. IEEE Trans. Pow. Electron. **37**(2), 2226–2236 (2021)
5. Yang, F., Song, X., Dong, G., Tsui, K.L.: A coulombic efficiency-based model for prognostics and health estimation of lithium-ion batteries. Energy **171**, 1173–1182 (2019)
6. Cheng, G., Wang, X., He, Y.: Remaining useful life and state of health prediction for lithium batteries based on empirical mode decomposition and a long and short memory neural network. Energy **232**, 121022 (2021)
7. Bai, G., Wang, P., Hu, C., Pecht, M.: A generic model-free approach for lithium-ion battery health management. Appl. Energy **135**, 247–260 (2014)
8. Klass, V., Behm, M., Lindbergh, G.: A support vector machine-based state-of-health estimation method for lithium-ion batteries under electric vehicle operation. J. Pow. Sources **270**, 262–272 (2014)
9. Liu, K., Tang, X., Teodorescu, R., Gao, F., Meng, J.: Future ageing trajectory prediction for lithium-ion battery considering the knee point effect. IEEE Trans. Energy Convers. **37**(2), 1282–1291 (2021)

10. Xu, H., Wu, L., Xiong, S., Li, W., Garg, A., Gao, L.: An improved CNN-LSTM model-based state-of-health estimation approach for lithium-ion batteries. Energy **276**, 127585 (2023)
11. Ma, G., et al.: Estimating the state of health for lithium-ion batteries: a particle swarm optimization-assisted deep domain adaptation approach. IEEE/CAA J. Automatica Sinica **10**(7), 1530–1543 (2023)
12. Guo, X., Wang, K., Yao, S., Fu, G., Ning, Y.: RUL prediction of lithium ion battery based on CEEMDAN-CNN BiLSTM model. Energy Rep. **9**, 1299–1306 (2023)
13. Fan, Y., Xiao, F., Li, C., Yang, G., Tang, X.: A novel deep learning framework for state of health estimation of lithium-ion battery. J. Energy Storage **32**, 101741 (2020)
14. Branke, J., Deb, K., Dierolf, H., Osswald, M.: Finding knees in multi-objective optimization. In: Proceedings of the Parallel Problem Solving from Nature-PPSN VIII: 8th International Conference, Birmingham, UK, September 18-22, 2004, pp. 722–731. Springer (2004)
15. Ma, Y., Shan, C., Gao, J., Chen, H.: Multiple health indicators fusion-based health prognostic for lithium-ion battery using transfer learning and hybrid deep learning method. Reliab. Eng. Syst. Saf. **229**, 108818 (2023)
16. Du, Y., et al.: AdaRNN: adaptive learning and forecasting of time series. In: Proceedings of the 30th ACM International Conference on Information and Knowledge Management, pp. 402–411 (2021)
17. Saha, B., Goebel, K.: Battery data set, NASA AMES prognostics data repository (2007)

Research on the Inconsistence and Equalization Technology of Lithium-Ion Battery in Energy Storage Device

Juan Yan[1(✉)], Chuanfeng Zhu[1], Cheng Zhang[2], Cungang Hu[1], and Wenping Cao[1]

[1] School of Electrical Engineering and Automation, Anhui University, Hefei 230601, China
1906504759@qq.com

[2] Institute for Future Transport and Cities, CALPS, Coventry University, Coventry, UK

Abstract. The energy storage device holds a pivotal position in renewable energy systems, facilitating the efficient accumulation and subsequent deployment of energy. However, the inconsistency of lithium-ion batteries has been a key factor limiting their performance and lifespan. This paper examines the inconsistencies of Li-ion batteries and discusses the application of various equalization techniques to improve overall performance and extend battery life. It also compares and analyzes the advantages and disadvantages of different equalization techniques, demonstrating their application in lithium battery packs through simulation. This paper provides important theoretical guidance for enhancing the overall performance of energy storage devices and prolonging battery life.

Keywords: Energy Storage · Lithium-ion Battery · Inconsistent · Balancing Strategy

1 Introduction

With the transformation and upgrading of the global energy structure, the academic community pays more and more attention to new energy and electric vehicles, the significance of battery storage becomes even more important [1]. Energy storage technology as a support large-scale renewable energy access and the key technology of efficient utilization, are increasingly being attention [2]. Lithium-ion batteries have emerged as a highly prevalent energy storage solution, attributed to their noteworthy energy density, extended lifecycle, and minimal self-discharge rate. However, when lithium-ion batteries form energy storage systems, inconsistencies between single batteries will inevitably occur due to the influence of many factors such as production process, material properties and use environment. This inconsistency not only affects the overall performance of the energy storage system, but can also lead to premature failure of some batteries and even lead to safety accidents. Therefore, it is of great significance to conduct in-depth research on the inconsistency of lithium-ion batteries and develop effective equalization technology for improving the safety and reliability of energy storage systems. The inconsistency of lithium-ion batteries is mainly manifested in voltage, capacity and internal

K. Li et al. (Eds.): LSMS/ICSEE 2024, CCIS 2218, pp. 331–342, 2025.
https://doi.org/10.1007/978-981-96-0232-2_26

resistance [3]. Voltage inconsistency may lead to overcharge or overdischarge of the battery in the process of charging and discharging, thus shortening the service life of the battery. The presence of capacity variations in lithium-ion batteries when operated in parallel arrangements can lead to uneven current distribution, compromising the system's overall performance. Furthermore, inconsistencies in internal resistance can cause temperature disparities during operation, affecting the thermal stability of the system. These inconsistencies not only constrain the usage of lithium-ion batteries in energy storage applications but also elevate maintenance costs and safety concerns. To address this, researchers have put forward various balancing techniques [4] to mitigate the issue of battery inconsistency. These techniques can be divided into passive and active equalization. Passive equalization technology mainly through the introduction of resistance or other energy-consuming components in the battery pack, the excess energy in the higher energy battery is consumed to achieve the purpose of balance. Although this technology is simple and easy, but the energy utilization rate is low, and it is easy to cause the problem of heat accumulation. Active equalization technology transfers energy from high energy battery to low energy battery through energy conversion and transfer to achieve efficient use of energy. This kind of technology usually includes capacitive equalization, transformer equalization and inductance equalization, etc., which has high energy utilization rate and better equalization effect. However, the existing equalization technology still faces some challenges in practical application. First of all, different equalization technologies have significant differences in performance, cost and complexity, and need to be selected and optimized according to specific application scenarios and needs. Secondly, with the expansion of the scale and complexity of energy storage systems [2], how to achieve efficient and accurate balance of a large number of single batteries has become an urgent problem to be solved. In view of the above problems, this paper aims to review the inconsistency and equalization technology of lithium-ion batteries in energy storage devices. In this paper, the causes and effects of the inconsistencies of lithium-ion batteries are analyzed in detail, and then the existing equalization strategies and technologies are introduced from the passive and active equalization aspects.

2 Lithium-Ion Battery Inconsistency Problem

Lithium-ion batteries, being a cornerstone of contemporary energy storage, are extensively utilized in electric vehicles, portable gadgets, energy storage setups, and numerous other domains [5]. However, with the expansion of its application scope and the increase of complexity, the inconsistency problem of lithium-ion batteries has gradually become prominent, which has become a key factor restricting its performance improvement and application expansion. The inconsistency of lithium-ion batteries is mainly reflected in its voltage, capacity, internal resistance, self-discharge rate and other key parameters. The inconsistency of these parameters is caused, on the one hand, by various factors such as production process, material difference and use environment [6]. On the other hand, as the battery is used and aged, these inconsistencies will be further amplified, resulting in rapid attenuation of battery performance and shortening of life.

As depicted in Fig. 1, the charging capacity of the battery pack is constrained by the lithium battery possessing the least charging capacity within the pack. Moreover, if any

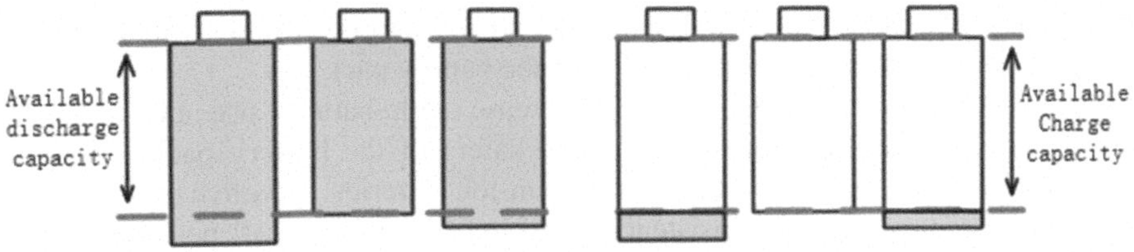

Fig. 1. The left side is the available discharge energy of the lithium battery, and the right side is the available charging energy of the lithium battery.

battery within the pack attains the charging or discharging terminus, the entire battery pack halts its charging and discharging process. Consequently, the pack's inconsistency significantly impacts the available charge and discharge capacity.

2.1 The Forms of Inconsistency

There are various forms of inconsistency in lithium-ion batteries [7], which are mainly reflected in the following aspects:

Battery Capacity Variation: The capacity of a lithium-ion battery, which represents the amount of power it can store, may vary among batteries of the same or different batches due to factors such as variations in manufacturing processes and materials. This discrepancy in capacity can lead to inconsistencies during charging and discharging, where batteries with smaller capacities may reach full charge or discharge sooner, while larger-capacity batteries may not achieve a complete charge or discharge cycle, ultimately affecting the overall energy utilization efficiency and performance of the battery system.

Internal resistance inconsistency: Internal resistance is the internal resistance of the battery, which affects the charging and discharging efficiency and heating of the battery [3]. Due to differences in factors such as material, manufacturing technology, lithium ion battery internal resistance may exist inconsistency. Inconsistent internal resistance will cause the battery to produce heat difference during the charging and discharging process, which may affect the safety and life of the battery.

Varying Self-Discharge Rates: The self-discharge rate signifies the varying degrees of natural discharge a battery experiences due to internal chemical reactions while in storage. The inconsistency of the self-discharge rate will lead to significant differences in the remaining power of each cell after long-term storage, which will affect the overall performance of the battery pack.

Voltage Disparity: The voltage of a lithium-ion battery is a crucial indicator of its operational state. However, variations in the performance of individual batteries can lead to inconsistencies in their operating voltage. This disparity in voltage can result in imbalances during charging and discharging, ultimately affecting the overall performance and safety of the battery pack.

State of Charge (SOC) inconsistency: SOC is the ratio of the remaining capacity of the battery to the rated capacity, which reflects the state of charge of the battery. Due to the inconsistency of battery capacity, internal resistance and other parameters, the SOC of the battery may also be inconsistent. This inconsistency may lead to inconsistent depth of charge and discharge during the charging and discharging process of the battery, affecting the energy utilization rate and life of the battery pack.

These inconsistencies may gradually increase with battery use and aging [8], adversely affecting the performance, life, and safety of the battery pack. Therefore, in the production, use and maintenance of lithium-ion batteries, effective measures need to be taken to reduce these inconsistencies and improve the overall performance and safety of the battery pack.

2.2 Cause Analysis of Inconsistency Problem

The emergence of lithium-ion battery inconsistencies can broadly be attributed to two phases: production and usage [9]. During production, variations in raw materials, manufacturing equipment, and processes can lead to inconsistencies in LIB parameters. In the usage phase, environmental differences during battery operation and storage primarily contribute to inconsistencies. Both these factors can impact the overall performance of the battery pack. Inconsistencies can occur during production. When there are differences in the raw materials of the battery, such as positive and negative electrode materials, electrolyte and diaphragm, etc., even from the same supplier, there may be small differences. These differences affect the battery's performance, such as capacity, internal resistance, and energy density. During battery production, the precision of critical process parameters, including winding and liquid injection, is essential for ensuring battery performance consistency. Even minor fluctuations in these parameters can result in significant variations in battery performance. The inspection and sorting standards for batteries on the production line can also affect the consistency of batteries. If the detection accuracy is not high or the separation standard is not strict, it may lead to batteries with large performance differences being mixed into the same batch.

Inconsistencies can also arise during battery usage. Specifically, variations in charging and discharging conditions, such as the charging current and discharge depth, can lead to imbalances in the chemical reactions within the battery, thereby causing differences in performance. Operating temperature: The operating temperature of a battery has a great effect on its performance. Differences in the operating temperature of different cells can lead to inconsistencies in their performance. For example, high temperatures expedite battery aging, whereas low temperatures hamper battery discharge performance. Operating environment: Battery operating environment, such as vibration, shock, and humidity, will also affect battery performance. These environmental factors can lead to small changes in the internal structure of the battery, which can affect the consistency of its performance.

3 Battery Equalization Strategy

As shown in Fig. 2, the variables of the equalization strategy mainly include battery voltage, state of charge (SOC), capacity and other indicators. These variables play a key role in the battery balancing process [10]. First, voltage is an easy indicator to measure, so it is often used in equalization control strategies. Cell polarization, however, may cause the voltage error, causing the balanced or unbalanced problem. Secondly, SOC as a equalization variable is more accurate than based on voltage variables. The SOC of the battery cannot be measured directly and must be estimated online or offline, which increases the cost and complicates the control algorithm. Capacity is another important equalization variable, but the actual capacity of the battery is difficult to determine accurately due to the size of the current and the temperature.

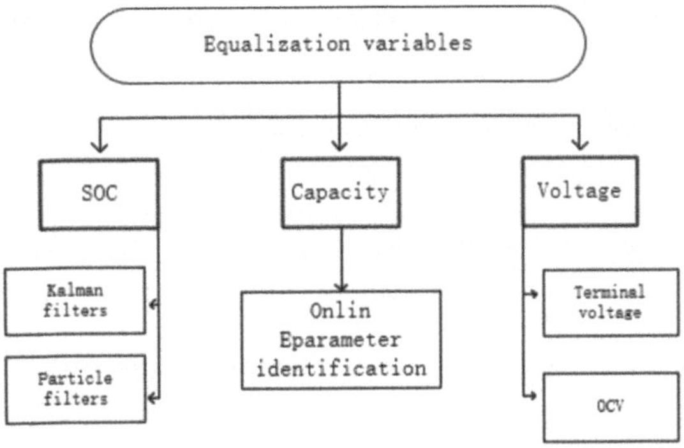

Fig. 2. Classification of balanced variables.

3.1 SOC-Based Equalization Strategy (SBES)

SOC based balancing strategy is a key technology in battery management system, its goal is to achieve SOC (State of Charge) balance among each cell in the battery pack. Small differences that may exist in the manufacturing and use of battery cells, as well as different charging and discharging conditions, the SOC between battery cells is prone to imbalance, which will affect the overall performance and life of the battery pack. The SOC-based equalization strategy ensures that the SOC of each cell is consistent by monitoring and adjusting the charge level between cells. In the process of strategy implementation, the battery management system will calculate the difference between the SOC of each battery cell and the average value according to the real-time data, and fine-tune the charge and discharge speed of each battery according to these differences.

When adopting SOC-based balancing strategies, it's important to recognize that directly measuring the SOC of a battery is challenging and often relies on intricate algorithms for estimation. Thus, in real-world scenarios, it's essential to strike a balance between the accuracy, real-time responsiveness, and computational costs of the algorithm to ensure effective SOC balancing control. Chen and colleagues [12] devised a feedforward neural network (NN) model for high-precision SOC estimation. Experimental outcomes demonstrate that this approach can converge even with an incorrect initial SOC value, yielding an estimation error of less than 2%. To enhance the SOC estimation accuracy of lithium batteries, a method leveraging the parallel fusion of the Residual Shrinkage Network (RSN) and Gated Recurrent Unit (GRU) has been introduced for SOC estimation. Literature [13] proposes a lithium battery SOC estimation method based on the parallel fusion of residual shrinkage network (RSN) and gated cycle unit (GRU) in order to improve the accuracy of state of charge (SOC) estimation of lithium batteries. Literature [14] proposes an online SOC estimation method for lead-acid batteries based on DKF-Bi-LSTM. Based on dual Kalman filter algorithm of secondary structure, model estimation and state estimation are respectively performed to improve SOC estimation accuracy. The state estimation accuracy of SOC is relatively important.

3.2 Voltage-Based Equalization Strategy (VBES)

The voltage equalization strategy is mainly used in battery management systems (BMS), especially in lithium-ion battery packs. Battery management system based on voltage equalization strategy is of great significance to improve the overall performance and service life of battery pack.

Battery voltage measurement during equalization has garnered attention from numerous researchers. Meng and his team [15] explored a rapid OCV estimation technique using time delay, achieving higher accuracy within a shorter resting period. Song et al. [16] introduced a method rooted in the charging voltage curve, offering high accuracy while minimizing computational costs. This strategy has been successfully integrated into the BMS system, and experimental results validate its applicability and precision.

3.3 Capacity-Based Equalization Strategy (CBES)

CBES offers a means to balance the actual battery capacity, yet determining this capacity accurately can be challenging due to variations in current, temperature, and other factors. This poses difficulties in ensuring consistent battery capacity across a pack. Ma et al. [17] explored balancing criteria based on capacity utilization and introduced an online active balancing strategy aimed at maximizing capacity. Similarly, Cui et al. [18] devised a balancing approach centered on rechargeable and discharge capacities to tackle inhomogeneity in battery packs. Wang et al. [19] put forth an active balancing method rooted in battery remaining capacity, and experimental findings validated the effectiveness of this strategy in energy storage systems.

4 Balanced Management System

The battery balance management system is crucial in minimizing inconsistencies among batteries within a battery pack. Its primary objective is to harmonize voltage, current, and temperature parameters across individual cells in a lithium-ion battery pack. This equilibrium enhances the performance, safety, and longevity of the entire battery pack. The lithium battery balance management system achieves this through a comprehensive set of technological approaches, including active balance and passive balance strategy, to achieve the balance between battery cells. Active equalization adjusts the state of charge between battery cells by controlling the charging and discharging process to keep them consistent. Passive equalization uses components such as resistors or capacitors to transfer excess charge from one cell to another cell to achieve electricity balance.

4.1 Passive Equalization Control

Passive balance controllers (PBCs) commonly utilize parallel resistors in conjunction with batteries to dissipate excess energy from batteries exhibiting higher voltages or states of charge (SOC).

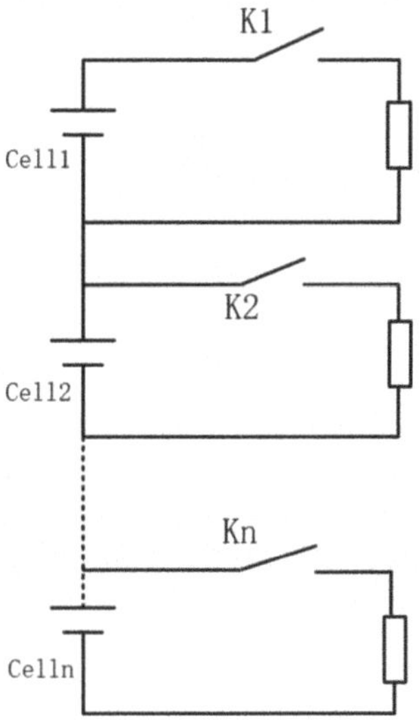

Fig. 3. Passive balanced circuit topology. The diagram shows a structure diagram composed of n batteries.

As shown in Fig. 3, a passive equalization controller can control switches (K1, K2 ..... Kn) to dissipate excess energy. If the energy of Cell1 is higher than that of other cells, the switch connected to Cell1 can be modulated to close, so that Cell1 can discharge, and the switch can work continuously or intermittently depending on the control strategy.

4.2 Active Equalization Control

Lithium-ion active balancing control is a crucial aspect of the lithium battery management system. Its core aim is to adjust the power difference among individual cells in the battery pack through active means, achieving a balanced power distribution. This ultimately enhances the performance and lifespan of the entire battery pack. Active equalization control mainly relies on an energy transfer mechanism, which is able to transfer the excess energy in the cell with higher charge to the cell with lower charge. This energy transfer is usually implemented through specific circuit and the control strategy, can ensure that the energy efficient and accurate transfer between battery monomer. Compared with passive equalization control, active equalization control has higher efficiency and more accurate control ability. Passive equalization control mainly achieves equalization by consuming the excess energy in the single battery with higher power, which will cause a waste of energy. While active balancing control can realize the effective utilization of energy, avoid the waste of energy.

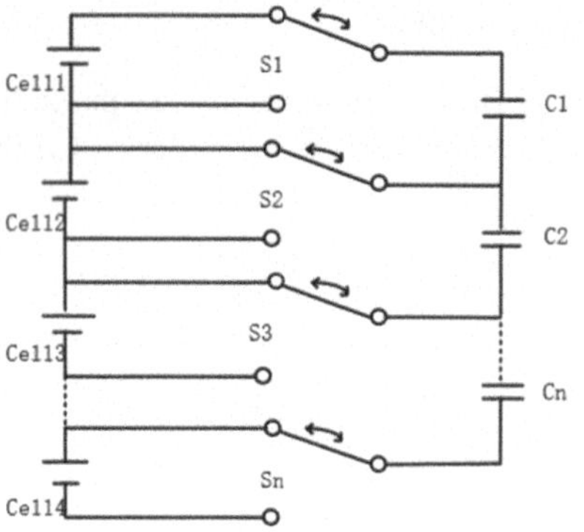

Fig. 4. Switched capacitance topology. Structure diagram of n batteries, n bidirectional switches, and n capacitors.

A typical switched capacitor topology is shown in Fig. 4, which consists of several capacitors and bidirectional switches that transfer energy between continuous cells via a control switch. The switched capacitor topology boasts simplicity and cost-effectiveness, yet demands sophisticated control when dealing with minute voltage disparities between adjacent cells. This design facilitates energy transfer among cells via capacitors.

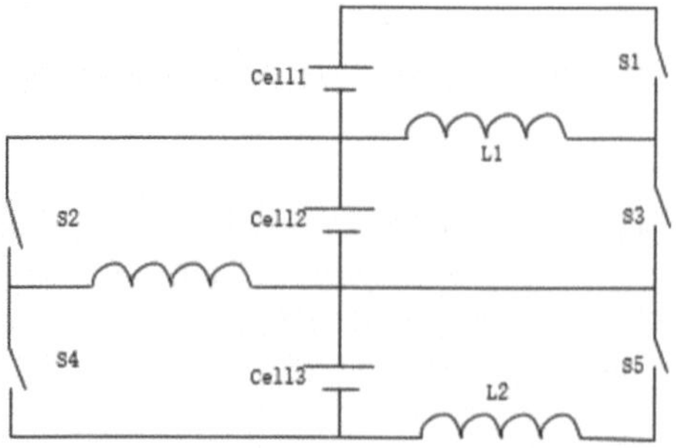

Fig. 5. Multi-inductance topology. An example diagram consisting of 3 battery cells and 3 inductors.

Figure 5 depicts a multi-inductance topology comprising inductors and switches. The MOSFET regulates energy transfer between cells. If Cell1 has excess energy compared to Cell2, S1 is initially activated to charge inductor L1 from Cell1. Subsequently, S1 is deactivated and S3 is switched on, allowing energy to transfer from Cell1 to Cell2. However, this topology's equalization speed is influenced by battery pack size, and since energy transfer is limited to adjacent cells, the speed tends to be slow.

Additionally, transformer-based balance controllers (TBBCs) offer an alternative approach, utilizing external transformers to facilitate energy exchange between cells and modules. TBBCs can be classified as single-winding or multi-winding transformer topologies. A DC/DC converter, or power converter, enables bidirectional energy conversion from one voltage to another, with PBBCs boasting high efficiency and precision in this regard.

5 Simulation Verification of Battery Balance

This paper simulates the battery inconsistency equalization under MATLAB environment, and verifies the effectiveness of equalization operation. Taking active equalization as an example, this paper sets up the equalization simulation circuit for 4 series battery packs with SOC as the target, and verifies the effectiveness of equalization measures.

Figure 6 shows the establishment of a simulation circuit for a single battery, which mainly includes a single battery and an SOC calculation module. The circuit can calculate the SOC of a single battery in order to find inconsistencies between batteries and facilitate subsequent balancing operations.

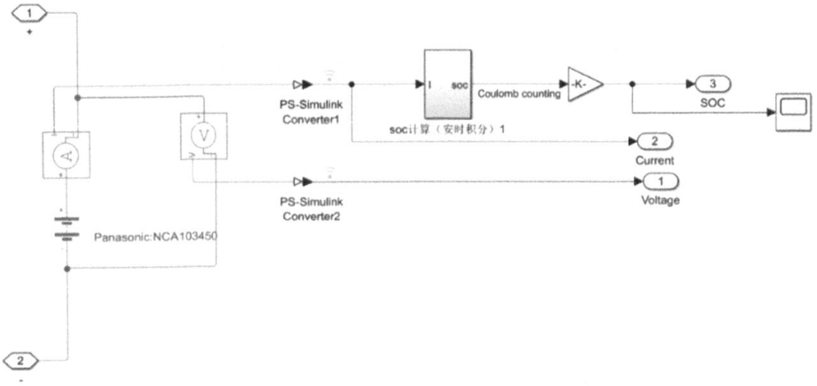

Fig. 6. Cell simulation circuit. It mainly includes batteries and SOC computing modules.

By combining the four cells together and adding the active balancing module, the main balancing circuit shown in Fig. 7 is formed, which can verify the effectiveness of the balancing operation mentioned in this article.

The simulation results are shown in Fig. 8. It can be seen that the initial SOC values of the four batteries are 70%, 65%, 60% and 40% respectively. It can be seen that the SOC values of the batteries in the initial stage are inconsistent to some extent. After subsequent balancing operations, the SOC of the batteries with higher energy decreases while that of the batteries with lower energy increases. You can see that the inconsistency of the battery SOC has been slowly decreasing, and eventually the difference is negligible compared to the initial state. It can be verified that the equalization control can arbitrarily reduce battery inconsistencies.

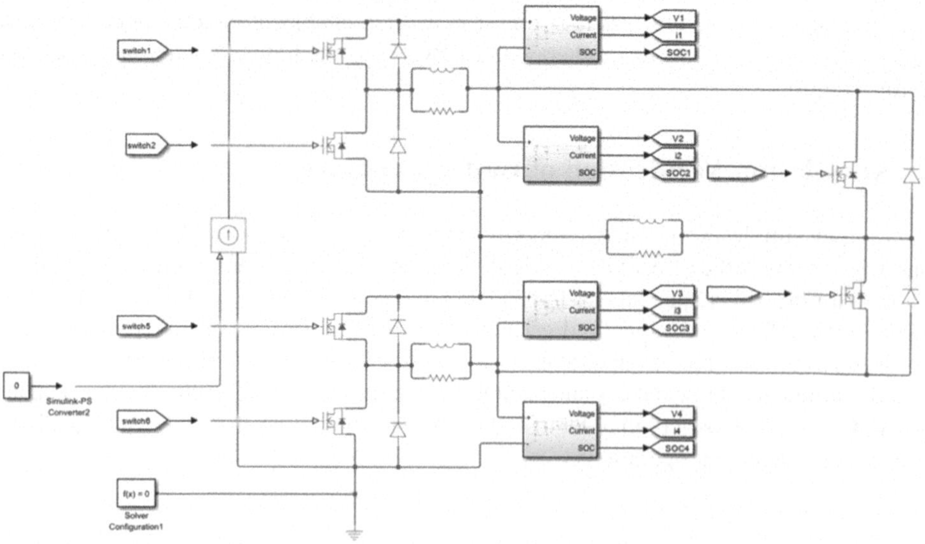

Fig. 7. Battery equalization main circuit. It consists of four battery modules and a balancing module.

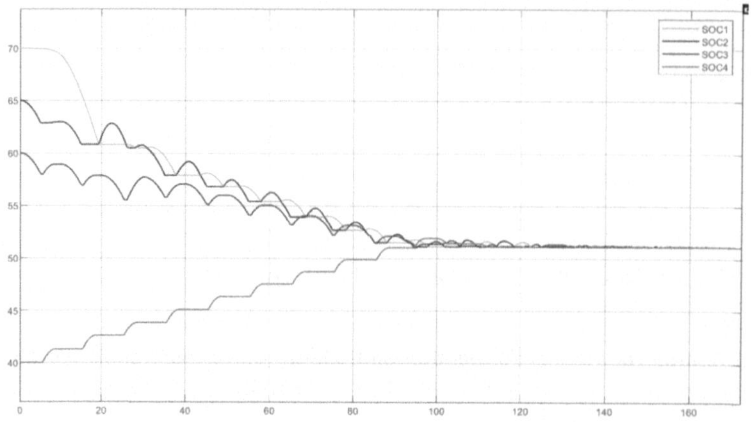

Fig. 8. Battery balance results. There are four separate curves representing the battery SOC

6 Conclusion

The inconsistency of lithium-ion battery energy storage device is the key factor affecting its performance and life, and the equalization technology is an effective way to solve this problem. This paper first gives an overview of the inconsistencies of the battery, analyzes the causes of the inconsistencies in the production and use process, and then expounds the balance control, analyzes the categories of the balance variables, and introduces the active balance and passive balance, and introduces the common balance topology.

Finally, the effectiveness of the equalization controller in dealing with the inconsistency problem is verified by simulation.

References

1. Wang, H., Shao, Y., Zhou, L.: Reliability analysis of battery energy storage systems: an overview. In: 2022 IEEE/IAS Industrial and Commercial Power System Asia (I&CPS Asia), Shanghai, China, pp. 2036–2040. IEEE (2022)
2. Reniers, J.M., Howey, D.A.: Digital twin of a MWh-scale grid battery system for efficiency and degradation analysis. Appl. Energy **336**, 120774 (2023)
3. Hua, Y., Zhou, S., Cui, H., Liu, X., Zhang, C., Xu, X., et al.: A comprehensive review on inconsistency and equalization technology of lithium-ion battery for electric vehicles. Int. J. Energy Res. **44**(14), 11059–11087 (2020)
4. Chetri, C., Huynh, A., Williamson, S.S.: A comprehensive review of active EV battery cell voltage balancing systems: current issues and prospective solutions. In: 2022 IEEE 1st Industrial Electronics Society Annual On-Line Conference (ONCON), Chengdu, China, pp. 1–6. IEEE (2022)
5. Bairwa, B., Hampannavar, S., Vishal, K.S., Bhargavi, K.M.: Modeling and simulation for capacity fade prediction of lithium-ion battery. In: IEEE EUROCON 2021 - 19th International Conference on Smart Technologies, Lviv, Ukraine, pp. 537–541. IEEE (2021)
6. Yang, Q., Cao, L., Shao, X., Yang, P., Gong, Y.: Aging simulation of lead-acid battery based on numerical electrochemical model. In: 2022 2nd International Conference on Electrical Engineering and Mechatronics Technology (ICEEMT), Hangzhou, China, pp. 348–351. IEEE (2022)
7. Xu, J., Mei, X., Wang, J.: A high power low-cost balancing system for battery strings. Energy Procedia **158**, 2948–2953 (2019)
8. Liu, V.T., Lu, H.Y., Wang, S.K.: New lithium battery balancing circuit design using isolated converter. In: 2019 IEEE Eurasia Conference on IOT Communication and Engineering (ECICE), Yunlin, Taiwan, pp. 346–349. IEEE (2019)
9. Yeoh, S.H., Pok, C.Y., Lum, K.Y., Yiauw, K.H.: Active cell balancing with DC/DC converter for electric vehicle. In: 2022 10th International Conference on Smart Grid and Clean Energy Technologies (ICSGCE), Kuala Lumpur, Malaysia, pp. 39–45. IEEE (2022)
10. Ahmad, A., Singh, V., Teja, A.V.R., Payami, S.: Parallel active charge balance technique of li-ion batteries using combined phase shifted and boost DC-DC Converter. In: IECON 2023-49th Annual Conference of the IEEE Industrial Electronics Society, Singapore, Singapore, pp. 1–6. IEEE (2023)
11. Tyrpekl, M., Zavrel, M., Kindl, V.: Comparison of active and passive battery balancing. In: 2022 20th International Conference on Mechatronics - Mechatronika (ME), Pilsen, Czech Republic, pp. 1–6. IEEE (2022)
12. Chen, C., Xiong, R., Yang, R., Shen, W., Sun, F.: State-of-chargeestimation of lithium-ion battery using an improved neuralnetwork model and extended Kalman filter. J. Clean. Prod. **234**, 1153–1164 (2019)
13. Quan, R., Liu, P., Zhang, J.: State of charge estimation of lithium battery based on RSN-GRU fusion network. J. Huazhong Univ. Sci. Technol. **52**, 1–7 (2024)
14. Lianbing, L., Yanjie, L., Hailiang, W.: Online SOC estimation method for valve-controlled lead-acid batteries based on DKF-Bi-LSTM. China Test **50**, 28–37 (2024)
15. Meng, J., Stroe, D., Ricco, M., Luo, G., Swierczynski, M., Teodorescu, R.: A novel multiple correction approach for fast open circuit voltage prediction of Lithium-ion battery. IEEE Trans. Energy Convers. **34**, 1115–1123 (2019)

16. Song, L., Liang, T., Lu, L., Ouyang, M.: Lithium-ion battery pack equalization based on charging voltage curves. Int. J. Electr. Power **115**, 105516 (2020)
17. Ma, Z., Jiang, J., Wen, F., Guo, H., Shi, F.: Design of equilibrium strategy of echelon use li-ion battery pack for energy storage system. Autom. Electr. Power Syst. **38**(3), 106–111 (2014)
18. Cui, X., Shen, W., Zhang, Y., Hu, C., Zheng, J.: Novel active LiFePO4 battery balancing method based on chargeable and dischargeable capacity. Comput. Chem. Eng. **97**, 27–35 (2017)
19. Wang, Y., Zhang, C., Chen, Z., Xie, J., Zhang, X.: A novel active equalization method for lithium-ion batteries in electric vehicles. Appl. Energy **145**, 36–42 (2015)
20. Ning, B., Xu, J., Cao, B., Wang, B., Xu, G.: A sliding mode observer SOC estimation method based on parameter adaptive battery model. Energy Procedia **88**, 619–626 (2016)

Multi-constraint Optimization for Energy Scheduling of Plug-In Electric Vehicles

Linxin Zhang[1,2], Zhile Yang[2(✉)], Yuanjun Guo[2], Qinge Xiao[2], and Juncheng Zhu[1,2]

[1] Faculty of Data Science, City University of Macau, Taipa 999078, Macau
[2] Shenzhen Institute of Advanced Technology, Chinese Academy of Sciences, Shenzhen 518055, Guangdong, China
zyang07@qub.ac.uk

Abstract. The integration of a substantial number of plug-in electric vehicles (PEVs) into power grid scheduling introduces complexities due to stochastic charging and discharging behaviors, which pose significant challenges to the stability and reliability of the power system. The PEV scheduling optimization extends the traditional combinatorial optimization problem, adding constraints and variables, and becomes a multi-constraint unit commitment (MCUC)problem. In this paper, a novel PEV scheduling optimization framework is proposed to address MCUC problem, taking economic cost and carbon emissions as the objectives, a multi-constraint mathematical model is established. Meanwhile, a new binary level-based learning optimization algorithm is proposed to solve flexible scheduling of PEVs in power systems. Finally, considering different numbers of units, the superiority of the proposed optimization algorithm is verified by comparison with different algorithms. The feasibility and effectiveness of the proposed solution are also analyzed before and after scheduling to verify its potential in solving the MCUC problem.

Keywords: grid scheduling · plug-in electric vehicle · swarm optimization · unit commitment

1 Introduction

The escalating adoption of plug-in electric vehicles (PEVs), driven by their environmental, energy-saving, and high-efficiency benefits, aligns closely with global initiatives such as the 3060 carbon neutrality goal [11]. This increase in the use of PEVs has attracted significant attention from scholars and social enterprises worldwide. Nevertheless, a critical challenge arises from the large-scale and random charging behaviors of these vehicles, which pose substantial risks to the planning and operational stability of power grids [5].

The integration of PEVs into the electrical grid represents a typical multi-constraint unit commitment (MCUC) problem [7,9,17]. Indeed, various researchers have proposed innovative solutions addressing the challenges, e.g.

K. Li et al. (Eds.): LSMS/ICSEE 2024, CCIS 2218, pp. 343–356, 2025.
https://doi.org/10.1007/978-981-96-0232-2_27

a constrained optimization framework aimed at achieving cost-effective battery charging strategies for PEV users was proposed in [8]. In [22], a power system active rescheduling method was considered to solve MCUC problems. However, the uncontrolled nature of the charging activities can exacerbate peak-valley load disparities, and diminish the economic efficiency of power networks, potentially compromising the safety and stability of the entire power system [16]. It is crucial to implement optimization technologies between PEVs and the power grid to address these challenges. simultaneously, the dynamic nature of PEV charging and discharging behaviors adds complexity to the MCUC problem, necessitating sophisticated strategies for optimal scheduling, load balancing, and ensuring grid stability [20]. Such measures are imperative not only for advancing the shift towards electric energy as a cornerstone of the energy consumption revolution and environmental pollution reduction but also for ensuring the sustainable integration of PEVs into the existing power system.

Presently, Meta-heuristics (MAs) optimization algorithms constitute a primary methodological approach for addressing MCUC problems. Genetic algorithm (GA) [1], differential evolution (DE) [13], particle swarm optimization (PSO) [12], and teaching learning based optimization (TLBO) [10] are favored for the robustness and adaptability in dealing with the complex decision spaces characteristic of MCUC problems, enabling the efficient identification of global optimal or local optimal solutions. Furthermore, to tackle MCUC binary conversion problems, where the start-stop state of generating units is represented by binary decision variables, various binary-coded variants of MAs have been introduced. Binary particle swarm optimization (BPSO) [17], binary differential evolution (BDE) [19], biogeography-based learning particle swarm optimization (BLPSO) [2], and binary whale optimization algorithm (BWOA) [6], are modified to adapt to the binary search space, that is, the position of each particle is represented as 0 and 1, making the algorithms suitable for optimization problems with discrete decision variables. With the large-scale popularization of PEVs, the dimensions of the systems have expanded exponentially. In response to this significant increase in complexity, researchers have incorporated various strategies into the algorithms to enhance performance. Notably, the competitive learning strategy [4] draws inspiration from human societal competition and cognition, offering a unique approach to enhance algorithmic efficiency. The social learning strategy [3] has been developed, further exemplifying the influence of human social dynamics on algorithmic design. These enhancements are tailored to efficiently handle the enlarged scale and intricacy of the problems arising from the extensive integration of PEVs into power systems, ensuring that the algorithms remain effective and robust in these more demanding scenarios. Subsequently, BCSO algorithm [14] and BSLPSO algorithm [23] leverage the knowledge of superior particles within the current swarm to guide the update process of other particles.

Different from BCSO and BSLPSO, the Level-Based Learning Swarm Optimizer (LLSO) [15] categorizes the particles into various levels, and the position update of each particle is influenced by two superior particles, each from a distinct randomly selected level, while the BCSO and BSLPSO utilize only one superior particle to guide the update. This multi-level interaction strategy not only augments the diversity of solutions but also adeptly strikes a balance between the exploration and exploitation phases in the optimization process.

In this paper, a novel PEV scheduling optimization framework is proposed to balance the integration of PEVs into the power system. Since the switch state of units is binary variables, a novel binary level-based learning optimization algorithm is proposed for solving large-scale MCUC problems. The structure of this paper is as follows: Sect. 2 primarily focuses on presenting the MCUC problem model. In Sect. 3, we introduce and detail the algorithm that has been developed. Section 4 is dedicated to the presentation and discussion of experimental results, serving to validate the effectiveness of the proposed algorithm. The paper summarizes with Sect. 5.

2 Problem Formulation

In response to the requirements of load demand, the optimization of the MCUC problems plays a pivotal role in maximizing the economic efficiency of the grid infrastructure. This encompasses both the minimization of operational expenses and the reduction of emissions harmful to the environment. The MCUC problem's objective function is developed, and a thorough analysis of the crucial constraints is conducted.

2.1 The Objective Function of MCUC Problem

In the proposed MCUC model, the objective function accounts for the power generation cost and carbon emissions of the unit and PEV under normal conditions within 24 h. It is worth noting that the cost of generating electricity is independent of the hour. The functions of f_{GC} and f_{CE} are given below:

$$f_{GC} = \min \sum_{t=1}^{T} \sum_{i=1}^{n} (F_i^{uni}(P_{i,t})u_{i,t} + S_{i,t}(1 - u_{i,t-1})u_{i,t}) \tag{1}$$

$$f_{CE} = \min \left(\sum_{t=1}^{T} \sum_{i=1}^{n} F_i^{emi}(P_{i,t})u_{i,t} \times pr_e + \sum_{t=1}^{T} \sum_{p=1}^{nv} E_v^{PEV} \times pr_e \right) \tag{2}$$

In formulas (1) and (2), the function f_{GC} signifies the operational cost for power generation in n units, including both fossil fuel expenses and start-stop costs. The f_{CE} represents the carbon emissions of thermal units and PEVs. Incorporating start-stop costs into the analysis enables a reduction in the expenses and losses stemming from recurrent activations and deactivations of operations.

E_v^{PEV} denotes the capacity of the vth electric vehicle. The $P_{i,t}$ and $u_{i,t}$, respectively, denote the electrical capacity produced by the ith unit during the time interval t and the switch state of the units. pr_e is the average trade price of emissions, as recommended by [21].

To reduce the complexity of the problem, this study offers a weight to strike a balance between economic costs and environmental implications, enhance solution efficiency, and respond promptly to power demand. The objective function is shown below:

$$F = \omega f_{GC} + (1 - \omega) f_{CE} \tag{3}$$

The ω of the formula 3 is the weight coefficient of balancing generation and carbon emissions and is set to 0.3 in this paper.

The fossil fuel costs and start-stop costs of the formula (1) are the costs required to start the unit from a shutdown state and connect it to the grid. The $F_i^{emi}(P_{i,t})$ represents the carbon emission costs. Modeled as follows:

$$F_i^{uni}(P_{i,t}) = a_i + b_i P_{i,t} + c_i P_{i,t}^2 \tag{4}$$

$$F_{i,t}^{emi}(P_{i,t}) = \alpha_i + \beta_i P_{i,t} + \gamma_i P_{i,t}^2 \tag{5}$$

$$S_{i,t} = \begin{cases} S_{H,i}, \ if \ MDT_i \leq TS_{i,t} \leq MDT_i + T_{cold,i} \\ S_{C,i}, \ if \ TS_{i,t} > MDT_i + T_{cold,i} \end{cases} \tag{6}$$

The cost coefficients for the fuel cost function $F_{i,t}(P_{i,t})$ are denoted by a_i, b_i, and c_i, respectively. α_i, β_i and γ_i are emission factors of CO_2 of the ith unit, respectively. The start-stop cost function $S_{i,t}$ consists of the hot start cost $S_{H,i}$ and the cold start cost $S_{C,i}$. $TS_{i,t}$ represents the continuous shutdown time of the ith unit. A hot start is defined as the downtime of a unit that exceeds the minimum downtime MDT_i but is less than the total of the minimum downtime and the cold start threshold $T_{cold,i}$, and vice versa.

2.2 Constraints of MCUC Problem

Within the practical domain, the UC model typically incorporates various constraints, facilitating the scheduling of power generation units to preserve the equilibrium between electricity provision and consumption. This paper considers several constraints, including limits on power generation capacity, minimum up/downtime requirements, PEV constraints, and so on.

1) *Generation constraint of thermal units*

In accordance with the specific generating capability of each unit, the output power is confined within a predetermined range to guarantee the regular operation of each unit. The maximum and minimum thresholds for the units are delineated in Eq. 7:

$$u_{i,t} P_{i,min} \leq P_{i,t} \leq u_{i,t} P_{i,max} \tag{7}$$

where $P_{i,max}$ and $P_{i,min}$ denote the maximum and minimum generating capacity of the ith unit at hour t, respectively.

2) *Power balance constraint*

The integration of electric vehicles increases the complexity of grid scheduling, to ensure the stable operation of the electric grid, and reduce fluctuations in grid frequency and voltage, it is essential to consider power balance constraints during grid scheduling. This involves the rational coordination of charging and discharging of electric vehicles, as well as the scheduling of other loads.

$$\sum_{i=1}^{n} P_{i,t}u_{i,t} = P_{D,t} + P_{PEV,t} \tag{8}$$

$P_{D,t}$ denotes the hourly power demand of the load, while $P_{PEV,t}$ signifies the load power attributed to PEVs. It is pertinent to note that within the power balance constraint model, the losses incurred during power transmission are not accounted for.

3) *Spinning reserve constraints*

Spinning reserve capacity represents a critical safeguarding measure, allocated specifically to uphold the stability and balance of the power system in the event of emergency scenarios. This constraint is essential for mitigating the impact of sudden demand surges, and is shown in the following formula:

$$\sum_{i=1}^{n} P_{i,max}u_{i,t} \geq P_{D,t} + SR_t \tag{9}$$

$$SR_t = m \times P_{D,t}. \tag{10}$$

where SR_t denotes the spinning reserve at hour t, as specified in Eq. 10. This study adopts a spinning reserve value equivalent to 10% of the conventional load, recommended by [?] for maintaining system reliability and operational security.

4) *Minimum up/down time of units*

The operational status of the unit is represented by a binary switch variable, manifesting in two discrete states: "0" for shutdown and "1" for operational. Each state is intrinsically linked with minimum up-time and down-time constraints to ensure optimal system reliability and efficiency. The formulation of this constraint is delineated as follows:

$$u_{i,t} = \begin{cases} 1, & if\ 1 \leq TON_{i,t-1} < MUT_i \\ 0, & if\ 1 \leq TOFF_{i,t-1} < MDT_i \\ 0\ or\ 1, & otherwise \end{cases} \tag{11}$$

where $TON_{i,t-1}$ represents the consecutive start-up time of the ith unit, and $TOFF_{i,t-1}$ represents the continuous shutdown period of the same unit. In the operational dynamics, if the operational duration of a unit falls short of the minimum up-time threshold MUT_i within the $t-1$ period, it is mandated to remain operational into the subsequent t interval, thereby being assigned a state of "1". Conversely, if the unit's downtime has not yet satisfied the minimum downtime requirement MDT_i, it is precluded from reactivation in the ensuing period.

5) *Constraints of PEVs*

In this study, the integration of PEVs into the grid is considered, necessitating the inclusion of specific constraints related to electric vehicles.

$$P_{PEV,t,max} \leq P_{PEV,t} \leq P_{PEV,t,max} \tag{12}$$

$$\sum_{t=1}^{T} P_{PEV,t} = P_{PEV,total} \tag{13}$$

The constraints are defined in Eq. 12 for the maximum charge and discharge power, and 13 for the charge load balance, respectively. These constraints ensure that the operation of PEVs within the electrical grid is both efficient and sustainable, aligning with the overall energy management objectives.

3 Methodology

The system model incorporating PEVs is marked by its high dimensionality, strong interdependencies, and significant nonlinearity. These features pose substantial challenges for conventional optimization techniques. To address these complexities, this study introduces a binary level-based learning swarm optimization framework, specifically designed to solve this problem effectively.

3.1 Original LLSO Optimization Algorithm

The SLPSO and CSO algorithms enhance diversity in high-dimensional problems by updating individuals with elite members, yet their reliance on a singular superior for guidance and population averaging curtails diversity growth.

LLSO employs a level-based learning approach, as shown in Fig. 1, categorizing particles into levels and utilizing the knowledge from the top two individuals for collective advancement, enhancing global search efficacy. The level-based learning mechanism of LLSO is defined as follows:

$$x_{i,j}^{d}(t+1) = x_{i,j}^{d} + v_{i,j}^{d}(t+1) \tag{14}$$

$$v_{i,j}^{d}(t+1) = r_1(t)v_{i,j}^{d}(t) + r_2(t)(x_{k_1,l_1}^{d}(t) - x_{i,j}^{d}(t)) \quad +\phi r_3(t)(x_{k_2,l_2}^{d}(t) - x_{i,j}^{d}(t)) \tag{15}$$

In Eq. 14, $x_{i,j}^{d}(t)$ denotes the position of the ith individual of the dth dimension from the jth level. In Eq. 15, $x_{k_1,l_1}^{d}(t)$ and $x_{k_2,l_2}^{d}(t)$ symbolize individuals chosen from two distinct levels, L_{l_1} and L_{l_2}, where l_1 and l_2 denote the indices of these levels, each selected from a range of higher levels [1, i-1]. The k_1 and k_2 represent the indices that selected two different individuals from selected levels within [1, S]. S represents the number of particles in each level. ϕ is the influence factor to control the update speed of the second selected individual. It is important to highlight that the index l_1 is smaller than l_2, both smaller than l_j, the x_{l_1} surpasses that of x_{l_2}, with both being superior to the x_{l_j}.

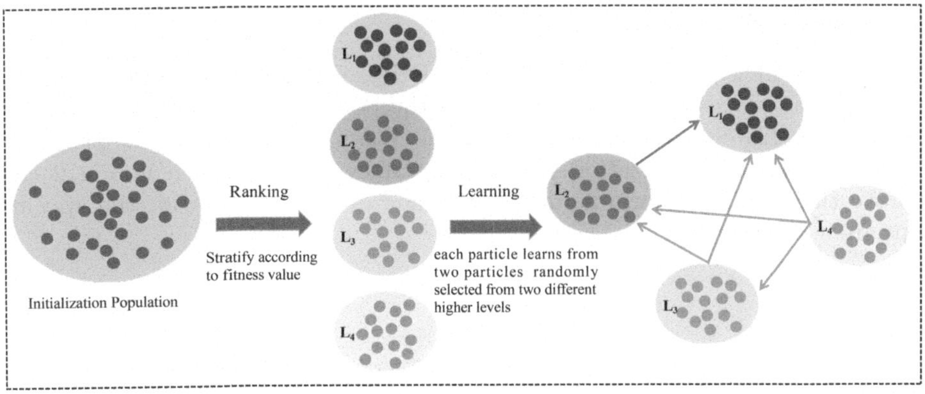

Fig. 1. The level-based learning mechanism of LLSO.

3.2 Improved LLSO for MCUC Problems

In the process of optimizing the MCUC problem, given the binary nature of unit status and the decimal characteristic of particle velocity, it is necessary to design the algorithm update strategy to ensure uniform encoding across particle attributes. This study adopts the V-shaped function for discretization, the improved LLSO update strategy is designed as follows:

$$\mathbf{X}^d(t+1) = \begin{bmatrix} x_{1,1}^d + B_{1,1}^d(t+1) & x_{1,2}^d + B_{1,2}^d(t+1) & \cdots & x_{1,M}^d + B_{1,M}^d(t+1) \\ x_{2,1}^d + B_{2,1}^d(t+1) & x_{2,2}^d + B_{2,2}^d(t+1) & \cdots & x_{2,M}^d + B_{2,M}^d(t+1) \\ \vdots & & \ddots & \\ x_{S,1}^d + B_{S,1}^d(t+1) & x_{S,2}^d + B_{S,2}^d(t+1) & \cdots & x_{S,M}^d + B_{S,M}^d(t+1) \end{bmatrix} \tag{16}$$

$$B_{i,j}^d(t) = 2 * |\frac{1}{1+e^{-v_{i,j}^d(t)}} - 0.5| \tag{17}$$

$$x_{i,j}^d(t) = \begin{cases} 1, & \text{if } rand < B_{i,j}^d(t) \\ 0, & \text{otherwise} \end{cases} \tag{18}$$

$\mathbf{X}^d(t+1)$ represents the update population of the dth dimension, where M denotes the number of levels. $B_{i,j}^d(t)$ represents the speed of the ith individual in the jth level in the dth dimension after binary conversion. The Eq. 18 represents the position of the ith individual after binary update. In addition, it is critical to emphasize that, to avoid erroneous updates of the most promising individuals, those within L_1 are preserved without modification, ensuring their direct progression to the subsequent generation.

3.3 Specific Steps of BLLSO

As electric vehicles continue to gain popularity, increased demand for large-scale charging places potential pressure on the power grid. Therefore, effective management strategies are needed to maintain energy balance, and the proposed binary level-based optimization scheduling framework, as shown in Fig. 2 can improve the reliability and operating efficiency of the power grid, the specific steps are as follows:

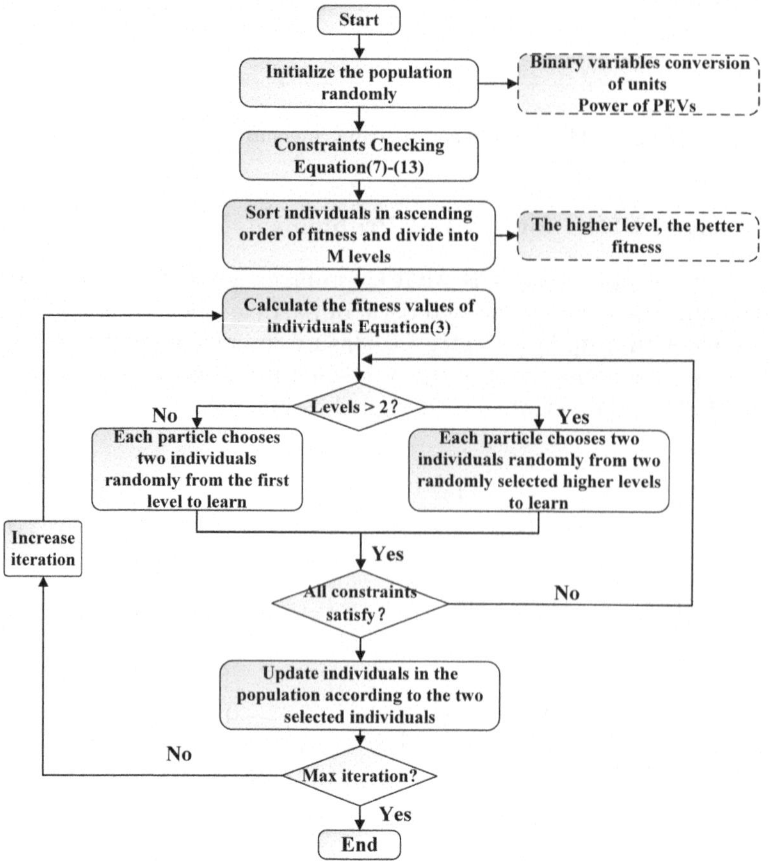

Fig. 2. The proposed BLLSO scheduling framework.

– Data Initialization: First, input and initialize unit data and PEVs' 24-hour charging details, including essential parameters such as max/min generation capacity, fuel coefficients, and unit initial states. Additionally, initialize the range of particle velocities and algorithm parameters for the optimization process.

- Constraint Handling: Initialized unit data must meet minimum up/down time limits or adjust on-off states accordingly. PEV data should comply with constraints 12 and 13.
- Individual stratification: After fitness evaluation, particles are ordered by ascending fitness values for stratification. Importantly, L_1 level individuals bypass updates in subsequent iterations to protect highly promising individuals from erroneous modifications. Those at the L_2 level update based on two leading individuals from L_1.
- Population update: In the BLLSO optimization process, individual positions are refined based on fitness evaluations. Subsequently, updated particles undergo verification to ascertain compliance with constraints, ensuring the optimization satisfies all system limitations.
- Termination evaluation: The process assesses whether the current iteration count is below the preset termination threshold. If not, it concludes by outputting the optimized results and halting the operation.

4 Experimental Results and Analysis

This study conducts a comparative performance analysis to assess the effectiveness and feasibility of the BLLSO algorithm in managing MCUC problems. Through this comparison with multiple algorithms under diverse scenarios, the research also explores how charge-discharge management of PEVs influences the power system's demand fluctuations, particularly in terms of peak and valley.

4.1 Optimize MCUC Problem for the Proposed Algorithm

This section compares the proposed algorithm with four PSO variant algorithms, namely BLPSO, BPSO, BCSO, and NBPSO [18]. The comparative analysis of the economic costs associated with various algorithms applied to the MCUC problem is shown in Table 1. A notable observation is the variance in cost savings achieved by BLLSO against other algorithms, ranging from a minimum of 1137.64 $/day$ to a maximum of 161921.29 $/day$ as the unit increases to 100. This significant gap underscores the robust competitiveness of the BLLSO in addressing large-scale, high-dimensional MCUC challenges.

Additionally, as depicted in Fig. 3, the superiority of BLLSO is evident not only in cost optimization but also in its convergence efficiency and overall optimization quality, particularly in scenarios involving up to 100 units. This enhanced performance is attributed to the level-based learning strategy, which fosters a diverse solution space, effectively circumventing the pitfalls of premature convergence and stagnation commonly encountered in optimization processes. In summary, BLLSO's algorithmic design strikes an optimal balance between exploration and exploitation capabilities, rendering it an advantageous solution for combinatorial optimization problems.

Table 1. The Economic benefits comparison between different algorithms.

Unit	Different Algorithms				
	LLSO	*BLPSO*	*BPSO*	*BCSO*	*NBPSO*
10	647844.07	655582.95	658265.96	659221.71	659057.60
20	1323127.22	1354725.62	1346231.97	1349804.99	1348629.85
40	3017330.34	3031596.98	3020052.99	3027087.41	3043436.99
60	5156405.68	5185173.48	5179857.19	5196686.28	5187541.25
80	7681862.95	7765107.48	7748591.58	7753115.20	7767889.58
100	10562180.04	10758322.03	10686916.41	10702019.62	10724101.33

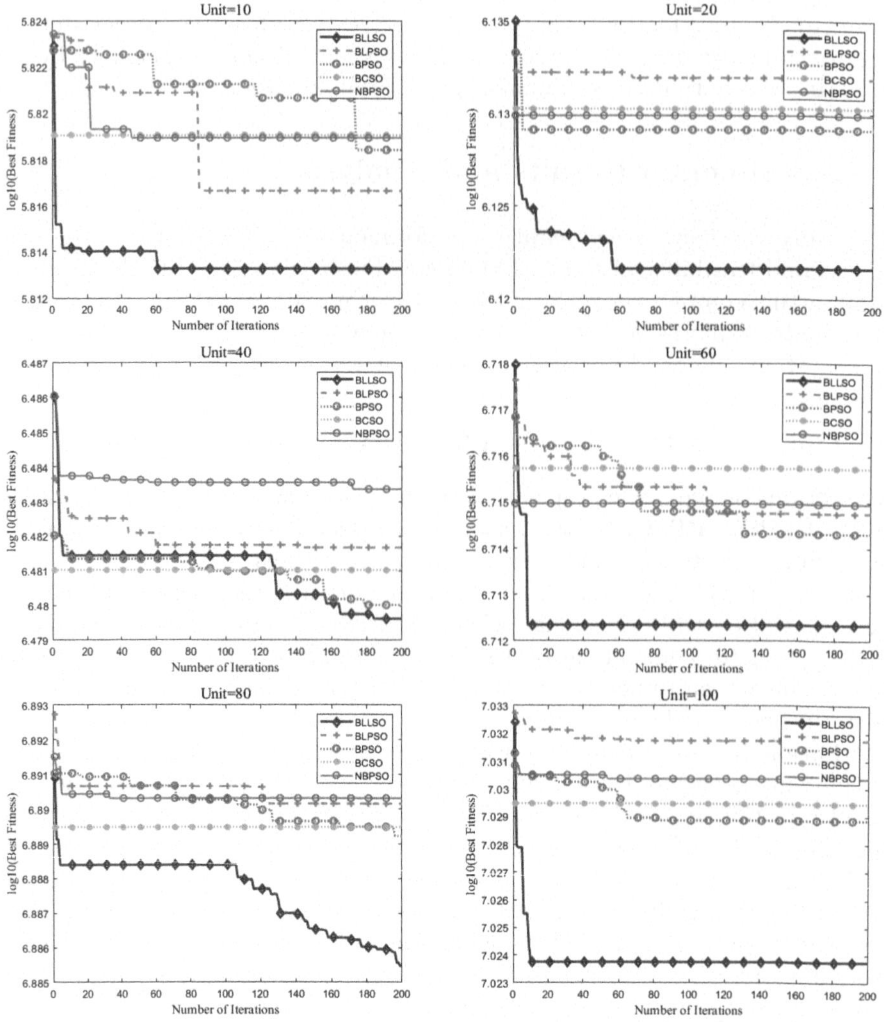

Fig. 3. Convergence results of different algorithms with different units.

4.2 The Scheduling Results of PEV Charge and Discharge on MCUC Problem

To explore the integration of charge-discharge management of PEVs in the electrical grid, this section hypothesizes a setup integrating 10 units with 150,000 PEVs, each PEV having a 30 KWh battery capacity, and the total PEV load approximates 1.114 GWh. The optimized charge-discharge management scheduling outcomes for PEVs are presented in Table 2 electric system. A positive PEV load indicates that the electric vehicle is charging, and a negative PEV load indicates that the electric vehicle is discharging. Peak power demand is observed from 9:00 a.m. to 14:00 and from 20:00 to 21:00, aligning with the peak generation times of units. During these peaks, PEVs discharge to meet demand while charging in low-demand periods, showcasing the algorithm's effectiveness in achieving demand stabilization through valley-filling and peak-shaving.

Table 2. The Optimization results of charge-discharge management of PEVs on MCUC problems

Hour (h)	Unit1 (MW)	Unit2 (MW)	Unit3 (MW)	Unit4 (MW)	Unit5 (MW)	Unit6 (MW)	Unit7 (MW)	Unit8 (MW)	Unit9 (MW)	Unit10 (MW)	Demand (MW)	PEV Load (MW)
1	321.36	407.15	0.00	0.00	124.49	0.00	0.00	0.00	0.00	0.00	700.00	153.00
2	324.41	436.77	0.00	0.00	141.82	0.00	0.00	0.00	0.00	0.00	750.00	153.00
3	408.05	377.95	0.00	0.00	162.00	0.00	0.00	55.00	0.00	0.00	850.00	153.00
4	272.07	412.20	128.60	129.60	160.52	0.00	0.00	0.00	0.00	0.00	950.00	153.00
5	435.17	400.21	125.32	21.18	134.42	0.00	0.00	0.00	0.00	0.00	1000.00	116.31
6	153.00	426.36	130.00	118.16	60.87	0.00	25.00	0.00	0.00	0.00	1100.00	115.07
7	153.00	455.00	109.55	130.00	82.52	20.00	25.00	0.00	0.00	0.00	1150.00	127.06
8	340.61	455.00	102.21	130.00	162.00	43.53	85.00	0.00	0.00	0.00	1200.00	118.35
9	265.67	454.39	89.64	105.42	111.39	41.47	25.00	46.59	40.88	0.00	1300.00	-119.55
10	400.39	397.97	130.00	48.97	98.28	79.61	40.80	10.14	37.54	54.52	1400.00	-101.78
11	454.05	181.23	86.38	129.54	157.98	79.55	43.75	55.00	55.00	54.52	1450.00	-153
12	280.23	421.88	114.64	115.02	112.07	72.57	75.90	55.00	45.02	54.67	1500.00	-153
13	329.87	292.47	102.91	102.21	160.94	79.77	75.29	41.13	43.63	55.00	1400.00	-116.78
14	455.00	254.39	60.27	48.48	162.00	64.84	85.00	0.00	55.00	10.00	1300.00	-105.01
15	438.82	347.00	129.44	130.00	162.00	20.00	85.00	0.00	0.00	0.00	1200.00	112.25
16	455.00	265.67	130.00	130.00	162.00	0.00	0.00	0.00	0.00	0.00	1050.00	92.67
17	307.45	455.00	117.22	130.00	128.53	0.00	0.00	0.00	0.00	0.00	1000.00	138.20
18	412.61	448.69	130.00	111.59	110.73	0.00	0.00	10.00	0.00	0.00	1100.00	123.62
19	455.00	299.20	130.00	130.00	162.00	70.81	84.08	0.00	0.00	0.00	1200.00	131.09
20	279.36	286.16	130.00	80.92	161.19	79.35	85.00	55.00	35.08	54.94	1400.00	-153
21	402.51	332.44	129.05	121.79	36.62	64.25	79.50	21.85	10.00	0.00	1300.00	-101.98
22	455.00	455.00	45.85	82.82	162.00	0.00	25.00	0.00	0.00	0.00	1100.00	125.67
23	455.00	341.56	130.00	0.00	126.44	0.00	0.00	0.00	0.00	0.00	900.00	153
24	455.00	455.00	33.00	0.00	0.00	0.00	0.00	10.00	0.00	0.00	800.00	153

Figure 4 provides a comparative depiction of unit output variations in the pre- and post-PEV scheduling situations. Observations from Fig. 4(a) reveal that, in the absence of PEV integration, Units 1 and 2 are disproportionately strained with excessive output, leading to an uneven load distribution across the 24 h. Conversely, Fig. 4(b) highlights that integrating PEVs into the grid markedly alleviates this output imbalance, distributing the load more equitably among units. This scheduling strategy not only eases the strain on specific units but also contributes to the overall stability and efficiency of the power system.

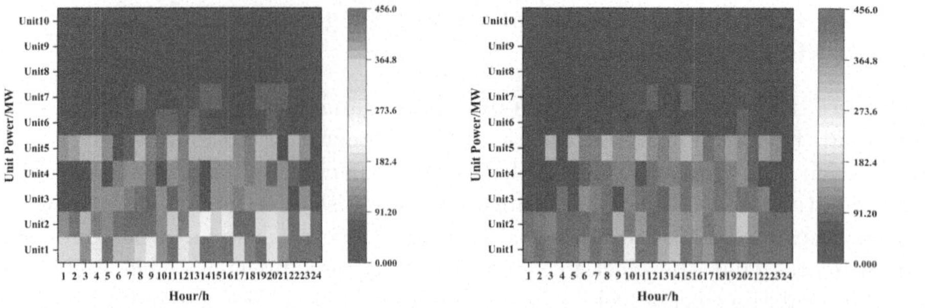

(a) The output power of 10 units integrated with PEVs

(b) The output power of 10 units without PEVs

Fig. 4. Heat map comparison of unit output under different scenarios.

5 Conclusion

This paper introduces a level-based learning optimization scheduling framework, designed to optimize the integration of PEVs into the power grid. This integration is targeted at enhancing grid stability, managing peak loads, and optimizing grid infrastructure, thereby posing a complex, large-scale optimization challenge. The proposed solution employs a novel binary level-based learning swarm optimization algorithm, tailored to address the specificities of unit switching states in generating units. The algorithm optimizes both fossil fuel costs and carbon emissions. Efficiently managing the charging and discharging of PEVs into the grid is central to this framework. The paper validates the effectiveness of the proposed scheduling approach through different scenarios, demonstrating not only its feasibility but also its economic advantages. The study underscores the significance of charge and discharge management of PEVs in future power systems, emphasizing its contribution to low carbon emissions, and economic benefits.

Acknowledgments. This study was financially supported by National Natural Science Foundation of China under grants 52077213 and 62003332, and Youth Innovation Promotion Association CAS 2021358, and Shenzhen Science and Technology Research and Development Fund JCYJ20200109114839874.

Disclosure of Interests. The authors have no competing interests to declare that are relevant to the content of this article.

References

1. Arroyo, J.M., Conejo, A.J.: A parallel repair genetic algorithm to solve the unit commitment problem. IEEE Trans. Power Syst. **17**(4), 1216–1224 (2002)
2. Chen, X., Li, K., Xu, B., Yang, Z.: Biogeography-based learning particle swarm optimization for combined heat and power economic dispatch problem. Knowl.-Based Syst. **208**, 106463 (2020)

3. Cheng, R., Jin, Y.: A social learning particle swarm optimization algorithm for scalable optimization. Inf. Sci. **291**(C), 43–60 (2015). https://doi.org/10.1016/j.ins.2014.08.039. https://www.scopus.com/inward/record.uri?eid=2-s2.0-84923326534&doi=10.1016%2fj.ins.2014.08.039&partnerID=40&md5=827e926499f77385986fa4ae776919df
4. Cheng, R., Sun, C., Jin, Y.: A multi-swarm evolutionary framework based on a feedback mechanism. In: 2013 IEEE Congress on Evolutionary Computation, pp. 718–724. IEEE (2013)
5. Fernandez, L.P., San Román, T.G., Cossent, R., Domingo, C.M., Frias, P.: Assessment of the impact of plug-in electric vehicles on distribution networks. IEEE Trans. Power Syst. **26**(1), 206–213 (2010)
6. Hussien, A.G., Hassanien, A.E., Houssein, E.H., Amin, M., Azar, A.T.: New binary whale optimization algorithm for discrete optimization problems. Eng. Optim. **52**(6), 945–959 (2020)
7. Jiang, Q., Zhou, B., Zhang, M.: Parallel augment lagrangian relaxation method for transient stability constrained unit commitment. IEEE Trans. Power Syst. **28**(2), 1140–1148 (2013)
8. Lander, L., Kallitsis, E., Hales, A., Edge, J.S., Korre, A., Offer, G.: Cost and carbon footprint reduction of electric vehicle lithium-ion batteries through efficient thermal management. Appl. Energy **289**, 116737 (2021)
9. Li, X., Zhai, Q., Zhou, J., Guan, X.: A variable reduction method for large-scale unit commitment. IEEE Trans. Power Syst. **35**(1), 261–272 (2019)
10. Roy, P.K., Sarkar, R.: Solution of unit commitment problem using quasi-oppositional teaching learning based algorithm. Int. J. Electr. Power Energy Syst. **60**, 96–106 (2014)
11. Sun, J.: Charging network+ new energy vehicles is an effective path to achieve carbon neutrality. China Ind. Inf. Technol. **2021**(04), 010 (2021)
12. Ting, T., Rao, M., Loo, C.: A novel approach for unit commitment problem via an effective hybrid particle swarm optimization. IEEE Trans. Power Syst. **21**(1), 411–418 (2006)
13. Trivedi, A., Srinivasan, D., Biswas, S., Reindl, T.: A genetic algorithm-differential evolution based hybrid framework: case study on unit commitment scheduling problem. Inf. Sci. **354**, 275–300 (2016)
14. Wang, Y., Yang, Z., Mourshed, M., Guo, Y., Niu, Q., Zhu, X.: Demand side management of plug-in electric vehicles and coordinated unit commitment: a novel parallel competitive swarm optimization method. Energy Convers. Manag. **196**, 935–949 (2019)
15. Yang, Q., Chen, W.N., Da Deng, J., Li, Y., Gu, T., Zhang, J.: A level-based learning swarm optimizer for large-scale optimization. IEEE Trans. Evol. Comput. **22**(4), 578–594 (2017)
16. Yang, W., Guo, J., Vartosh, A.: Optimal economic-emission planning of multi-energy systems integrated electric vehicles with modified group search optimization. Appl. Energy **311**, 118634 (2022)
17. Yang, Z., et al.: A binary symmetric based hybrid meta-heuristic method for solving mixed integer unit commitment problem integrating with significant plug-in electric vehicles. Energy **170**, 889–905 (2019)
18. Yang, Z., Li, K., Niu, Q., Xue, Y.: A novel parallel-series hybrid meta-heuristic method for solving a hybrid unit commitment problem. Knowl.-Based Syst. **134**, 13–30 (2017)

19. Yuan, X., Su, A., Nie, H., Yuan, Y., Wang, L.: Application of enhanced discrete differential evolution approach to unit commitment problem. Energy Convers. Manag. **50**(9), 2449–2456 (2009)
20. Zhang, H., Moura, S.J., Hu, Z., Qi, W., Song, Y.: Joint pev charging network and distributed pv generation planning based on accelerated generalized benders decomposition. IEEE Trans. Transport. Electrificat. **4**(3), 789–803 (2018)
21. Zhang, N., Hu, Z., Dai, D., Dang, S., Yao, M., Zhou, Y.: Unit commitment model in smart grid environment considering carbon emissions trading. IEEE Trans. Smart Grid **7**(1), 420–427 (2015)
22. Zhao, S., Li, K., Yang, Z., Xu, X., Zhang, N.: A new power system active rescheduling method considering the dispatchable plug-in electric vehicles and intermittent renewable energies. Appl. Energy **314**, 118715 (2022)
23. Zhu, X., Zhao, S., Yang, Z., Zhang, N., Xu, X.: A parallel meta-heuristic method for solving large scale unit commitment considering the integration of new energy sectors. Energy **238**, 121829 (2022)

A Novel CNN-Transformer Capacity Estimation Model for Real-World Lithium-Ion Battery Pack

Yin-Yi Soo, Yujie Wang(✉), and Haoxiang Xiang

Department of Automation, University of Science and Technology of China, Hefei 230027, China
wangyujie@ustc.edu.cn

Abstract. Lithium-ion batteries (LIBs) are widely used in electric vehicles (EVs) due to their long cycle life, high energy density, and fast charging abilities. However, the natural decline in LIB performance over time, leading to reduced capacity and power, presents challenges. This is especially true in battery packs, where inconsistencies among individual cells and non-uniform aging patterns complicate the estimation of state of health (SOH). Capacity serves as a crucial parameter for assessing SOH, providing insights into remaining energy storage capability and maintenance needs. Many studies have focused on estimating the capacity of single LIB cells or experimental datasets, but applying LIB packs in real-world situations requires customized models that consider environmental factors such as temperature and road conditions. This research introduces a novel hybrid model named CNN-Transformer for LIB pack capacity estimation, leveraging a combination of 1D-Convolutional Neural Network (CNN) feature extraction and transformer networks. The proposed method is evaluated against traditional recurrent neural network (RNN)-based methods, revealing superior performance. This evaluation highlights the effectiveness of the CNN-Transformer in managing long-term dependencies and overcoming the limitations of attention mechanisms found in transformer networks. Finally, the proposed methods are validated using real-world EV dataset, demonstrating practical effectiveness in LIB pack capacity estimation. In conclusion, this research contributes innovative approaches to real-world LIB pack capacity estimation, addressing key challenges and showcasing superior performance compared to traditional methods. These advancements hold significant promise for advancing the energy storage industry and ensuring the safe and stable operation of LIB packs in diverse applications.

Keywords: Lithium-ion Battery Pack · Capacity Estimation · Transformer network · Convolution neural network

1 Introduction

Lithium-ion batteries (LIB) have become irreplaceable in portable electronic devices, electric vehicles (EV), and grid-scale energy storage systems due to

K. Li et al. (Eds.): LSMS/ICSEE 2024, CCIS 2218, pp. 357–370, 2025.
https://doi.org/10.1007/978-981-96-0232-2_28

long cycle life, high energy density, and fast charging speed [25]. In real-world applications, several hundred LIB cells are connected in series to form a battery pack in order to meet power requirements [15]. However, the degradation of performance LIB is inevitable in the form of capacity and power fade [2]. Besides, the battery pack of LIB encounters a unique set of challenges such as inconsistencies in parameters among individual cells, the extraction of features, and the underlying causes of parameter inconsistency [3]. The non-uniform thermal field or different locations of battery cells in the LIB pack exhibit different and non-uniform aging patterns [7]. This inconsistency in battery cells makes the state of health (SOH) estimation of the LIB pack more complicated.

Capacity is the common key parameter in quantifying the degradation degree of SOH. LIB capacity allows users to assess the remaining energy storage capability of the system, and provide meaningful insights into remaining useful life and potential maintenance needs [24]. Numerous research has been conducted on LIB capacity estimation or SOH for single cell and experimental dataset [9,19]. However, real-world applications of LIB often utilize the LIB pack and face environmental challenges such as temperature [21], vehicle mileage, and road condition. Therefore, to model the real-world operating system of the LIB pack and to ensure the safe and stable operation of the LIB pack, an effective capacity model is needed for LIB pack.

2 Literature Review

The LIB capacity estimation can mainly divided into two categories: model-based methods and data-driven methods. A model-based approach is to approximate the dynamic behavior inside the cell by building a specific physicochemical model, however, this approach is difficult to apply to real-world pack-level LIB due to the complexity of model construction, parameterization, and computation. Particularly during actual operational environments and the impact of aging, which involve narrow SOC intervals and high current rate (C-rate) [27]. Therefore, data-driven methods will be applied in this research for LIB pack capacity estimation. Feature extraction is the fundamental process in a data-driven model. This process involves transforming raw data into a more concise and representative form, capturing the essential relationships and patterns necessary for a data-driven model. Several feature extraction methods for the LIB pack will be discussed.

2.1 Feature Extraction of LIB Pack

Health Factor (HF) extraction is the fundamental process in estimating battery SOH. However, feature extraction of LIB pack is complicated due to the inconsistent coupling between single cell batteries in pack batteries [23]. Song et al. introduced a principal component analysis (PCA) model for feature extraction in LIB pack at the pack level. However, PCA assumes linearity among features and may not perform optimally with real-world data [17]. She et al. present a method of applying the incremental capacity analysis (ICA) method for pack-level battery feature extraction which realizes pack-level battery SOH estimation based on cell-level tests [16]. Liu et al. proposed a novel ICA curve determination method based on the Hermite interpolation (HI) method with total voltage [11]. Jorge et al. proposed a wrapper for feature subset selection [6]. This method effectively captures temporal dependencies but is very computationally expensive. Yang et al. introduce a generalized matrix consisting of pack-level aging trend features and cell-level inconsistency features for feature extraction [23]. Zou et al. proposed a one-dimensional convolutional neural network (CNN) for LIB pack feature extraction [29]. One-dimensional CNN has proven to be suitable for time-series modeling due to its ability to capture the spatial and temporal features from the inputs [28].

2.2 Data-Driven Method

Data-driven methods are popular in SOH models due to their advantages of flexibility and being model-free [5]. Data-driven methods for SOH estimations usually utilize the collected data, and investigate hidden useful information by various data analysis methods such as support vector machine [13], gaussian process regression [8], artificial neural network (ANN) [12,18]. Xiang et al. introduce a two-level battery health diagnosis model using relaxation voltage and present a novel gaussian mixture ensemble learning method in SOH estimation [22]. Change et al. proposed a gaussian process regression for parameter prediction in LIB pack to overcome the inconsistency in battery cells [1]. Song et al. apply feed-forward neural network and genetic algorithm for two real-world data of EV with a big data platform for SOH estimation [17]. LSTM is also one of the popular RNN-based methods in SOH estimation with its strong generalization capabilities and training stability [?]. Yayan et al. proposed a stacked Bi-LSTM model structure in batteries SOH prediction which enables a robust training phase [26]. Qu et al. proposed a new method that combines particle swarm optimization with LSTM to solve both SOH Monitoring and remaining useful life (RUL) prediction [14]. However, the conventional RNN-based approach struggled with lengthy input and output sequences and lacked flexibility in handling variable input sequence lengths. To resolve the current weakness of the RNN-based approach, the attention model transformer stands out as an intriguing data-driven network.

2.3 Transformer Network

A Transformer network is an attention model that replaces the recurrent layers in encoder-decoder architectures with multi-headed self-attention [20]. The innovation of the transformer network resolves the issue of handling scalable input sequence length and long input and output length. Transformer models introduce position feature embedding to indicate the absolute or relative position information of data in the sequence and to describe features by positional encoding. Gu et al. proposed a CNN-Transformer SOH model for NASA LIB [4].

This research utilizes CNN for parameter optimization on experimental data but does not extend the research to LIB pack and real-world data. However, despite the Transformer network emerging as a promising new RNN-based model for capacity estimation, its attention mechanism has limitations in capturing spatial relationships within datasets. This is because the attention mechanism employed in transformers assigns weights to all input positions, making it challenging to focus specifically on local variables or spatial relationships. Besides, transformers process the entire sequence simultaneously, making it less intuitive for the model to distinguish hierarchical relationships.

2.4 Main Research Gap

The current capacity estimation of LIB packs encounters two primary challenges: difficulties in feature extraction for LIB packs and the absence of a real-world degradation model for LIB packs. In order to address these challenges, a pioneering capacity estimation model based on CNN-Transformer is introduced and validated using real-world EV data. A transformer network is introduced to address the challenge of handling long sequences in time series data, to effectively model the capacity degradation of LIB pack operation in real-world scenarios. However, the transformer network suffered from extracting hierarchical relationships and spatial relationships within datasets. To overcome the weakness of the transformer network, a one-dimensional convolutions neural network (1D-CNN) feature extraction method is being implemented. The 1D-CNN is being used to extract feature from the pack-level LIB features. These extracted features are then incorporated into the transformer model as input. The implementation of 1D-CNN also helps to extract useful features from the LIB pack to overcome the challenges of difficulties in LIN pack feature extraction.

2.5 Motivation and Contribution of this Work

The motivation of this work is to achieve accurate and efficient battery pack capacity estimation under real-world conditions. The proposed capacity estimation method aims to resolve the long-term dependencies for sequential data and its application in real-world LIB packs. The innovations and contributions of this paper are:

(1) A new data-driven hybrid model named CNN-Transformer is proposed to achieve a capacity estimation model for the LIB pack. Unlike the traditional RNN-based methods, this research applied a new Transformer network combined with CNN feature extraction for real-world LIB pack capacity estimation. The proposed model is evaluated and compared with other ANN models such as LSTM, Bi-LSTM and GRU. The proposed model proved to show better results than the other ANN model and The evaluation metrics are mean absolute error (MAE), mean squared error (MSE), and root mean square error (RMSE). The results of the research prove that the proposed CNN-Transformer can deal with long-term dependencies.

(2) A one-dimensional convolutions neural network (1D-CNN) feature extraction from the pack-level LIB features is proposed. The CNN extraction method can capture local variables hierarchically and spatial relationships from the dataset, which resolves the attention mechanism limitations of a single transformer network. The features extracted by the CNN network are evaluated with linear and non-linear evaluators. The features extracted by the CNN are additionally included in the capacity estimation network. The experimental results show that the proposed method has great potential to handle real-world LIB pack capacity estimation.
(3) This research utilizes a real-world LIB pack dataset of EV bus which is composed of 608 cells in parallel and series. The dataset consists of internal and external factors including current, total voltage, environment temperature, and total mileage. The proposed CNN-Transformer method performs effectively on real-world datasets and has proven to be practical and beneficial to real-world LIB pack capacity estimation.

3 Data Description

The data utilized in this study was gathered from the battery module of a fully electric vehicle between 2012 and 2013. The electric vehicle, utilized for commuting purposes, operates in Hefei, Anhui, China. The battery module consists of 608 cells connected in both parallel and series configurations. Each cell is a $LiFePO_4$ battery with a capacity of 10Ah and a rated voltage of 3.2V. The battery module is structured into 152 series-connected groups, with each group comprising 4 cells in parallel.

The module's current and voltage are monitored by the battery management system (BMS) and recorded onto a secure digital memory card. The BMS employs a distributed architecture with a Controller Area Network (CAN) network comprising one master and seven slaves. Each cluster incorporates two battery management chips (LTC 6803), which collect data on 24 battery voltages. The bus current is measured by the LEM CAB300 sensor and transmitted to the master via the CAN network. Recorded data includes timestamps, mileage, total voltage, total current, pack voltage, state of charge (SOC), maximum/minimum voltage, and more. The sampling interval is set at 1 s, and data collection occurs from December to August. During operation, the electric vehicle operates in four states: acceleration, braking, parking, and constant speed driving, leading to three battery pack operation modes: discharging, charging, and standby. Due to the presence of an energy feedback device in the EV, braking generates back electromotive force directed to the battery pack. The dataset used is shown in Table 1.

3.1 Capacity Degradation

The charging and discharging data for real-world LIB pack operation is random and incomplete. To address this challenge, a capacity calculation method

Table 1. BMS Data

Time	Mile/kM	Voltage/V	Current/A	SOC	Vmax/mV	Vmin	Tmax/°C	Tmin/°C
8:51:56	70	561.2	0.9	99.6	3344	3338	25	25
8:51:57	70	561.1	2.3	99.6	3343	3337	25	25
8:51:58	70	560.8	7.5	99.6	3341	3335	25	25
8:51:59	70	560.5	8.1	99.6	3340	3333	25	24
8:52:00	70	560.5	8.2	99.6	3339	3333	25	24
8:52:01	70	560.4	7.9	99.6	3339	3333	25	25

based on SOC is adopted. The charging data are extracted from the dataset, and their current, time and SOC are being used to calculate the capacity. The mathematical formula 1 is used to calculate the capacity per cycle Q_c for the LIB pack.

$$Q_c = \frac{\int_{t_0}^{t} I(t)dt}{SOC(t) - SOC(t_o)} = \frac{Q}{\Delta SOC} \tag{1}$$

where $SOC(t)$ and $SOC(t_0)$ represent the SOCs of the battery pack at the time steps t, respectively. $I(t)$ represents the current of the battery pack at the time step t and Q_c is the maximum available capacity of the battery pack in cycle c. The capacity calculated for each cycle is subsequently smoothed using a Kalman filter.

4 Proposed Method

The Fig. 1 shows the research flow of this research. The process begins with data cleaning and charging data extraction. Subsequently, this charging data is grouped by cycle to analyze key metrics such as average voltage, temperature, current, and mileage. The LIB pack's capacity is then calculated using a mathematical formula referenced as 1 and smoothed by a kalman filter. Following this, the aggregated data undergoes 1D-CNN feature extraction to uncover latent patterns correlated with capacity. All features, including those from CNN and aggregated sources, are evaluated for linear and non-linear relationships with capacity. Features failing to meet correlation criteria are eliminated from the capacity estimation model. Finally, the remaining correlated features are input into the CNN-Transformer network for capacity estimation. The parameter of the 1D-CNN and transformer network is illustrated in Table 2.

The raw data will be split into 80% train and 20% test data for the 1D-CNN feature extraction model and capacity estimation model. The training data will undergo 1D-CNN feature extraction, and then feed into the capacity estimation model. The remaining 20% test data will be used later for evaluating the capacity estimation model. All the network training in this research will be trained by Laptop Legion Y530 with 8 GB RAM, 64bit operating system, Windows 10 home edition and intel core i7-8750H.

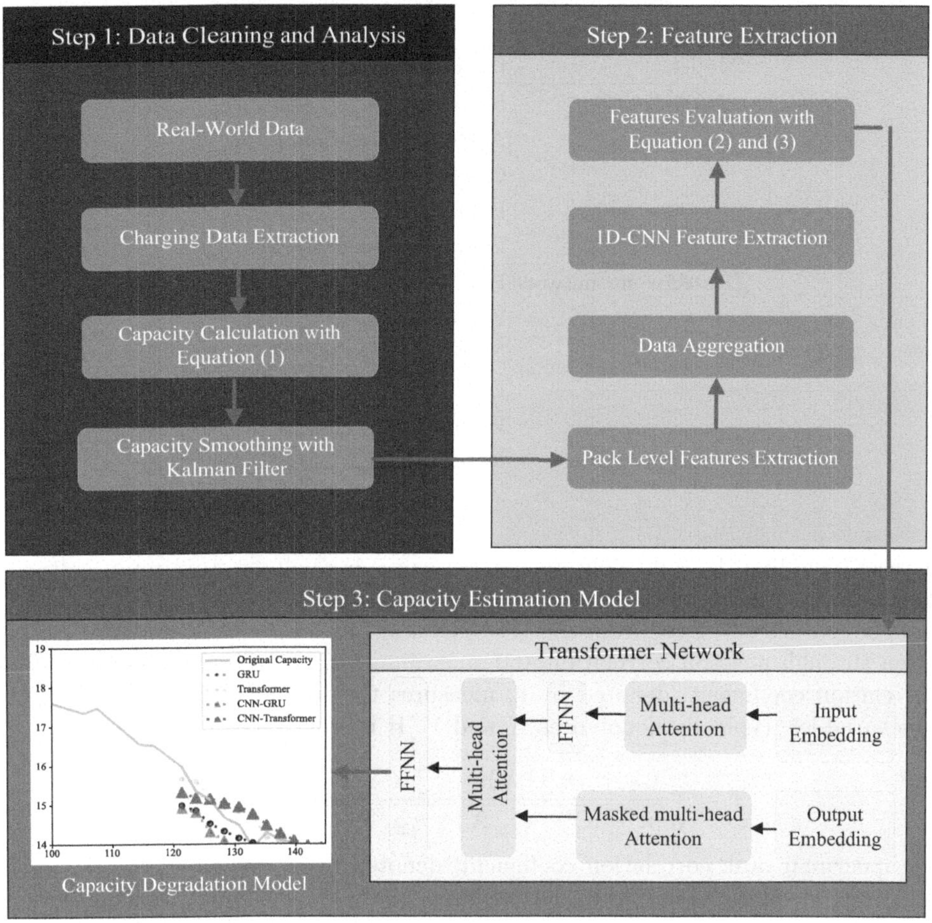

Fig. 1. Research Flow

4.1 Convolutional Neural Network (CNN)

Convolutional neural network (CNN) draws inspiration from feed-forward ANNs, with 1D-CNN being particularly suitable for time series modeling [10]. A CNN consists of three main layers: input layers, hidden layers, and output layers. The input layer facilitates the transfer of original data to the first hidden layer. The output layer generates the desired outputs for subsequent processing. The hidden layers consist of the fully connected layer, the max-pooling layer, and the convolutional layer. The parameter of 1D-CNN used in this research is illustrated in Table 2. This method proves highly effective in discerning simple patterns and generating more complex patterns at an elevated level when the data fragment exhibits a low local correlation. One feature will be extracted from the raw

Table 2. Model Parameter

Model	Parameter	Range
1D-CNN	Kernel size	3
	Activation function	Relu
	Convolution filter	32
	Pooling size	2
	Dense layer neurons	64
Transformer network	Batch size	32
	Epoach	100
	Number of heads	8
	Embedding dim	16
	Ffn units	8
	learning rate	0.001

dataset including the mile, voltage, current, soc, max voltage, minimum voltage, maximum and minimum temperature with 1D-CNN.

The feature extracted from the 1D-CNN and raw feature will undergo evaluation through pearson correlation and spearman rank correlation. The Pearson correlation coefficient, denoted by r, measures the linear relationship between two variables, typically denoted as X and Y. It is shown as the formula 2:

$$r = \frac{n\sum XY - \sum X \sum Y}{\sqrt{[n\sum X^2 - (\sum X)^2][n\sum Y^2 - (\sum Y)^2]}} \quad (2)$$

Spearman rank correlation coefficient, denoted by ρ, measures the strength and direction of the monotonic relationship between two variables. It is computed based on the ranks of the observations rather than the actual values. The formula for spearman rank correlation is:

$$\rho = 1 - \frac{6\sum d_i^2}{n(n^2 - 1)} \quad (3)$$

The correlation between the features in the raw dataset and the 1D-CNN features should exceed 0.05 in Pearson correlation coefficient and have a p-value of less than 0.05 in spearman rank correlation.

4.2 Transformer

The Transformer model adheres to a structured design including pointwise, stacked self-attention, and fully connected layers for both the encoder and decoder components. By eliminating the need for recurrence and convolutions, the Transformer model has innovated the implementation of attention mechanisms, relying exclusively on the self-attention mechanism. This mechanism comprises scaled-dot-product attention and multi-head attention, marking a significant departure from traditional approaches.

The scaled dot-product attention firstly computes a dot product for each query, q, with all the keys, k. It subsequently divides each result by $\sqrt{d_k}$ and proceeds to apply a soft-max function:

$$Attention(Q, K, V) = softmax(\frac{QK^T}{\sqrt{d_k}})V \tag{4}$$

The equation for the multi-head attention mechanism is built on the above-scaled dot-product attention mechanism, as shown below:

$$Multihead(Q, K, V) = Concat(head_1, ..., head_h)W^o \tag{5}$$

here each $head_i$, i = 1, ... h, implements a single attention function characterized by its own learned projection matrices as:

$$head_i = Attention(QW_i^Q, KW_i^k, VW_i^V) \tag{6}$$

where q and k represent vectors of dimensions, and d_k contains the queries and keys, respectively. v denotes a vector of dimensions and d_v contains the values. Q,K, and V which denote matrices that pack together sets of queries, keys, and values, respectively. W_Q, W_K, and W_V represent projection matrices employed to create distinct subspace representations of the query, key, and value matrices, respectively. W_O indicates a projection matrix for the multi-head output.

The attention mechanism processes the entire sequence equally with Q,K, and V, but does not focus specifically on local variables. This poses challenges in capturing local variables hierarchically and spatial relationships for transformer. To resolve the constraints inherent in the attention mechanism of the transformer, the initial raw data will undergo 1D-CNN feature extraction before being fed into the transformer model. The data will be split into 80% train and 20% test data. The proposed method is being evaluated with other ANNs models with MAE, MAPE, RMSE, and R^2, which are given as:

$$\text{MAE} = \frac{1}{n}\sum_{i=1}^{n}|y_i - \hat{y}_i| \tag{7}$$

$$\text{MAPE} = \frac{100}{n}\sum_{i=1}^{n}\left|\frac{y_i - \hat{y}_i}{y_i}\right| \tag{8}$$

$$\text{RMSE} = \sqrt{\frac{1}{n}\sum_{i=1}^{n}(y_i - \hat{y}_i)^2} \tag{9}$$

$$R^2 = 1 - \frac{\sum_{i=1}^{n}(y_i - \hat{y}_i)^2}{\sum_{i=1}^{n}(y_i - \bar{y})^2} \tag{10}$$

where y_i is the actual values, $\hat{y}_i$ are the predicted values. $\bar{y}$ is the mean of y_i and n is the number of testing samples.

5 Result and Discussion

5.1 Feature Extraction with 1D CNN

The below Table 3 shows the pearson correlation and spearman's ranks for the extracted features from CNN and aggregation. Both average maximum and minimum voltages exhibit no correlation with capacity in either Pearson correlation or Spearman's rank and, therefore will be excluded from the model input. The 1D-CNN extracted feature demonstrates a significant correlation with capacity in both linear and non-linear relationships. Therefore, it will be incorporated into the capacity estimation model.

Table 3. Model Parameter

Features	Pearson correlation		Spearman's rank	
Average Maximum Voltage	0.025988	Remove	0.610221660	Remove
Average Minimum Voltage	0.082887	Remove	0.015264745	Remove
Average Total Voltage	−0.670812	Keep	0.00013686	Keep
Average Current	0.586954	Keep	3.167139e-09	Keep
Average Maximum Temperature	−0.769161	Keep	2.1221388e-13	Keep
Average Minimum Temperature	−0.558013	Keep	5.274858e-09	Keep
CNN Feature	−0.723544	Keep	6.20671e-10	Keep

5.2 Capacity Estimation with Transformer Network

The Table 4 illustrates the performance of the proposed CNN-Transformer capacity estimation model compared with other ANN models. The proposed CNN-Transformer model exhibits the best performance as compare to the other ANNs models. The degradation model by each ANN model is illustrated in Fig. 2. The transformer model outperforms the other ANN models due to its ability to capture long-range dependencies in the real-world LIB pack data. The enhanced CNN-Transformer performs better than the transformer alone, exhibiting improvements of 21% for MAE, 37% for MSE, 21% for RMSE, and 15% for r-square. This suggests that the proposed method has surpassed the limitations of the attention mechanism and effectively captures hierarchical local variables and spatial relationships from the dataset.

Table 4. Evaluation of different ANNs

ANNs	MAE	MSE	RMSE	R^2
LSTM	0.090833	0.013705	0.11707	0.32418
Bi-LSTM	0.085828	0.01434	0.11974	0.29293
GRU	0.096624	0.012806	0.11316	0.36851
Transformer	0.06205	0.00592	0.07696	0.70794
CNN-LSTM	0.10893	0.02413	0.15534	−0.18997
CNN-BiLSTM	0.13706	0.03113	0.17643	−0.53487
CNN-GRU	0.06641	0.00671	0.08195	0.66879
CNN-Transformer	0.04886	0.00369	0.06078	0.81779

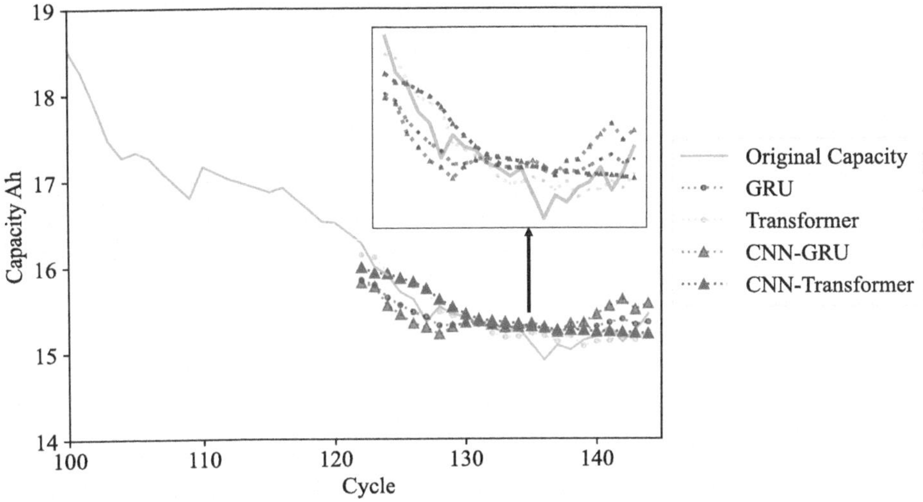

Fig. 2. Capacity Estimation Model of LIB pack with ANN Models

6 Conclusion

Real-world LIB pack capacity estimation plays a vital role in developing the energy storage industry. In this research, a novel CNN-Transformer method has been proposed. The proposed method leverages 1D-CNN to capture hierarchical local variables and spatial relationships. This research utilizes real-world EV operated from the years 2012 to 2013. The original dataset undergoes extraction of charging data and aggregation at the pack level. Then, capacity is calculated for each cycle, followed by the implementation of 1D-CNN to capture hidden hierarchical features associated with capacity. The extracted 1D-CNN feature is subjected to evaluation for linear and non-linear correlation, demonstrating a strong correlation with capacity. Various ANN models such as LSTM, GRU, transformer, CNN-GRU and CNN-Transformer are employed and compared in predicting the LIB pack capacity degradation model. The proposed CNN-Transformer outperforms other ANN models and achieves more competitive performance. The enhancement of 1D-CNN to the transformer also exhibited improvements as compared to the original transformer by 21% for MAE, 37% for MSE, 21% for RMSE, and 15% for r-square. The experimental result for real-world EV proves that (1) The proposed CNN-Transformer effectively addresses long-term dependencies, manages scalable input sequence lengths, and performs well in the real-world LIB pack degradation model. (2) The combination of CNN and Transformer is able to surpass the constraints of the attention mechanism from the original transformer, leading to the enhancement of model performance. (3) The proposed 1D-CNN feature extraction method effectively retrieves informative features from the LIB pack level, facilitating their application in capacity estimation models. Future research will concentrate on feature extraction from both pack and cell level of real-world LIB packs, and formulating a comprehensive capacity degradation model for both pack and cell levels.

References

1. Chang, C., et al.: Prognostics of the state of health for lithium-ion battery packs in energy storage applications. Energy **239**, 122189 (2022)
2. Edge, J.S., et al.: Lithium ion battery degradation: what you need to know. Phys. Chem. Chem. Phys. **23**(14), 8200–8221 (2021)
3. Feng, F., Hu, X., Hu, L., Hu, F., Li, Y., Zhang, L.: Propagation mechanisms and diagnosis of parameter inconsistency within li-ion battery packs. Renew. Sustain. Energy Rev. **112**, 102–113 (2019)
4. Gu, X., See, K.W., Li, P., Shan, K., Wang, Y., Zhao, L., Zhang, N.: A novel state-of-health estimation for the lithium-ion battery using a convolutional neural network and transformer model. Energy **262**, 125501 (2023)
5. Hu, X., Jiang, J., Cao, D., Egardt, B.: Battery health prognosis for electric vehicles using sample entropy and sparse Bayesian predictive modeling. IEEE Trans. Ind. Electron. **63**(4), 2645–2656 (2015)
6. Jorge, I., Mesbahi, T., Samet, A., Boné, R.: Time series feature extraction for lithium-ion batteries state-of-health prediction. J. Energy Storage **59**, 106436 (2023)
7. Klein, M., Tong, S., Park, J.W.: In-plane nonuniform temperature effects on the performance of a large-format lithium-ion pouch cell. Appl. Energy **165**, 639–647 (2016)
8. Kong, J.Z., Yang, F., Zhang, X., Pan, E., Peng, Z., Wang, D.: Voltage-temperature health feature extraction to improve prognostics and health management of lithium-ion batteries. Energy **223**, 120114 (2021)
9. Li, K., Wang, Y., Chen, Z.: A comparative study of battery state-of-health estimation based on empirical mode decomposition and neural network. J. Energy Storage **54**, 105333 (2022)
10. Liu, H., Mi, X., Li, Y.: Smart deep learning based wind speed prediction model using wavelet packet decomposition, convolutional neural network and convolutional long short term memory network. Energy Convers. Manage. **166**, 120–131 (2018)

11. Liu, P., Wu, Y., She, C., Wang, Z., Zhang, Z.: Comparative study of incremental capacity curve determination methods for lithium-ion batteries considering the real-world situation. IEEE Trans. Pow. Electron. **37**(10), 12563–12576 (2022)
12. Lu, J., Xiong, R., Tian, J., Wang, C., Sun, F.: Deep learning to estimate lithium-ion battery state of health without additional degradation experiments. Nat. Commun. **14**(1), 2760 (2023)
13. Pang, X., et al.: An interval prediction approach based on fuzzy information granulation and linguistic description for remaining useful life of lithium-ion batteries. J. Pow. Sources **542**, 231750 (2022)
14. Qu, J., Liu, F., Ma, Y., Fan, J.: A neural-network-based method for RUL prediction and soh monitoring of lithium-ion battery. IEEE Access **7**, 87178–87191 (2019)
15. Schuster, S.F., Brand, M.J., Berg, P., Gleissenberger, M., Jossen, A.: Lithium-ion cell-to-cell variation during battery electric vehicle operation. J. Pow. Sources **297**, 242–251 (2015)
16. She, C., Zhang, L., Wang, Z., Sun, F., Liu, P., Song, C.: Battery state-of-health estimation based on incremental capacity analysis method: synthesizing from cell-level test to real-world application. IEEE J. Emerg. Sel. Top. Pow. Electron. **11**(1), 214–223 (2021)
17. Song, L., Zhang, K., Liang, T., Han, X., Zhang, Y.: Intelligent state of health estimation for lithium-ion battery pack based on big data analysis. J. Energy Storage **32**, 101836 (2020)
18. Tian, Y., Dong, Q., Tian, J., Li, X., Li, G., Mehran, K.: Capacity estimation of lithium-ion batteries based on optimized charging voltage section and virtual sample generation. Appl. Energy **332**, 120516 (2023)
19. Topan, P.A., Ramadan, M.N., Fathoni, G., Cahyadi, A.I., Wahyunggoro, O.: State of charge (SoC) and state of health (SoH) estimation on lithium polymer battery via Kalman filter. In: 2016 2nd International Conference on Science and Technology-Computer (ICST), pp. 93–96. IEEE (2016)
20. Vaswani, A., et al.: Attention is all you need. In: Advances in Neural Information Processing Systems, vol. 30 (2017)
21. Wang, Y., et al.: A comprehensive review of battery modeling and state estimation approaches for advanced battery management systems. Renew. Sustain. Energy Rev. **131**, 110015 (2020)
22. Xiang, H., Wang, Y., Zhang, X., Chen, Z.: Two-level battery health diagnosis using encoder-decoder framework and gaussian mixture ensemble learning based on relaxation voltage. In: IEEE Transactions on Transportation Electrification (2023)
23. Yang, S., et al.: A hierarchical enhanced data-driven battery pack capacity estimation framework for real-world operating conditions with fewer labeled data. J. Energy Chem. **91**, 417–432 (2024)
24. Yang, S., Zhang, C., Jiang, J., Zhang, W., Zhang, L., Wang, Y.: Review on state-of-health of lithium-ion batteries: characterizations, estimations and applications. J. Clean. Prod. **314**, 128015 (2021)
25. Yang, Z., Huang, H., Lin, F.: Sustainable electric vehicle batteries for a sustainable world: perspectives on battery cathodes, environment, supply chain, manufacturing, life cycle, and policy. Adv. Energy Mater. **12**(26), 2200383 (2022)
26. Yayan, U., Arslan, A.T., Yucel, H.: A novel method for soh prediction of batteries based on stacked LSTM with quick charge data. Appl. Artif. Intell. **35**(6), 421–439 (2021)
27. Zhang, X., Zou, Y., Fan, J., Guo, H.: Usage pattern analysis of Beijing private electric vehicles based on real-world data. Energy **167**, 1074–1085 (2019)

28. Zhao, X., Jiang, N., Liu, J., Yu, D., Chang, J.: Short-term average wind speed and turbulent standard deviation forecasts based on one-dimensional convolutional neural network and the integrate method for probabilistic framework. Energy Convers. Manage. **203**, 112239 (2020)
29. Zou, R., Duan, Y., Wang, Y., Pang, J., Liu, F., Sheikh, S.R.: A novel convolutional informer network for deterministic and probabilistic state-of-charge estimation of lithium-ion batteries. J. Energy Storage **57**, 106298 (2023)

Thermal Monitoring of Li-Ion Batteries Using Convolutional Neural Networks and Fibre Bragg Grating Sensors

Li Zhang[1], Kang Li[2], Jingjie Ma[1](✉), Xingyu Jia[3], and Shiyun Liu[2]

[1] School of Electrical and Electronic Engineering, Shanghai Institute of Technology, Shanghai 201418, China
mjj_staff@sit.edu.cn

[2] School of Electronic and Electrical Engineering, University of Leeds, Leeds LS2 9JT, England

[3] School of Industrial and Information Engineering, Politecnico di Milano, Milan, Italy 20133

Abstract. Thermal safety of Li-ion Batteries is an important issue in their applications in electric vehicles and grid-tied energy storage. Thermal monitoring of Li-ion Batteries is crucial in preventing thermal failures and prolong lifetime. However, the existing temperature monitoring method are sensitive to corrosion and electromagnetic radiation, without considering the aging factor and non-electric features leading to low accuracy for full life cycle thermal monitoring of batteries under different operating conditions. To tackle the challenges, this paper presents a thermal model using convolutional neural networks incorporating FBG sensors. In order to more accurate, the attribute information fusion of non-electric variables (wavelengths) and the electric variables (voltage, current, state of charge) is developed for battery temperature monitoring in the proposed method. The experimental results based on different cells indicate that the proposed model has desirable accuracy which achieves the root mean squared error around 0.2°C on the validation data.

Keywords: Li-ion Batteries · Thermal monitoring · Convolutional Neural Networks · Fiber Bragg grating sensors

1 Introduction

Lithium-ion batteries (LIBs) are commonly used in power station [1,2] and in electric vehicles (EVs) [3,4]. Temperature is a critical factor which effect the degradation of characteristics for a lithium-ion battery. Generally, the usable capacity increases with the rise of the operating temperature until it reaches a specific value and it decreases rapidly when the temperature falls below room temperature [5]. In addition to the impacts on the usable capacity, temperature also influences battery internal resistance which is more sensitive to the variation of SOC at the lower operating temperature since the active particles travel

K. Li et al. (Eds.): LSMS/ICSEE 2024, CCIS 2218, pp. 371–385, 2025.
https://doi.org/10.1007/978-981-96-0232-2_29

slower. Therefore, temperature monitoring of the Li-ion batteries has an important role to estimate the internal states, prevent thermal failures and prolong battery use life. To guarantee the thermal safety and maintain battery performance, thermal monitoring should be accurate, reliable and timely. Moreover, accurate thermal monitoring can be used to develop advanced charging and discharging strategies. Due to the internal temperature of li-ion batteries is often not directly measurable, the surface temperature of cells is often used for thermal monitoring because its variation tendency is in connection with the changes of the internal temperature. The existing methods of thermal monitoring based on the surface temperature are classified into two categories including direct measurement using thermal sensors and state of temperature estimation using model-based method [6].

Different thermal sensors [7,8], such as thermocouples, resistance temperature detectors (RTD) and infrared sensors, have been used to directly measure the surface temperature of batteries. Among them, the traditional thermal sensor, such as thermocouples and thermistors, are often sensitive to corrosion and electromagnetic radiation which can only measure the temperature of a single location using one probe [9]. However, the temperature profile of single location is non-uniform inside and outside, hence single location measurement is insufficient. For large scale battery pack, a great deal of sensors and other supporting facilities should be deployed resulting in challenges to the wiring management, which causes signal drifts and corruptions on account of strong electromagnetic inferences. Infrared sensors are non-contact temperature sensors to avoid short circuits, while their accuracy is greatly reduced because of the scattering and reflection of infrared rays affected by operating condition [10]. The fiber bragg grating (FBG) sensors which can be inscribed with multiple FBG sensors have been widely used for battery temperature measurements [11] owing to the merits of smaller size, corrosion resistance, nonconductivity and anti-electromagnetic interference [12,13]. Numbers of the surface temperatures of commercial cells have been measured using the FBG sensors [11,14]. However, the above direct methods have their limitations for battery temperature monitoring that the layout of sensors would have significant impact on the measurement accuracy [15] and the temperature of cells is affected by a range of factors, making it difficult to acquire with high precision [16].

To address these problems, the model-based method is another solution which estimates the battery state of temperature (SOT) by establishing a suitable battery model. To this end, various models have been developed to estimate the temperature of batteries, including physical models [17], equivalent circuit models (ECMs) [16] and data-driven models. The physical thermal model are usually complex to achieve high accuracy, while the equivalent circuit model has the difficulties in estimating model parameters which often are identified offline [18]. The accuracy of these parameters tends to be lower in different working conditions. Thus, the data-driven model is attracting increasing interests to estimate surface temperatures of li-ion batteries which is more user-friendly and more independent of of physical or off-line trained ECM models. In [19], a feed-forward neural network (FNN) model is proposed to estimate battery temperatures. However, the temperature of cells is estimated using few charge and discharge cycles which is not suitable in later life cycles because of aging. In [20], the long short term memory (LSTM) method is proposed to estimate battery temperature changes induced by the increased irreversible heat due to battery ageing and the average temperature of each cycle is added into the model. However, measurements from the same battery are usually adopted both for training and testing in the above methods.

Although the thermal monitoring based on data-driven models does not require the knowledge of explicit complex physical models, some issues still exist. Firstly, most temperature estimation methods do not consider the aging factor leading to the lower accuracy for the cell temperature estimation in a full life cycle. Secondly, to process enormous data is highly time-consuming and requires huge storage space. Thirdly, the electronic sensor signals are susceptible to electro magnetic interferences (EMIs), leading to degraded performance. To overcome the shortcomings of thermal monitoring methods based on data-driven models, a novel thermal monitoring method based on convolutional neural networks (CNN) integrated with multi-fiber bragg grating (multi-FBG) sensor signals is proposed. The main works of this paper are given as follows:

- Different from existing methods, the partial charging curves of voltage, current and wavelength shift from FBG sensors are employed for inputs in developing the CNN model in order to achieve higher accuracy.
- The adaptive moment estimation (Adam) optimizer is used when training the CNN model to ensure fast online estimation of battery temperature. Besides, the data of whole life cycles in each cell is employed to train the CNN model to reflect the battery aging effect.
- Experimental results indicate that the proposed method has the ability of producing accurate and comparable temperature estimation for different battery cells.

The remainder of the paper is structured as follows. Section 2 briefly introduces the CNN and Adam algorithm. The framework of the thermal estimation method using convolutional neural networks (CNN) fusing the information of multi-fiber bragg grating sensors is explained in Sect. 3. In Sect. 4, the performance of experimental simulations is analysed to evaluate the excellent performance of the proposed method. Section 5 is a conclusion.

2 Preliminaries

2.1 Convolutional Neural Networks

Due to the characteristics of sparse connectivity, sharing weights, and pooling layers, convolutional neural networks (CNNs) has less number of parameters to maintain the accuracy with the same number of layers. Moreover, the CNN method is good at extract features automagically and reducing the risk of overfitting which can be used to process large scale time series signals, so that it is widely used in battery state estimation [21]. Since the size of filter is smaller than that of the input, a partially connected pattern for neurons between adjacent layers is developed to ensure the sparse connectivity. To illustrate the structure of CNNs, a simple CNN composed by a input layer, a convolution layer, a pooling layer, a fully connected layer and a output layer is given in Fig. 1.

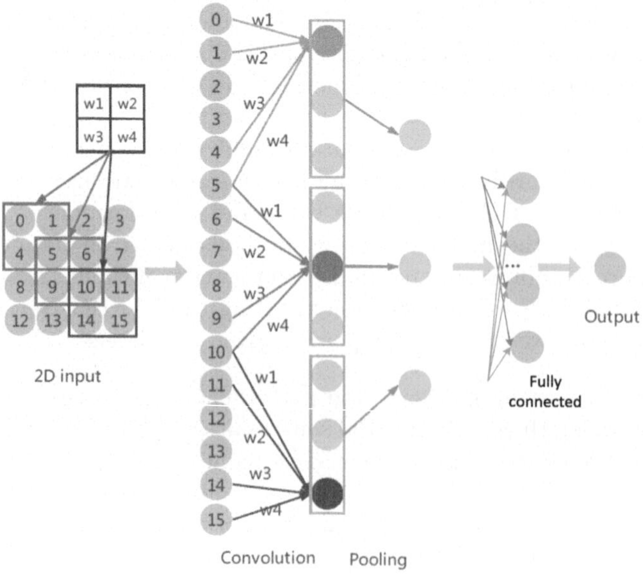

Fig. 1. A convolutional neural network [21]

In Fig. 1, firstly, several 2-dimensional matrices are often employed at the input layer to make the processing of features extraction possible. Secondly, the convolutional operation is employed to the input with the filter (2×2) in a convolutional layer. After reformulating the input from the 2-dimensional image shape into time series shape, the output node of a convolutional layer is obtained by connecting the convolution filter to the small subset from the inputs, contributing to the advantage of the sparse connectivity using the replacement of the matrix multiplication with convolutions. It is worth noting that the output nodes are generated by the filter convolutional sliding over the input subsets. Thus, all the output nodes share the weights using the filter value in the convolving map to calculate outputs. The size of filter determines the number of inputs related to one output node. For the rth convolving map in layer l, its output node $\boldsymbol{C}_{r,l}^{a,t}$ at a^{th} row and t^{th} column is given as following:

$$\boldsymbol{C}_{r,l}^{a,t} = f(\sum_{v} \sum_{i=0}^{m_l-1} \sum_{j=0}^{n_l-1} \boldsymbol{x}_{v,l}^{a \cdot s_l+i, tc \cdot d_l+j} \boldsymbol{w}_{v,r,l}^{i,j} + \boldsymbol{b}_{r,l}) \quad (1)$$

where $\boldsymbol{x}_{v,l} \in \boldsymbol{R}^{p\times q\times v}$ is the input of the lth layer at the vth channel (also called depth) with the size of $p\times q$. Except that it is the initial input in the first layer, it is the output of the previous layer. v denotes the number of convolving maps of the $(l-1)^{th}$ layer. $\boldsymbol{w}_{v,r,l} \in \boldsymbol{R}^{m_l\times n_l}$ denotes the vth channel of the rth filter in layer l, the size of which is $m_l \times n_l$. The sliding stride of the filter is set as $(s_l,\ d_l)$. $\boldsymbol{b}_{r,l}$ is the bias at the rth convolving map. $f(\cdot)$ is the nonlinear activation function which is used to improve the processing ability of the nonlinear relationships for a complex nonlinear system. Exponential linear unit (elu) is selected as the activation function in this paper for the experiments.

Unit (elu), which is given by

$$f(x) = \begin{cases} \alpha * (exp(x) - 1), & x <= 0 \\ x, \quad x > 0 \end{cases} \tag{2}$$

Thirdly, the pooling layer is used to cut down the size of the feature maps after the previous convolution layer and the number of the corresponding parameters for the next layer. Generally speaking, the down-sampling method is adopt for the pooling layer. The max pooling strategy which is commonly is given as

$$\boldsymbol{P}^{a,t}_{r,l+1} = \max_{0\leq i\leq m_{(l+1)}-1, 0\leq j\leq n_{(l+1)}-1} \{\boldsymbol{C}^{a\cdot s_{(l+1)}+i, t\cdot d_{(l+1)}+j}_{r,l+1}\} \tag{3}$$

and an another commonly strategy called average pooling is given as follows

$$\boldsymbol{P}^{a,t}_{r,l+1} = \frac{\sum_{i=0}^{m_{(l+1)}-1} \sum_{j=0}^{n_{(l+1)}-1} (\boldsymbol{C}^{a\cdot s_{(l+1)}+i, t\cdot d_{(l+1)}+j}_{r,l+1})}{m_{(l+1)} \times n_{(l+1)}} \tag{4}$$

where $(m_{(l+1)},\ n_{(l+1)})$ denotes the size of the pooling region, while the sliding stride of the pooling filter at $l+1^{th}$ layer is set as $(s_{(l+1)},\ d_{(l+1)})$.

The CNN in Fig. 1 just contains one filter with a size of 2×2 so that the number of the corresponding input subset is 4 for one output. Assuming the sliding stride as (1, 1)and the pooling filter is a 3-by-3 matrix, the size of the pooling output would be 1×3. Besides, additional convolution layers or pooling layers or flatten layers or fully connected layers can be selected for complex systems. Then, the final output is calculated by

$$\boldsymbol{O} = f(\sum_{j=1}^{z} \boldsymbol{x}(j)\boldsymbol{w}(j) + \boldsymbol{b}) \tag{5}$$

where $\boldsymbol{x}$ is the input matrix of the output layer. The corresponding linear weights and the bias of the output layer are $\boldsymbol{w}$ and $\boldsymbol{b}$ respectively. In this paper, the output $\boldsymbol{O}$ denotes the battery temperature. The statistics between the estimated temperature and the reference temperature $\boldsymbol{y}(i)$, such as root mean squared error (RMSE) E_{rmse}, mean squared error (MSE) E_{mse} and mean absolute error (MAE) E_{mae} are used to evaluate the accuracy of the CNN model.

2.2 Adaptive Moment Estimation

To build an accurate convolutional neural network, the optimization of parameters between two connected layers is also essential. Due to the advantages of high computing efficiency, small memory occupation and fast descending speed for parameter adjustment, the Adam algorithm has been widely adopted [22]. Taking the weights optimization of the fully-connected layer for CNN model as an example, the cost function is given as follows according to Eq. (5):

$$f_t(\boldsymbol{\Theta}) = (\boldsymbol{O}_t - \boldsymbol{y}(t))^2 \tag{6}$$

where $\boldsymbol{y}(t)$ and $\boldsymbol{x}(t)$ are output and input at time instant t respectively. $\boldsymbol{\Theta} = [\theta_1, \cdots, \theta_n]^T \in \Re^{n\times 1}$ denotes the linear weights between the fully-connected layer and the output layer.

Then, the gradients are given by calculating the partial derivative of the cost function f_t with respect to the variable $\boldsymbol{\Theta}$

$$g_t = \nabla_{\theta_\bullet} f_t(\bullet) = (\boldsymbol{O}_t - \boldsymbol{y}(t)) \cdot \nabla_{\theta_\bullet} \boldsymbol{O}_t \tag{7}$$

where $\boldsymbol{O}_t$ is the output estimation at time instant t. g_t is the gradient with respect to the cost function at t^{th} training iteration.

Thus, the biased 1st and 2nd moments of the gradient is:

$$\begin{aligned} m_t &= \beta_1 m_{t-1} + (1-\beta_1) g_t \\ v_t &= \beta_2 v_{t-1} + (1-\beta_2) g_t^2 \end{aligned} \tag{8}$$

where β_1 and β_2 are exponential decay rates within $[0,1)$. e is a small value to maintain numerical stability. α is the stepsize. m_t and v_t are the biased 1st moment (the mean) and 2nd moment (the uncentered variance) of the gradient respectively.

And the unbiased estimation of the 1st and 2nd moments is obtained by substituting the biased 1st and 2nd moments into the following equation:

$$\begin{aligned} \hat{m}_t &= m_t/(1-\beta_1^t) \\ \hat{v}_t &= v_t/(1-\beta_2^t) \end{aligned} \tag{9}$$

where β_1^t and β_2^t are β_1 and β_2 to the power t respectively. $\hat{m}_t$ and $\hat{v}_t$ are the bias-corrected values of m_t and v_t respectively.

Finally, the weights are updated as following:

$$(\theta_\bullet)_t = (\theta_\bullet)_{t-1} - \alpha \hat{m}_t / (\sqrt[2]{\hat{v}_t} - e) \tag{10}$$

Therefore, the steps of nonlinear parameters estimation using Adam algorithm are shown in Algorithm 1.

Algorithm 1. Parameter optimization based on the Adam algorithm

Require: Cost function with parameters $f_t(\boldsymbol{\Theta})$.
Ensure: Parameters for CNN $\boldsymbol{\Theta}$.
Initialization: Set exponential decay rates for the moment estimates, minimum and stepsize : $\beta_1 = 0.9$, $\beta_2 = 0.999$, $e = 1e - 8$, $\alpha = 0.001$; the 1st and 2nd moment vector are set as: $m_0 = 0$, $v_0 = 0$; the maximum number of iterations: iterations;
for $t = 1$ to $iterations$ **do**
1. Get gradients of parameters θ (the element of $\boldsymbol{\Theta}$) to the cost function $\hat{\boldsymbol{y}}_t$ at timestep t based on (7);
2. Update biased 1st moment estimation based on (8);
3. Update biased 2nd raw moment estimate based on (9);
5. Update parameters based on (10);
end for
return $\boldsymbol{\Theta}$.

3 Methodology

Consider the noises and corruptions induced into the sensor signals, a machine learning based method is applied to estimate the temperature in this paper. However, the internal dynamics of li-ion batteries is a complex and time-varying nonlinear electro-thermal-mechanical process, therefore the information including electric signals, temperature and strain measurements are all to reflect the coupling relations. To capture the strongly nonlinear dynamics, CNN which can automatically extract the local and global features using multi-layer convolution operation is applied in this paper.

3.1 Multiple Sensing Data Fusion

FBG Measurements

FBG Working Principle. FBG is generally formed by exposing a small section of light sensitive fiber to the intensity period distribution of laser light using the holographic interferometry or phase mask method. The refractive index of the optical fiber which is increased through the exposure will changes permanently because of the irradiated light intensity. Then small amounts of light would be reflected in every position of the refraction change which are mixed coherently to create a large reflection at a specific wavelength [23]. Thus the Bragg wavelength (λ_B) is defined as the wavelength at the position of the maximum reflection. It is computed as follows

$$\lambda_B = 2n_e\Lambda \tag{11}$$

where n_e is the effective refractive index at the grating location and Λ is the constant refractive index modulation.

When the incident light enters the FBG, selective reflexes will occur once the Bragg condition is satisfied. Since the grating pitch of a fiber grating is along with the axially distributed, the axial strain and refractive of a fiber grating will change following the external temperature and strain changes, so as to the wavelength will also change. Specifically, n_e and Λ in Eq. (11) will be influenced by temperature changes due to the thermo-optic effect and the thermal expansion effect respectively. While n_e and Λ in Eq. (11) will also be influenced by strain changes due to the stress-optic effect. Therefore, the measurement signal λ_B can be used to calculate the corresponding the corresponding temperature and strain variation.

Thus the relationship between the wavelength shift ($\Delta\lambda_B$), the temperature variation (ΔT) and the strain variation ($\Delta\varepsilon$) is given as follows:

$$\frac{\Delta\lambda_B}{\lambda_B} = (\alpha + \xi)\Delta T + (1 - P_e)\Delta\varepsilon \tag{12}$$

where $\Delta\lambda_B$ is the wavelength change. λ_B is the initial wavelength. P_e, α and ξ denote the elastic-optic coefficient, the thermal-expansion coefficient and the thermo-optical coefficient [24], respectively. ΔT and $\Delta\varepsilon$ represents the changes in strain and temperature, respectively. Based three known coefficients, the changes in temperature and strain can be decoupled using more than two Eq. (12) which should be not related.

Non-electric Signals Using FBG Sensors. According to the FBG working principle, FBG sensors are very sensitive to both temperature and strain. The temperature is affected by thermo-optical effect and thermal expansion effect, while strain is affected by elastic-optical effect and the change of fiber grating period [25]. To this end, the wavelength shift is usually caused by both the temperature changes and the strain changes. Thus, the temperature should be represented simultaneously using the wavelengths of FBG sensors and the strain of battery. Refer to [13,14], three wavelengths of different position are applied in this paper.

Electric Signals Using Battery Tester. To collected the electrical signal of the above li-ion batteries, a battery tester, named NEWARE BTS-4000 was applied to provide the charging/discharging strategy during the charging/discharging cycles and acquire the battery data including terminal voltage, circuit current and capacity. The sampling frequency was set to 1 Hz for all the tests performed in this study. To keep the constant temperature and adjust the ambient temperature, all the tests in this study were performed inside a BINDER thermal chamber. The host computer installed software BTS-7.5 was employed to program the test strategy and save the test data.

CNN Based Data Fusion. In order to verify the accuracy of the developed model, standard cycle profile was employed to the examined cells which includes a Constant-Current Constant-Voltage (CC-CV) charging process and a Constant-Current (CC) discharging process. The upper cut-off voltage of the charging process is 3.6 V, while the lower cut-off voltage of the discharging process is 2 V. During the charging/discharging cycling, the current rate of the constant current process is all set to 1C. The voltage, charging/discharging current, capacity and the wavelengths in one cycle are given in Fig. 2.

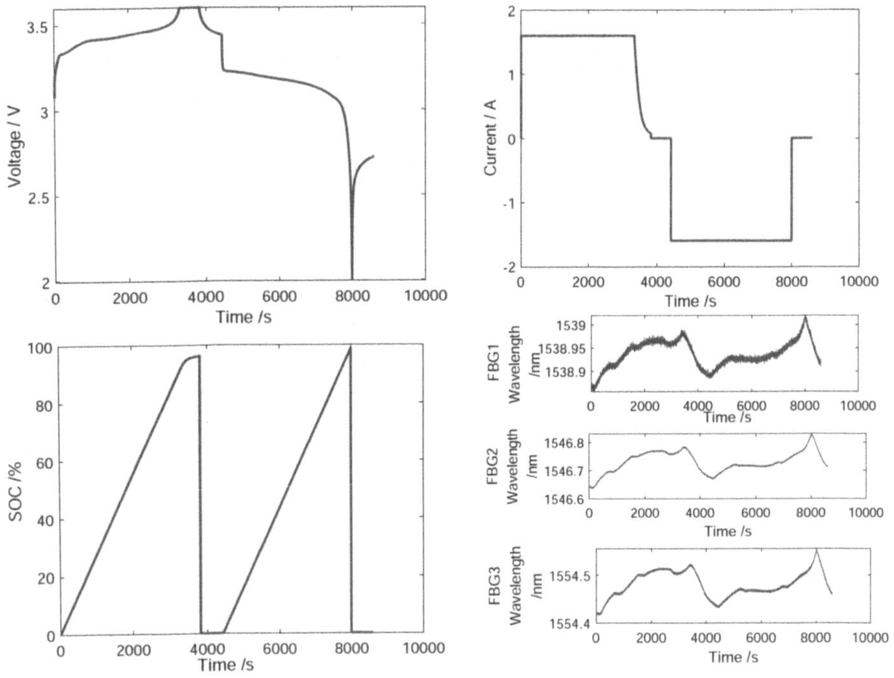

Fig. 2. The signal measured during the charging and discharging process

From Fig. 2, it is obviously that the measured wavelengths using FBG1, FBG2 and FBG3 have the exactly similar trends. The crests and troughs of these curves also occur in the similar position. However, the noises of these wavelength curves are in different shapes due to the uncertainties including random fluctuation, the sudden temperature changes and the internal/external forces, etc. The wavelengths should be processed to guarantee the performance of the temperature model at first. Thus, the temperature monitoring model is built using CNN method because the metrics of feature extraction.

3.2 Fused Solution Based on CNN

Since the data collected using different sensors while mutual coupling, it is essential to fuse them for the full exploitation. The CNN method are effective tools for extracting high-level features from data that exhibits spatial or temporal correlations using convolution operations. Thus, the CNN method is applied in this study.

Time Series Signal Transformation. The CNN method has a wide range of tasks in image processing, natural language processing, and speech recognition which often consist of high-dimensional signals. Due to the fact that the data collected in this study is time series, it is necessary to research how to extract effective features for time series signals using CNN methods. In [21], a time series-to-image transformation method is proposed for the battery capacity estimation using CNN which is detailed below.

Firstly, to make full use of historic measurements, n continuous data points are selected as one input from each measured variable of one cycle recursively. The operation is executed until the number of remaining data less than n so as to M data chunks of each variable are generated. Thus,

$$M = \text{floor}(\frac{L-n}{n-c}) + 1 \tag{13}$$

where floor$(\cdot)$ is the function to obtain the greatest integer that is not larger than the input parameter. M is the number of input chunks of one cycle. L is the number of data in one cycle in total. n and c are the length of a segment and the number of the overlap data between two adjacent chunks respectively.

Secondly, the n samples in each chunk are converted into a 2-dimensional matrix with the size of $\sqrt{n} \times \sqrt{n}$. The number of signal is namely as the dimension of the converted images. Therefore the input samples for the CNN method can be obtained.

Furthermore, these measurements using different sensors with different scale should be normalized to ensure the training speed and the estimation accuracy. The min-max normalization strategy as a commonly method is adopted in this paper. Assuming the value z falling within $[-1, 1]$ after normalized, they can be obtained from the data of measurement $\boldsymbol{x}^k = x_1^k, x_2^k, \cdots, x_n^k$ by

$$z_i^k = \frac{x_i^k - \min(\boldsymbol{x})}{\max(\boldsymbol{x}) - \min(\boldsymbol{x})} \times 2 - 1 \tag{14}$$

where $\boldsymbol{x}^k$ is the measured value in the kth data chunk. $\boldsymbol{x}$ is a set of all training data of each variable. n is the length of training data of each chunk.

In summary, the data transformation method has the advantages of easily implementation and no predefined parameters. It made processing time series signals based on CNNs is realized. The procedure is detailed in Algorithm 2.

Algorithm 2. Time Series Signal Transformation

Require: L Charging data samples $(\boldsymbol{data}_i, y_i)$ $\boldsymbol{data}_i \in R^m, y_i \in R^1, i = 1..., L$, the number of the variables is m, the selected interval c.
Ensure: 2-dimensional image samples $(\boldsymbol{input}_i, y_i)$ $\boldsymbol{input}_i \in R^{\sqrt{n}\times\sqrt{n}\times m}$, $y_i \in R^1$, $i = ..., M$, the number of the selected data at each iteration is n, the number of the input chunk is M.
1: **Initialization:** Calculate the number of the input chunk using Eq. (13).
2: **for** $t = 1$ to M **do**
3: a) Select n time series data from $(\boldsymbol{data}_i, y_i)$ begin with $(i-1)*(n-c)$:
4: b) Convert the n points into a $\sqrt{n} \times \sqrt{n}$ matrix.
5: **end for**
6: **Normalization**: Calculate the samples with 2-dimensional image represents using Eq. (14).

CNN Model for Temperature Estimation. Using the transformed high-dimensional data as inputs, the temperature estimation model can be then built using the CNN method shown in Fig. 3. This temperature estimation model based on CNN methods has less parameters because of shared weights in multi-activation functions and is insensitive to small translation of the input caused by the summary statistic using the pooling architecture, so as to avoid overfitting.

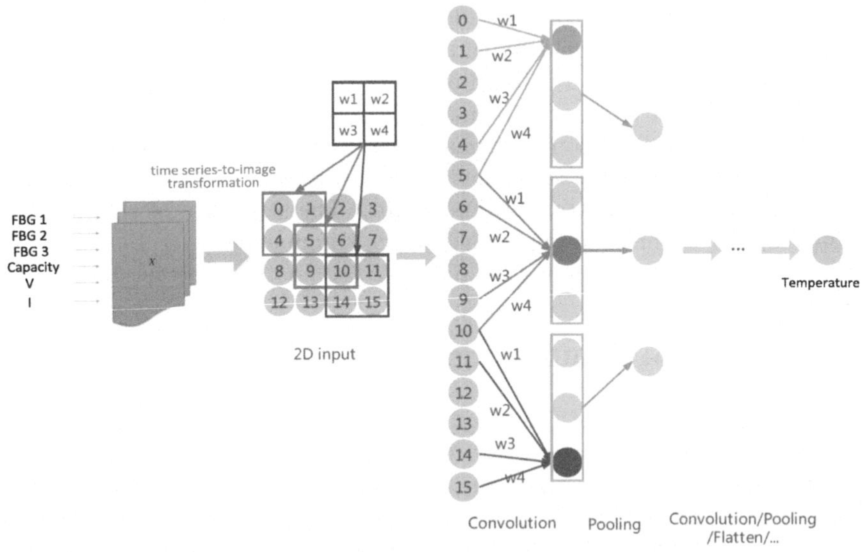

Fig. 3. The temperature estimation model using the CNN method

Due to the relatively small size of the input sample (15 × 15 × 6), a CNN structure listed in Table 1 is employed based on LeNet-5 [21]. As is shown in

Table 1, the proposed temperature estimation model using CNN includes two alternating convolutional, two pooling layers, one convolutional layers, a flatten layer and two dense layers (fully connected layers).

Table 1. The temperature estimation model structure using the CNN method

Model	Layer	Layer name	Stride	Output size	Parameters	Activation
AdamCNN	L1	Conv_1(16@5 × 5 × 6)	(1,1)	(15 × 15 × 16)	2416	elu
	L2	Maxpool_1(2 × 2)	(2,2)	(7 × 7 × 16)	0	-
	L3	Conv_2(16@3 × 3 × 16)	(1,1)	(7 × 7 × 16)	2320	elu
	L4	Maxpool_2(2 × 2)	(2,2)	(3 × 3 × 16)	0	-
	L5	Conv_3(8@3 × 3 × 16)	(1,1)	(3 × 3 × 16)	2320	elu
	L6	Flatten	-	144	0	-
	L7	Dense1	-	50	7250	elu
	L8	Dense2	-	1	51	

For the weight and height of the input image are set by the data number of each data chunk, while depth is set as the number of variables. Assuming $n = 225$, the weight and height of the input will both equal to 15, while the depth is 6 in this study (the 6 variables are current, voltage, capacity and the wavelengths of FBG1, FBG2, FBG3). The filter structure in convolutional layers is marked as $M@w \times h \times d$ which denotes that there are M filters and their size is $w \times h \times d$. Among them, d is the input depth of this layer. Then, the pooling layer obtained smaller feature maps using M feature maps with 2×2 pooling region. Following the final convolutional layer, the flatten layer transforms the $i \times j \times k$ features into a vector with $i \times j \times k$ neurons. In the end, the vector is fully connected to the output (temperature). The structure of the model are all manual set by experiment experience which summarized in Table 1 including the correspond output size and the number of parameters of every layer.

4 Experimental Simulations

To verify the performance of the novel thermal monitoring method, the experimental data of cell #1 is employed to training, while the data of cell #4 is applied to validation. The temperature measured using the thermocouple is as reference value. The performance of the proposed method in training and validation are list in Table 2. The RMSE is short for the Root Mean Squared Error, while MAE is short of the maximum absolute error.

The temperature estimated using the above proposed method based on cell #1 and the results of a cycle is shown in Figs. 4 and 5 respectively, while the temperature forecasted using the above novel method based on cell #4 is illustrated in Fig. 6. It is shown that the predicted temperature based on the proposed

Table 2. The training and validation statistics of the estimated temperature

	RMSE	MAE	MSE	MaxAE
Training using cell #1	0.2	1	0.03	6
Validation using cell #4	0.2	1	0.03	5

method can keep pace of the reference temperature. To improve the figure clarity, a cycle curve of the validation result based on the proposed method is given in Fig. 7. Most of the validation errors of Fig. 7 spread within ± 0.5°C, only apart from the points at the peak valley of SOC value that the validation errors of them mostly fall on ± 1°C. Both of them is allowable for the accuracy tolerance of the thermocouple.

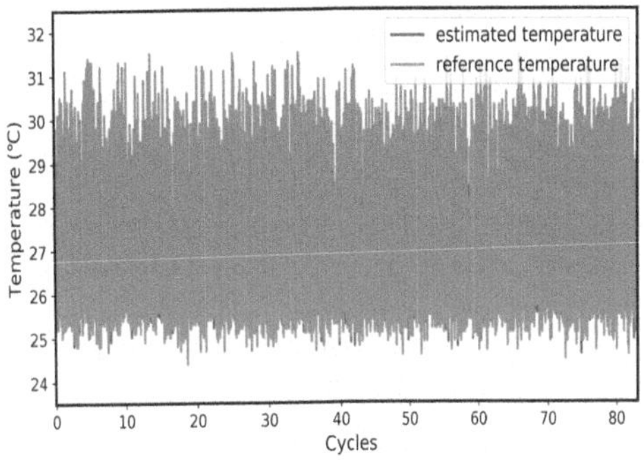

Fig. 4. Temperature fitting of cell #1

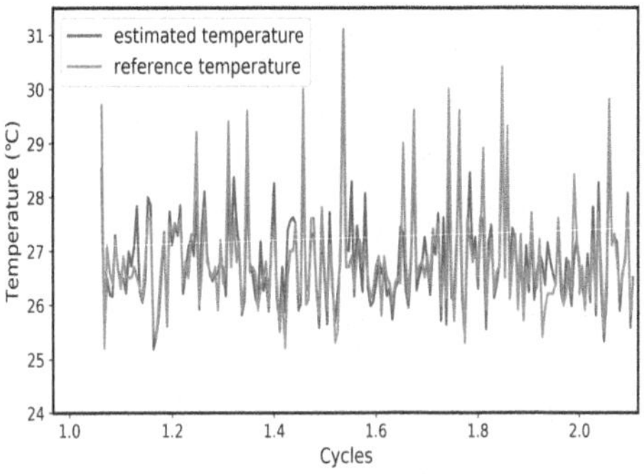

Fig. 5. Fitting results of Cycle #12 of cell #1

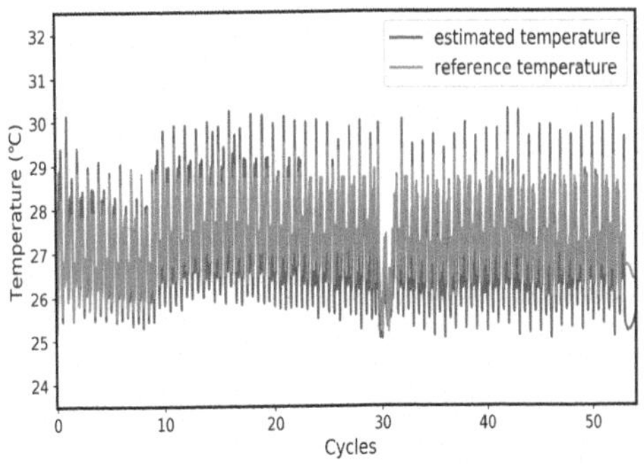

Fig. 6. Temperature validation of cell #4

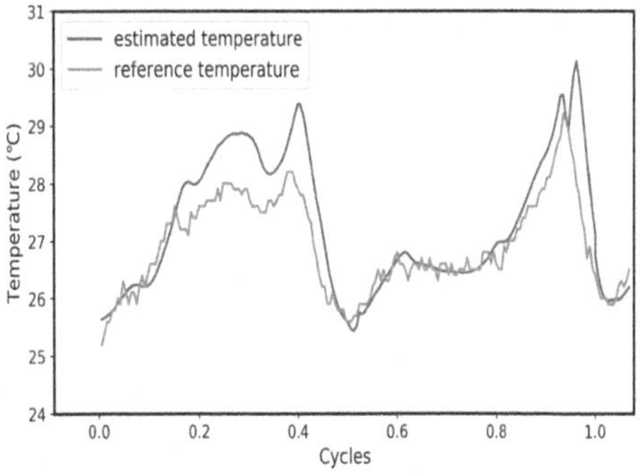

Fig. 7. Validation results of Cycle #1 of cell #4

5 Conclusions

A novel thermal monitoring method based on convolutional neural networks (CNN) integrating with measurements from multiple fiber bragg grating sensors has been presented in the paper, which reduces the negative impacts caused by sensors layout and the effect of measurements noise. Different temperature and strain features are incorporated into the CNN model to improve the estimation accuracy. Cell #1 and cell #4 which are made of the same materials are employed to analyse the performance of the proposed method. The experiment results confirm that the proposed thermal monitoring method has been capable of high estimation accuracy, and its root mean squared error on the validation data is 0.2°C. To prune the CNN model to reduce the computational complexity and save storage space will be studied in future.

Acknowledgments. The authors are would like to thanks the Shanghai Institute of Technology under Grant ZQ2023-2. This work was supported by the Smart Energy Systems School of Electronic and Electrical Engineering in university of Leeds.

References

1. Xu, B., Oudalov, A., Ulbig, A., Andersson, G., Kirschen, D.S.: Modeling of lithium-ion battery degradation for cell life assessment. IEEE Trans. Smart Grid **9**(2), 1131–1140 (2018)
2. Jin, Y., Zhao, Z., Miao, S., Wang, Q., Sun, L., Lu, H.: Explosion hazards study of grid-scale lithium-ion battery energy storage station. J. Energy Storage **42**, 102987 (2021)
3. Liu, K., Kang, L.I., Qiao, P., Cheng, Z.: A brief review on key technologies in the battery management system of electric vehicles. Front. Mech. Eng. **14**(1), 47–64 (2019)
4. Zhang, C., Li, K., Deng, J., Song, S.: Improved realtime state-of-charge estimation of LiFePO 4 battery based on a novel thermoelectric model. IEEE Trans. Ind. Electron. **64**(1), 654–663 (2017)
5. Waag, W., Käbitz, S., Sauer, D.U.: Experimental investigation of the lithium-ion battery impedance characteristic at various conditions and aging states and its influence on the application, special issue on advances in sustainable biofuel production and use - XIX international symposium on alcohol fuels - ISAF. Appl. Energy **102**, 885–897 (2013)
6. Xiong, R., Li, X., Li, H., Zhu, B., Avelin, A.: Neural network and physical enable one sensor to estimate the temperature for all cells in the battery pack. J. Energy Storage **80**, 110387 (2024)
7. Fowler, M., Dincer, I., Fraser, R., Agelin-Chaab, Panchal, S.: Experimental temperature distributions in a prismatic lithium-ion battery at varying conditions. Int. Commun. Heat Mass Transf. Rapid Commun. J. **71**, 35–43 (2016)
8. Mutyala, M.S.K., Zhao, J., Li, J., Pan, H., Yuan, C., Li, X.: In-situ temperature measurement in lithium ion battery by transferable flexible thin film thermocouples. J. Pow. Sources **260**, 43–49 (2014)
9. Raijmakers, L., Danilov, D., Eichel, R.A., Notten, P.: A review on various temperature-indication methods for li-ion batteries. Appl. Energy **240**, 918–945 (2019)

10. Huang, F., et al.: Real-time monitoring of temperature field distribution of three-element lib lithium battery using FBG arrays. IEEE Sens. J. **23**(24), 30473–30480 (2023)
11. Peng, J., Jia, S., Yu, H., Kang, X., Yang, S., Xu, S.: Design and experiment of FBG sensors for temperature monitoring on external electrode of lithium-ion batteries. IEEE Sens. J. **21**(4), 4628–4634 (2021)
12. Rente, B., et al.: Lithium-ion battery state-of-charge estimator based on FBG-based strain sensor and employing machine learning. IEEE Sens. J. **21**(2), 1453–1460 (2021)
13. Li, Y., et al.: A hybrid machine learning framework for joint SOC and SOH estimation of lithium-ion batteries assisted with fiber sensor measurements. Appl. Energy **325**, 119787 (2022)
14. Liu, S., Li, K.: Thermal monitoring of lithium-ion batteries based on machine learning and fibre Bragg grating sensors. Trans. Inst. Meas. Control. **45**(8), 1570–1578 (2023)
15. McTurk, E., Amietszajew, T., Fleming, J., Bhagat, R.: Thermo-electrochemical instrumentation of cylindrical li-ion cells. J. Pow. Sources **379**, 309–316 (2018)
16. Zhang, C., Li, K., Deng, J.: Real-time estimation of battery internal temperature based on a simplified thermoelectric model. J. Pow. Sources **302**, 146–154 (2016)
17. Shen, J., Zhang, Z., Shen, S., Zhang, Y., Chen, Z., Liu, Y.: Accurate state of temperature estimation for lithium-ion batteries based on square root cubature Kalman filter. Appl. Therm. Eng. **242**, 122452 (2024)
18. Wang, Y., et al.: Temperature state prediction for lithium-ion batteries based on improved physics informed neural networks. J. Energy Storage **73**, 108863 (2023)
19. Panchal, S., Dincer, I., Agelin-Chaab, M., Fraser, R., Fowler, M.: Thermal modeling and validation of temperature distributions in a prismatic lithium-ion battery at different discharge rates and varying boundary conditions. Appl. Therm. Eng. **96**, 190–199 (2016)
20. Zhu, S., He, C., Zhao, N., Sha, J.: Data-driven analysis on thermal effects and temperature changes of lithium-ion battery. J. Pow. Sources **482**, 228983 (2021)
21. Li, Y., Li, K., Liu, X., Zhang, L.: Fast battery capacity estimation using convolutional neural networks. In: Transactions of the Institute of Measurement and Control (2020)
22. Kingma, D., Ba, J.: Adam: a method for stochastic optimization. arXiv preprint arXiv:1412.6980 (2014)
23. Hong, C., Bao, C., Fei, J., Zhang, Y., Wang, X.: Application of FBG technology in additive manufacturing: monitoring real-time internal temperature of products. IEEE Sens. J. **21**(5), 6003–6011 (2021)
24. Zhang, L., et al.: Real-time battery temperature monitoring using FBG sensors: a data-driven calibration method. IEEE Sens. J. **22**(19), 18639–18648 (2022)
25. Zhao, Y., Liao, Y.: Discrimination methods and demodulation techniques for fiber Bragg grating sensors. Opt. Lasers Eng. **41**(1), 1–18 (2004)

Author Index

K. Li et al. (Eds.): LSMS/ICSEE 2024, CCIS 2218, pp. 387–388, 2025.
https://doi.org/10.1007/978-981-96-0232-2

SPRINGER NATURE

GPSR Compliance

The European Union's (EU) General Product Safety Regulation (GPSR) is a set of rules that requires consumer products to be safe and our obligations to ensure this.

If you have any concerns about our products, you can contact us on ProductSafety@springernature.com

In case Publisher is established outside the EU, the EU authorized representative is:

Springer Nature Customer Service Center GmbH
Europaplatz 3
69115 Heidelberg, Germany

The manufacturer's authorised representative in the EU is Springer
Nature Customer Service Centre GmbH, Europaplatz 3, 69115 Heidelberg,
Germany. If you have any concerns regarding our products, please
contact ProductSafety@springernature.com

Printed and bound by CPI Group (UK) Ltd, Croydon, CR0 4YY
01/10/2026
02225933-0002